W9-BCC-927

THE EARTH

THE EARTH

Edited by Peter J Smith

MACMILLAN PUBLISHING COMPANY
A Division of Macmillan, Inc.
New York

Editor Dougal Dixon
Art Editor John Ridgeway
Designer Ayala Kingsley
Picture Editor Linda Proud
Picture Researcher Caroline Lucus
Senior Editor Lawrence Clarke

Advisors
Dr Preston Cloud, University of California
Sir Kingsley Dunham FRS, University of Durham
Dr John Tuzo Wilson FRS, Ontario Science Center

Contributors
Dr Philip Allen (16)
Professor Jerry van Andel (12)
Dr John Catt (17)
John Downes (18)
Dr Stephen Drury
Professor Andrew Goudie (20)
Dr John Gribbin
Dr P M Kelly (11)
Professor J L Knill (21)
Dr Peter J Smith (Introduction, 1,2,3,4,5,6,7,8,9,10,14,15,22,23)

AN EQUINOX BOOK

Planned and produced by:
Equinox (Oxford) Ltd, Littlegate House, St Ebbe's Street, Oxford OX1 1SQ

Published by:
Macmillan Publishing Company
A Division of Macmillan, Inc.
886 Third Avenue
New York, N.Y. 10022
Collier Macmillan Canada, Inc.

Library of Congress Catalog Card Number 86-8506

ISBN 0-02-907441-X

Printed in the Netherlands by Royal Smeets Offset bv, Weert

Introductory pictures (pages 1-8)
1 Death Valley, California
2-3 Alaskan scene
Right Atoll near Tahiti
7 Night scene with horizon
8 Grand Canyon, Arizona

Contents

Introduction

In their wilder moments, some Earth scientists are liable to claim that geology is the oldest of the sciences. After all, did not the people of the Stone Age put at least one type of geological material to very good use? The Romans mined ores; the Greek, Pythagoras, was familiar with marine fossils on land during the sixth Century BC; and in the first century BC the Chinese invented a form of magnetic compass, an instrument that relies on one of the Earth's most fundamental geophysical properties. All this is true, but mere knowledge or use of a few random geological phenomena does not a science make.

The more sober members of the geological community will say that their subject only became a proper science in the late 18th or early 19th century – and there is more truth in that. Indeed, it is customary to see Edinburgh's James Hutton (1726-97) as "The Founder of Modern Geology" in recognition of the remarkable theory he first put forward in 1785.

The first revolution

Hutton saw the Earth as a giant machine that had run and run and would continue to do so. The continents are gradually eroded away, and the debris thus formed is washed down to the sea where it accumulates as sedimentary layers. The sediments become compacted and, through the expansive power of the Earth's heat, are uplifted and contorted. The cycle of erosion and uplift then begins again and repeats itself time after time. In addition, the Earth's heat is sufficient to melt rocks at depth; and these, intruded into the upper crust, are also subject to erosion.

This, in broad outline, is what we still believe today. Indeed, so familiar are the concepts of erosion, uplift and igneous activity that it is sometimes hard to believe that they have not always been perfectly obvious. In the late 18th century, however, they made for a revolution – the first revolution in the Earth sciences. Hutton was even denounced as an atheist, not least because his scheme required an extremely old Earth, an Earth that had, in fact, "no vestige of a beginning, – no prospect of an end". To Old Testament Christians this was anathema, for had not the Irish Prelate, James Ussher (1581-1656), calculated that Genesis had taken place in the year 4004 BC?

Nor was the conflict just over the Earth's age. No less important was the very nature of the planet's activity. Theology admitted divine intervention, exemplified by Genesis itself and the Flood (which Ussher claimed had taken place in 2349 BC!). Hutton, by contrast, was proposing the existence of only "natural" processes. The wearing down of mountains, the elevation of continents and the melting of rocks were inherent in the Earth's behavior. They had always taken place and always would; the geological phenomena of the past were no different from those of today; and divinely-inspired catastrophic events were necessary to explain neither the past nor the present.

The two faces of geology

It took Hutton's views many decades to become widely accepted; but once they had been, the science of geology expanded with amazing rapidity. Using fossils and the law of superposition (that the upper strata are usually younger than the lower), geologists were able to arrange in order of age many of the rock formations immediately accessible, particularly those in Europe. The geological time scale evolved. The existence of past ice ages was discovered. Dinosaur remains were dug up. It became clear that many continental rock formations had undergone huge tectonic movements. Geological maps appeared in ever-greater profusion.

On the institutional front, geological societies were formed locally, nationally and internationally. National Geological Surveys were set up. Geological journals multiplied. Geology books proliferated. Specimen collecting became a mania. It is no exaggeration to say that geology became the predominant science of much of the 19th century. Nor was it just a matter for the professionals. Never before had a science become so much a part of the popular culture (especially in Britain), and perhaps never again was one to do so. Geological textbooks became best-sellers, and the geology lecture became a popular form of entertainment and instruction, at least among the educated classes.

Why did all this happen? It would be nice to think that it was largely as a result of an urge for intellectual enlightenment, as indeed it was to a large extent. But it was much more. Geology, no less than machinery, was the key to the Industrial Revolution. Geological surveys and investigations revealed the raw materials that the new industrial system required – coal, iron, building stone, sand, clay, aggregates and water. Geological studies indicated where best to site the infrastructure of the industrial society – the canals, roads, railways and civil engineering structures. And the intellectual and economic facets of geology fed upon each other. Tunnels, cuttings and other excavations revealed geological sections never before seen, providing new geological information with industrial applications.

This duality – the "pure" and the "applied" – has been, is, and will always remain, crucial to geology's continued progress, although the proportions of the two aspects of the subject in the total mix have varied with time. During the first half of the 19th century, the intellectual and the economic ran together at a high level; but thereafter, until well into the 20th century, the intellectual excitement gradually died down. Geology remained, as always, a vital component of the industrial society; indeed, its economic role became ever more significant as population increased and new natural material, such as oil and the rarer metals, played an ever-increasing part in society. As far as the basic science was concerned, however, it sometimes looked as though little of fundamental interest remained to be discovered. The subject became routine and, to many, boring.

The second revolution

But then something quite remarkable happened. During the 1960s the second revolution in the Earth sciences came about, the result of the discovery of what is now known as plate tectonics. The continents, hitherto thought to be stationary, were found to be drifting. The ocean floors, believed to be old and inactive, were discovered to be young and in motion. The Earth's outer shell, hitherto regarded as a single entity (albeit comprising both oceanic and continental parts), was seen to be divided into a limited number of "plates" reacting with their neighbors along the edges.

The new global tectonics has completely changed the way we regard the Earth. Few aspects of geology have remained untouched by it. It has led to a much greater intellectual understanding of the Earth's behavior, just as it produced a much greater appreciation of how natural resources were formed and where they are now most likely to be found. Theoretical and practical interpretations of the Earth have continued to go hand in hand but at a greatly increased pace.

In this book we see the Earth in the exciting light of the plate-tectonic revolution. Not all the knowledge gained prior to the revolution was wrong; nor is that acquired since anywhere near complete. There is still much that geologists do not know, as there are no doubt some things they have misunderstood. The search for the full picture of the Earth will go on for a long time yet; but in the meantime, here is the planet as it appears to us today.

The Dynamic Earth

Pangea – the source of all our continents... Seafloor spreading – filling the gaps between continents...Plate tectonics...PERSPECTIVE...Early thoughts on continental drift...The proofs from rocks and fossils...The magnetically recorded evidence...The plates of the skin of the Earth...Investigating the birthplace of the plates...Where scientists disagree

As recently as the late 1950s, most geologists believed the Earth's crust to be horizontally immobile and that the continents were fixed in the positions in which they had formed during the the Earth's early history. The crust of the ocean was considered equally permanent. Since it was always underwater the weather could not erode it. Indeed, the ocean floors supposedly did nothing but collect the debris washed off the land. They were covered by billions of years' worth of sediments and fossils going back to the beginning of life.

But by the end of the 1960s, Earth scientists had proved every one of these well known, long established "facts" to be wrong. The continents move across the surface of the Earth with apparent ease. As for the oceanic crust, not only is it in constant motion, but the crust now in existence was produced only during the last 5 percent of the Earth's 4,600 billion years of existence (▶ page 78). Between 1960 and 1970 there had, in short, been a revolution in geological thought.

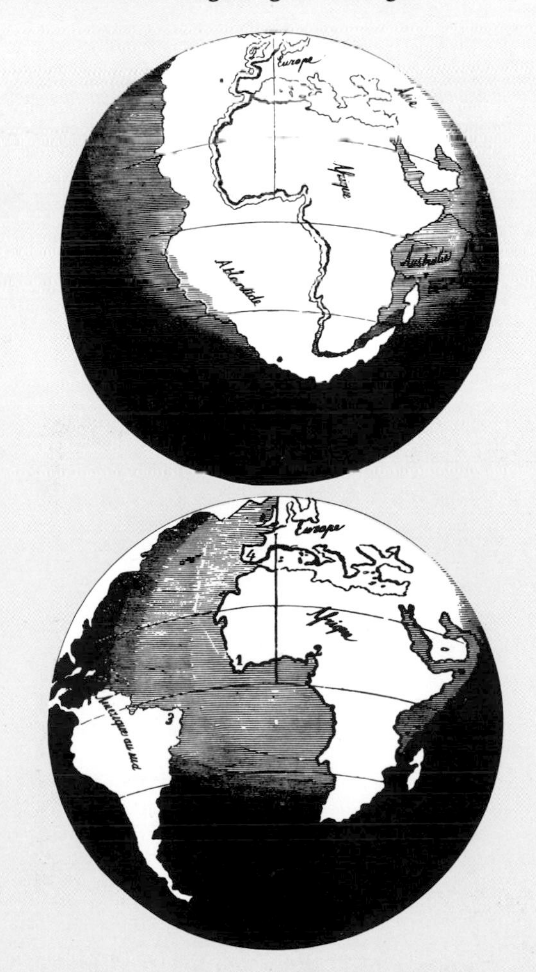

▶ Many of us have looked at a map of the world and been struck by the similarity in the shapes of the coastlines at each side of the Atlantic Ocean. The east coast of South America looks as if it could fit snugly into the west coast of Africa. It is almost as if these two continents were once part of the same landmass and have since been ripped apart along an irregular, Earth-sundering fissure. Earth science has come round to that opinion too. The notion is not new – philosophers have been suggesting it for centuries. The first person to draw up a map to give an impression of how the original combined continent might have looked was Antonio Snider in 1858. The upper globe shows the Earth at the time of Adam, shortly after the Creation. The lower globe shows the Earth at the time of Noah, after the Flood and associated divinely-induced earthquakes had done their damage.

The continental drift concept: Early hints

Although most Earth scientists refused to accept drift until the 1960s, the idea has a long history. Francis Bacon, the English statesman and natural philosopher, was the first to compare continental outlines, drawing attention in 1620 to the similarity in shape of the Atlantic coasts of Africa and South America. He regarded this likeness as an example of "conformable instances which are not to be neglected", but examined the matter no further. The idea of continental movement was introduced in 1668 by François Placet, the French moralist, who suggested that America might have been formed by "the conjunction of many floating islands which became connected with each other". Placet's aim was theological, however, and he adduced no real evidence for his proposal, which was nevertheless a curious pre-echo of a modern discovery (▶ pages 101-8).

Theodor Lilienthal, the German theologian, went further than Bacon, pointing out in 1756 that the coasts of Africa and South America were complementary. He noted that these coasts "would almost fill up each other in case they would stand side by side", even suggesting that the two continents were once actually joined. Alexander von Humboldt, the German scientist and traveler, also noticed the fit and first drew attention to the geological similarities of the two continents. However, his explanation, in 1801, was that the Atlantic had been gouged out by a huge current.

The first drift hypothesis

Antonio Snider, an American resident of Paris, was first to put forward a detailed hypothesis of continental breakup and separation. In 1858 he speculated that as the Earth cooled, the surface crystallized in a very irregular fashion, giving an uneven distribution of surface material. He suggested that a large continental mass had formed on one side of the globe, and that this arrangement was very unstable. Equilibrium was restored at the time of Noah's Flood, when volcanic material rose along fractures in the landmass, suddenly pushing America aside. In evidence, Snider cited the fit of the Atlantic coastlines, the existence of similar rock formations on the opposing edges of Africa and South America, and the similarity of fossils discovered on the two continents.

The continental movement envisioned by Snider was perhaps not so much a drift as a sprint, however, for it took place almost instantaneously in association with divine catastrophe (Noah's Flood). This ensured that it would not be taken seriously by geologists, most of whom had only recently rejected catastrophism – the idea that past geological events had taken place with great rapidity – in favor of uniformitarianism – the theory that suggests that the past geological events were similar to those that can be witnessed today. The latter philosophy ruled out sudden supernatural intervention, and claimed that any change in the surface of the Earth would have to be explained in terms of the processes that take place there at the present day. The processes of the Earth's interior were then, of course, unknown.

Plate tectonics ahead of its time

A flight from the poles

▲ ***Taylor surmised that there was once a continent over the North Pole. It broke up and fragments spread away from the pole because of the centrifugal force of the Earth's rotation. Their leading edges crumpled up forming fold mountains – the Alps, Rockies and Himalayas.***

▼ ***Earth movements were once attributed to divine acts, as illustrated in the dramatic painting 'The Flood', by Francis Darby (1793-1861).***

Drift becomes a science

The first truly scientific version of continental drift was proposed by Frank Bursley Taylor, an American geologist, in a lecture in 1908 and in an article published two years later. Taylor envisaged that there were originally two large landmasses at the North and South Poles, respectively. At sometime during the Tertiary Period these landmasses began creeping towards the equator. As they moved they broke up, and their leading edges were forced up into the Tertiary mountain ranges, such as the Alps and the Himalayas.

Taylor was largely ignored by the geological community of his day, however, for a number of reasons. First, few geologists were ready to accept an idea that was contrary to any theory that had existed since scientific geology had begun in the late 18th century. Second, Taylor's evidence lacked any reference to the traditional disciplines of paleontology and stratigraphy. The third reason was that Taylor was not part of the geological establishment in that he held no post at a university or research institution. He was a loner whose research was financed by his rich father. He was always in an uncertain state of health and traveled everywhere in the field with a doctor and later with his wife. Other geologists regarded him as an unknown quantity, even an eccentric.

And finally, any attention that Taylor might have been about to receive was soon to be diverted to Alfred Wegener, the German meteorologist and astronomer, an altogether more formidable proponent of the case for continental drift.

Alfred Wegener put continental drift on the map.

Wegener the outsider

Alfred Wegener (1880-1930) was the son of an evangelical preacher in Berlin. He obtained a higher degree in astronomy and a university post in meteorology. He spent most of the rest of his life teaching and researching the latter subject. He was killed in an accident during an expedition to Greenland and never lived to see his hypothesis validated. Wegener was not a geologist, and the geological establishment bitterly resented an outsider poaching upon their territory. In the United States, Wegener was even denounced as a charlatan by people who chose to forget that he was a serious scientist, albeit in a different field.

Francis Bacon noticed the apparent fit of Africa and South America barely 100 years after Columbus's voyage

The birth of the "supercontinent" theory

Earth scientists regard Alfred Wegener as the true progenitor of the continental drift concept, not because he was the first to think of it (which he was not) but because he was the first to provide a very detailed scientific case in its favor. His first article on the subject was published in 1912 and his book, "Die Entstehung der Kontinente und Ozeane" (The Origin of Continents and Oceans), appeared in 1915. Unfortunately, the first edition of the book remained largely unknown outside Germany because World War I prevented its being translated. The second edition of 1920 was not translated either, although it attracted interest in Europe. However, the third edition (1922) appeared in English, French, Spanish, Swedish and Russian in 1924 and immediately generated controversy, especially in Britain and America.

Wegener's main conclusion was that there was originally one supercontinent, which he referred to as "a pangæa" (all-Earth or all-land) but which has subsequently acquired the proper name Pangea. Sometime during the late Mesozoic Era, Pangea fragmented and the fragments – today's continents – slowly drifted to their present positions. To support his case, Wegener presented evidence from a wide variety of disciplines – geology, geophysics, paleontology, biology and paleoclimatology.

Mixed reception to a challenging theory

Many geologists, especially in Britain, were initially sympathetic to Wegener's ideas, although they felt the case to be unproven. In the United States, by contrast, the reaction was generally unsympathetic, often hostile, and sometimes even venomous. Every bit of Wegener's evidence was examined and found wanting. Geologists were able to put forward alternative interpretations that did not require drift.

The mechanism of drift was never clear. Wegener had suggested that the forces moving the continents came from the Earth's rotation, but this was unconvincing even to those sympathetic to the idea of drift. By the late 1920s, however, a number of geologists, most notably Arthur Holmes in Britain, had concluded that the upper part of the Earth's mantle, though solid, was not necessarily immobile. Perhaps it could slowly "creep" over long periods of time, allowing the crust above to move sideways. And if the upper mantle contained a source of energy, such as heat, perhaps convection currents could propel the continents.

Unfortunately, people still had difficulty in envisaging drift. However large the forces being exerted on the continents from below, how could they possibly plow their way through solid ocean floor? In 1929 Arthur Holmes suggested that mountains would be forced up where continents were pressing forward against the oceanic crust. He also claimed that, because of the high temperatures and pressures there, rocks would increase their density and sink away into the mantle. In this way continents would make slow progress in moving from their parent Pangea.

But whatever its merits or demerits, Holmes' 1929 paper was ignored, possibly because it was published in an obscure journal. Also largely ignored was "Our Wandering Continents", a book published in 1937 by the South African geologist Alexander du Toit. In this detailed volume du Toit enthusiastically followed Wegener, strengthening his hypothesis by making corrections to it and adding many new data in its support. Since the mass conversion to continental drift during the 1960s, secondhand copies of what has become a classic treatise on continental drift are rare.

After the initial stir created by the appearance of Wegener's work in the 1920s, and despite some subsequent support by a few influential geologists, continental drift was not to attract wide interest again until the late 1950s. Then new data began to arrive from an unexpected quarter. The revolution in the Earth sciences was about to begin, almost 50 years after Wegener's first words on the subject.

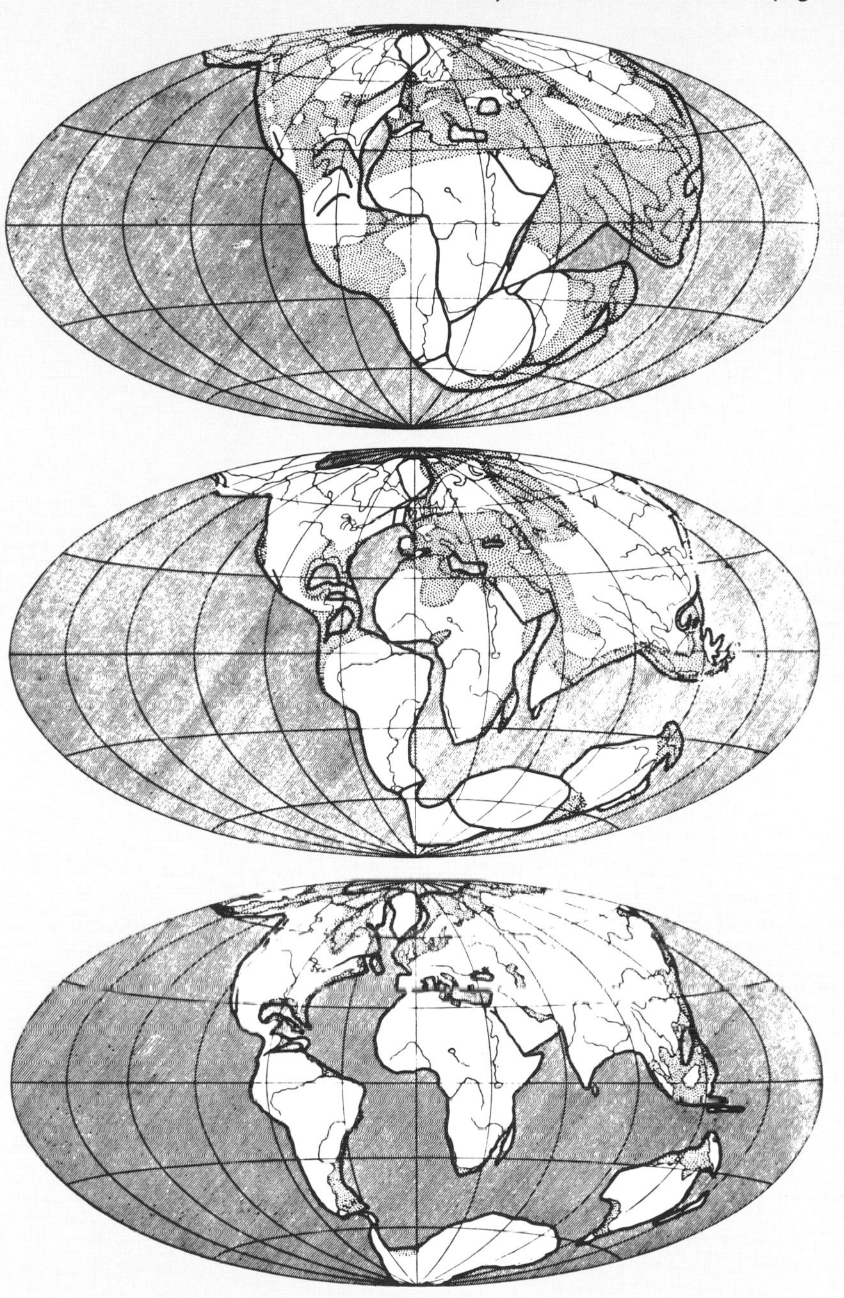

▲ Wegener's maps, produced in 1912, are remarkably similar to the paleogeographic maps produced today with our vastly increased knowledge of the subject. The top map shows the state of the world in the Upper Carboniferous (or Pennsylvanian) period. The single continent of Pangea existed then. The middle map is of Eocene times, when Pangea had begun to split up. The lower map is of the lower Quaternary – almost the present day. The stippled areas on the maps represent shallow continental seas.

Until the 1960s only surface features could be used to study the movements of continents

Gondwana fossils

Some of the most impressive of the paleontological evidence for continental drift comes from fossils of creatures that lived between Upper Carboniferous (or Pennsylvanian) and Triassic times on the Gondwana continents. Identical species of the seed fern "Glossopteris" occur on all five major landmasses as well as on Madagascar. Before continental drift was accepted, some paleontologists had made the astonishing suggestion that seeds of "Glossopteris" must have blown across thousands of kilometers of ocean. The reptile "Lystrosaurus" had long been known in South Africa and India when, in 1969, its remains were first found in Antarctica. The fossil reptiles "Mesosaurus" and "Cynognathus" are both found in Africa and South America. The ability of such reptiles to cross wide oceans was once incorrectly attributed to the existence of long-submerged "land bridges" between the continents.

Fossils prove continental drift

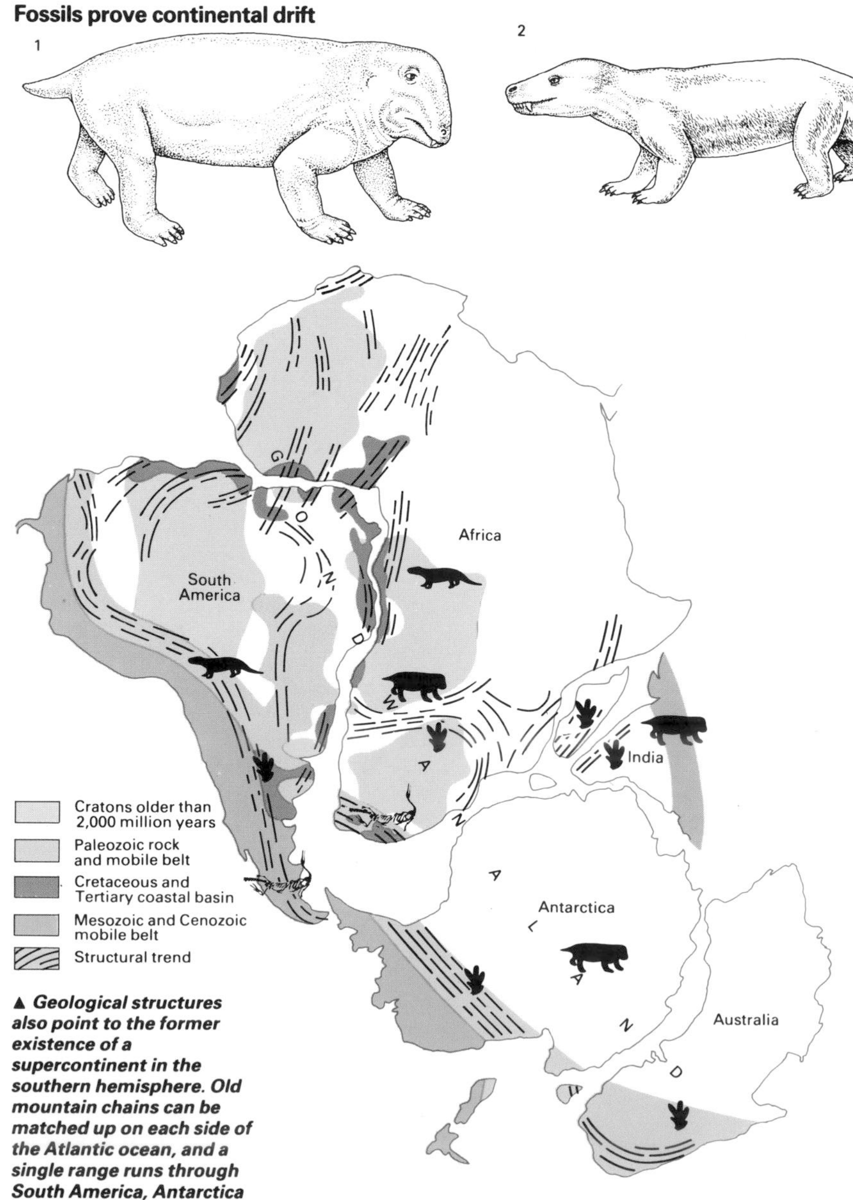

◄ ▲ Many different kinds of fossil organisms found scattered over the continents indicate that at least some of the continents were once jointed together. Fossils of the freshwater swimming reptile "Mesosaurus" (above) and the fox-like "Cynognathus" (2) have been found in Brazil and South Africa. The hippopotamus-like reptile "Lystrosaurus" (1) lies in Triassic rocks of Africa, India and Antarctica. The seed-fern "Glossopteris" (3) is found in rocks on all the southern continents.

▲ Geological structures also point to the former existence of a supercontinent in the southern hemisphere. Old mountain chains can be matched up on each side of the Atlantic ocean, and a single range runs through South America, Antarctica and Australia.

Geological correlation

Important geological evidence for continental drift comes from the matching of rock formations and structural features across continental edges. In both South America and Africa there are shield areas in which the rocks are older than 2 billion years, separated by sharp boundaries from rocks that are about 550 million years old. When the two continents are reconstructed in their pre-drift positions, the rocks and the boundaries correlate well. The structural trends, or "grains", of the continents also match. The opponents of drift once dismissed such correlations as coincidence, or attributed them to the existence of a block of land between the continents that had subsequently sunk out of sight.

The later geological histories of southwest Africa and southeast Brazil provide further evidence for drift. From about 550 million to about 100 million years ago when the two continents were together, the sequences of erosion, sedimentation, glaciation, flooding, coal formation, afforestation and volcanic activity were common to the two areas. They happened at the same time on both continents. Thereafter, Brazil entered a tropical regime while southwest Africa became desert. This divergence occurred after Africa and South America were separated and moved away as the continental drift hypothesis requires.

Magnetic rocks point the way

1

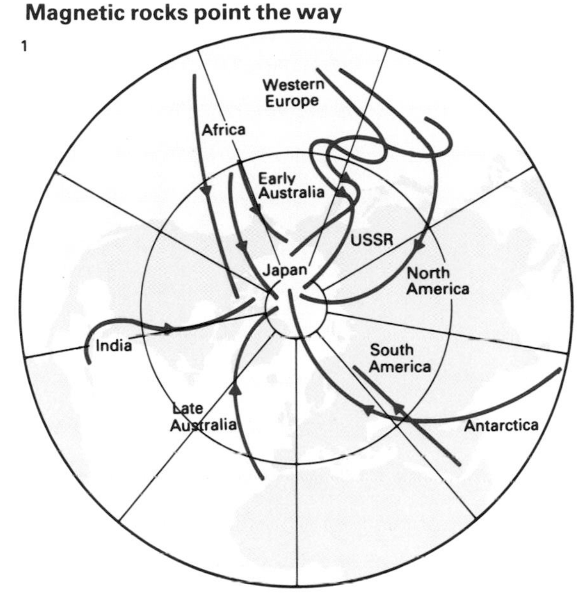

2

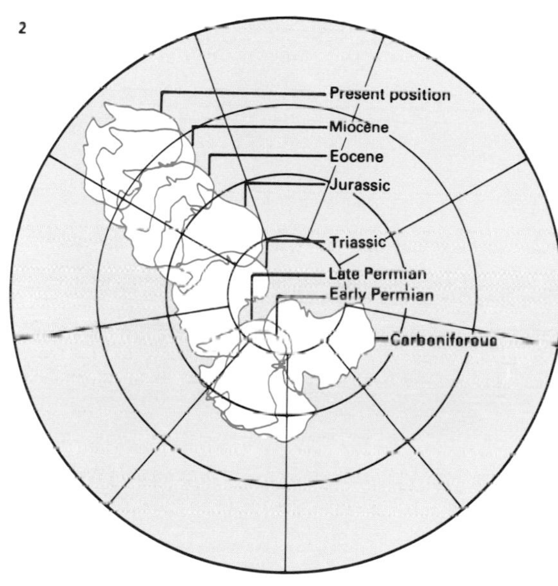

3

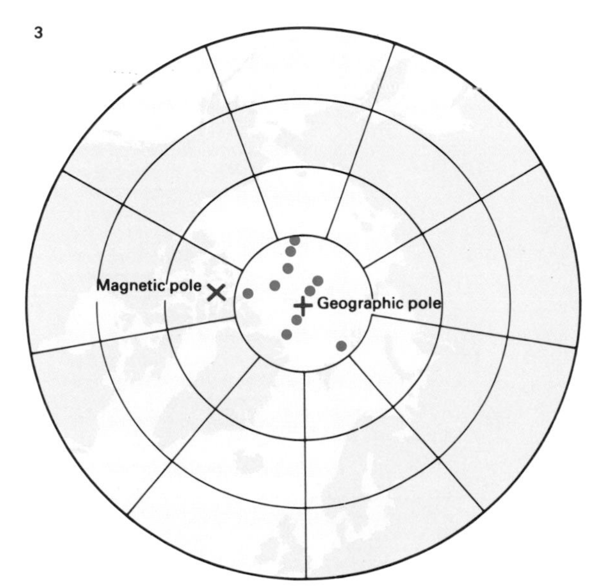

All of today's continents were once joined together in one supercontinent, Pangea, surrounded by a universal ocean, Panthalassa. Pangea was an awkwardly shaped landmass partly divided into two by a triangular arm of Panthalassa known as the Tethys Sea. Pangea began to disintegrate about 200 million years ago, having possibly remained intact for at least several hundred million years before that. The first split occurred at the apex of the Tethys where Pangea was particularly narrow, producing two lesser supercontinents, Laurasia and Gondwana. Laurasia, to the north, comprised what are now North America, Europe, and most of Asia. Gondwana, to the south, included South America, Africa, Australia, Antarctica and India.

By about 180 million years ago, Gondwana had begun to break up into South America-Africa, Australia-Antarctica, and India. By about 135 million years ago, the South Atlantic had begun to open between South America and Africa; and India had begun an unusually rapid northward journey towards Asia, with which it was to collide and cohere about 30 million years ago. Australia and Antarctica did not separate until about 45 million years ago, and North America did not finally break away from Europe until 5-10 million years later. Thereafter the separate continents slowly drifted to where they are now, having, since they left Pangea, moved northwards (with the exception of Antarctica) and undergone various degrees of rotation.

Africa, which has never lain far from Eurasia, has been moving northwards relative to Europe and Asia for some time. It has already joined its northern neighbor in the Middle East and is close to doing so at the tip of Spain. As this continues, the Mediterranean gets narrower and narrower and will ultimately disappear. But the classic case of continental collision is India, which is now welded to Asia. In its flight from Gondwana to its present home, India traveled at speeds of up to 17cm a year, although continental drift rates are more usually in the range 0-7cm a year.

◀ When a rock is formed its magnetic particles will be aligned along the magnetic field of the Earth. This alignment wil indicate the position of the Earth's magnetic pole at the time of formation. When this is done for many rocks of different ages it is found that the position of the pole has changed throughout time – it seems to have traced a route around the globe, called a "polar wander curve". Different continents have different polar wander curves (1). This shows that the different continents have moved in different directions relative to the pole and one another. Australia, for example, can be shown to have moved from the region of the South Pole to its present position in the last 300 million years (2). There is, inevitably, some error involved, but when a similar study is done on rocks formed recently (the last 7,000 years) the poles obtained lie close to the present pole (3).

Paleomagnetism: Proof at last

During the late 1950s, Earth scientists found that they could measure the weak magnetizations locked into rocks (▶ pages 45-56). When a rock forms, any magnetic particles in it will orientate themselves along the Earth's magnetic field. Once a physicist has measured the orientation of a rock's magnetic particles it is possible, using simple trigonometry, to determine the latitude at which the rock was formed and the past orientation of the continent upon which it lay. These readings can give the position of the North Magnetic Pole at the time. Paleomagnetic data can be plotted as ancient (north) pole positions – paleomagnetic poles – on maps of the world.

When earth scientists plotted the paleomagnetic north poles obtained from older and older rocks from a single continent, they found that the poles lay along a smooth curve – a polar wander curve – leading away from the present pole.

Two interpretations were possible. Either the magnetic pole itself had gradually moved to its present position, or the continent had moved instead.

When polar wander curves are constructed for other continents, however, they do not coincide. As there can only be one North Magnetic Pole at any instant, the diverging curves must indicate that the two continents have moved with respect to one another in the past.

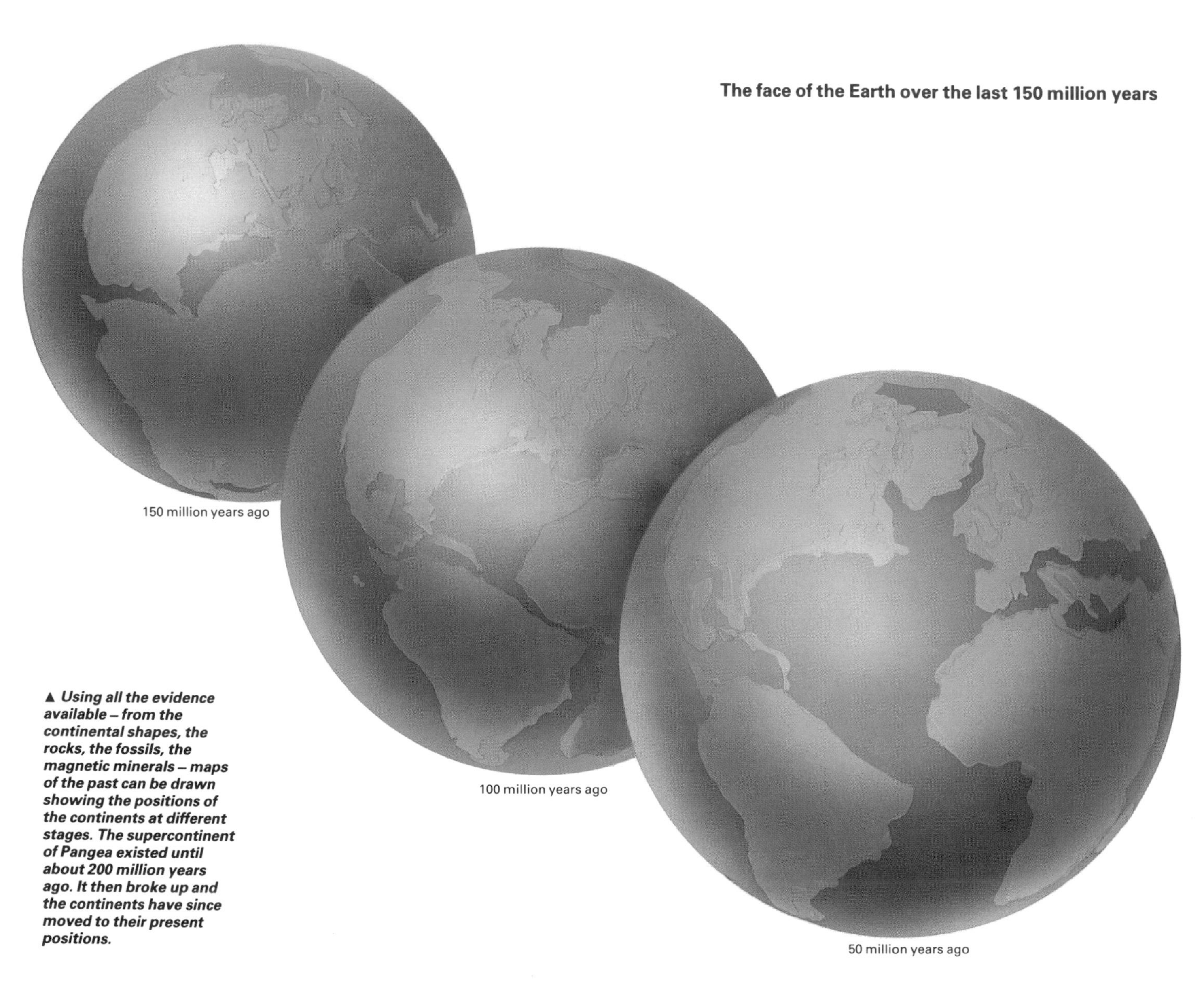

▲ Using all the evidence available – from the continental shapes, the rocks, the fossils, the magnetic minerals – maps of the past can be drawn showing the positions of the continents at different stages. The supercontinent of Pangea existed until about 200 million years ago. It then broke up and the continents have since moved to their present positions.

◄ ► When a supercontinent breaks up into smaller landmasses it does so along a series of cracks, or faults, at the weak points. The land bounded by the faults subsides, forming a rift valley. The Great Rift Valley of East Africa (left) is such a structure, stretching from East Africa, through the Red Sea to the Jordan Valley. It shows that continental drift is still continuing at the rate of a few centimeters per year. In 50 million years time it may be that the Great Rift Valley will have widened to a structure like the Red Sea, and that East Africa will have begun to split away into a separate continental mass, as Madagascar did 50 million years ago (right).

► A continent may pull apart in response to an upward movement of material below the continental crust. The rocks in the area are split by faults, and a rift valley forms. All this is accompanied by volcanoes, with molten material pushing its way up the faults (1). As the two parts of the continent move away from each other they take the fault and rift structures with them, and these are eventually covered by sediment as the sea floods into the gap between (2). The Atlantic coasts of Africa and South America show these features indicating that they originally formed along a rift valley (3).

The rifting of a continent

1

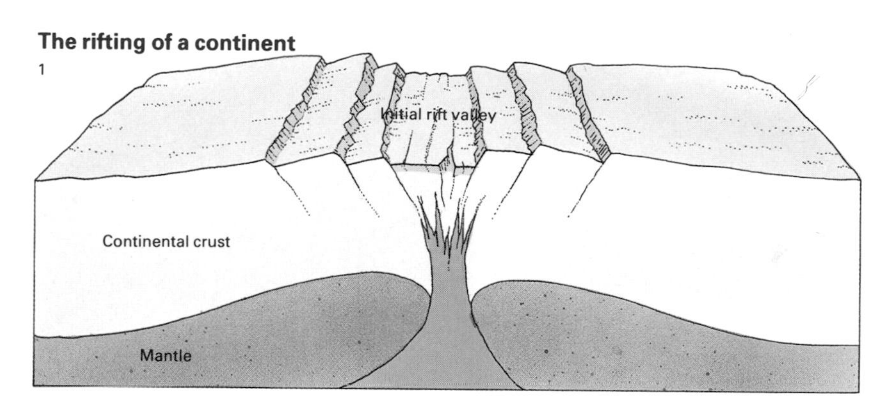

2

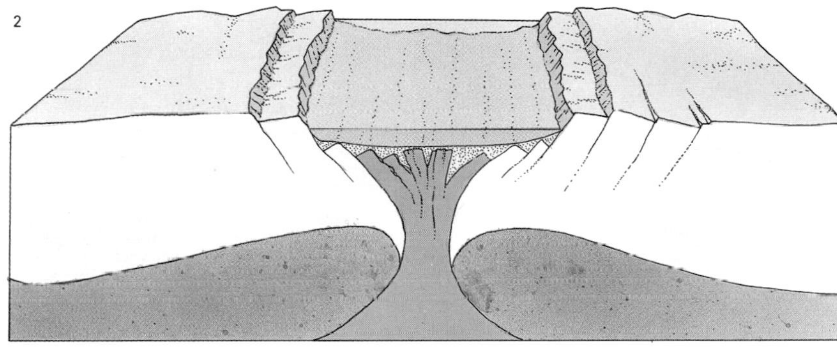

3

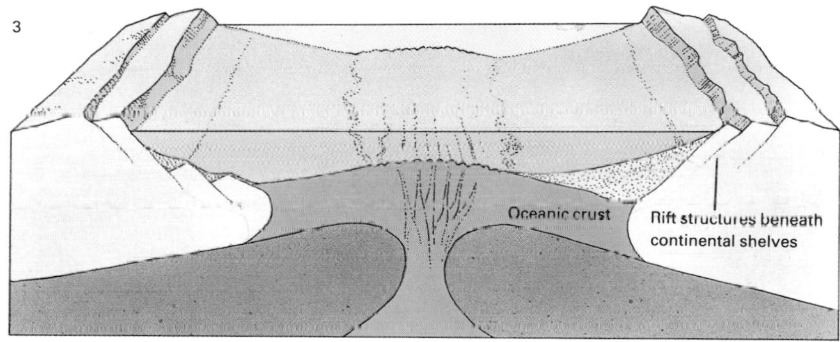

Today

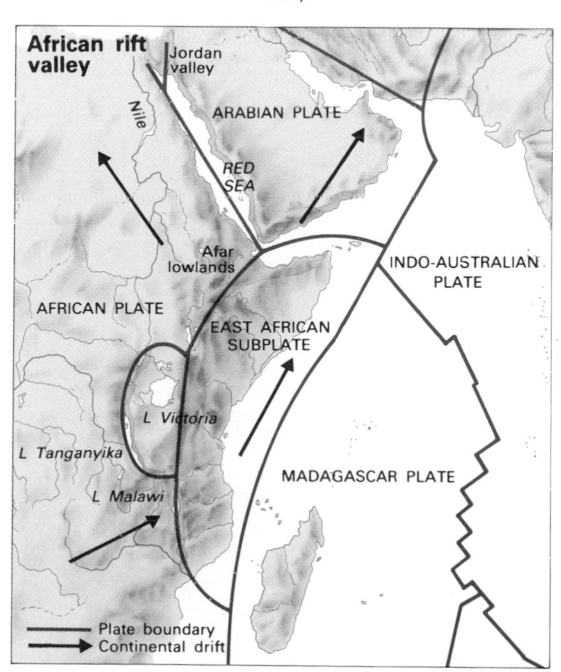

The evidence that proved beyond all reasonable doubt that the continents have drifted came during the late 1950s and early 1960s from paleomagnetism. Before that, all but the most pro-drift of geologists had regarded the more traditional data from geology, paleontology and paleoclimatology as, at best, equivocal and, at worst, negative. Once proof of drift had come from a field somewhat outside geology, however, geologists began to take seriously the idea of moving continents.

Thus continental drift puts beyond mere coincidence the good geographical matching of many pairs of coastlines (strictly, continental shelf edges) and the almost perfect fit of some. It explains why tectonic features and rock formations of particular ages are sometimes continuous between continents now divided by vast expanses of water. It resolves the problem of why certain flora and fauna which have undergone identical evolutionary development are now found in fossil form on widely separated landmasses. And it provides a reason for the distribution of different climate types in different parts of the world at different stages of geological time. On the whole continental drift provides a more self-consistent explanation for the geographical disposition of rocks, flora and fauna than does a set of stationary landmasses.

The new science of plate tectonics encompasses the old notion of continental drift, and the younger ideas of seafloor spreading

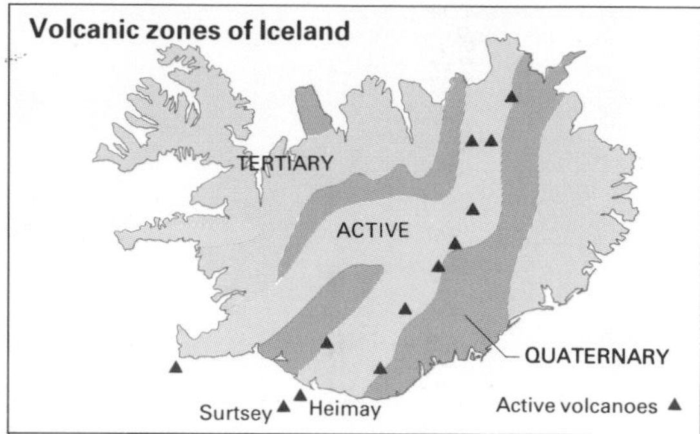

◄ ▼ Iceland is one of the few places in the world where an oceanic ridge reaches the surface of the sea. The North Atlantic constructive plate margin can be traced across the island by the zones of current volcanic activity and the rifting structures that are found there. The entire island is volcanic in origin.

Magnetic messages

Paleomagnetic studies soon led to the surprising discovery that, although rocks become magnetized parallel to the local direction of the Earth's magnetic field at the time of their formation, only 50 percent of rocks are magnetized in the same direction as the field. The other 50 percent are magnetized in precisely the opposite direction. These are called normally and reversely magnetized rocks respectively. Paleomagnetists therefore came to the conclusion – somewhat reluctantly at first, for they could hardly believe it – that at the times that the reversely magnetized rocks formed, the Earth's field itself must have been reversed. The field is continually switching from one polarity to the other. With the field in a reversed state, a compass needle of the type in use today will point approximately south rather than towards north.

When molten material from the mantle rises at oceanic ridges to form new oceanic lithosphere, it cools, solidifies and becomes magnetized in the Earth's field direction. It then spreads away from the ridges in both directions, allowing new material to rise between. Successive sections of the lithosphere therefore become magnetized first in one field direction and then the other; the lithosphere records continuously the alternations of the field. In short, the ocean floors are not only conveyor belts, they also act as tape recorders. The alternating sections of normally and reversely magnetized rock making up the lithosphere then act in effect as large magnets, producing their own weak fields at the Earth's surface. Sensitive instruments towed behind survey ships can detect these fields and plot the positions of the magnetized rocks.

The interpretation of oceanic magnetic anomalies in 1963 by the British geophysicists Fred Vine and Drummond Matthews soon convinced most Earth scientists of the reality of spreading.

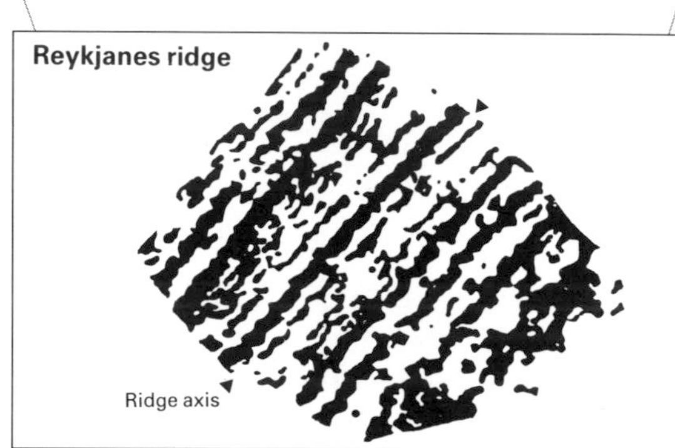

► We can see that new material is being produced on the North American and Eurasian plates at the mid-Atlantic ridge by the magnetic pattern of the ridge rocks. Magnetic surveys carried out south of Iceland show the rocks to be normally magnetized (black) and reversely magnetized (colored) in parallel stripes. Each ridge flank carries a pattern that is a mirror image of the other.

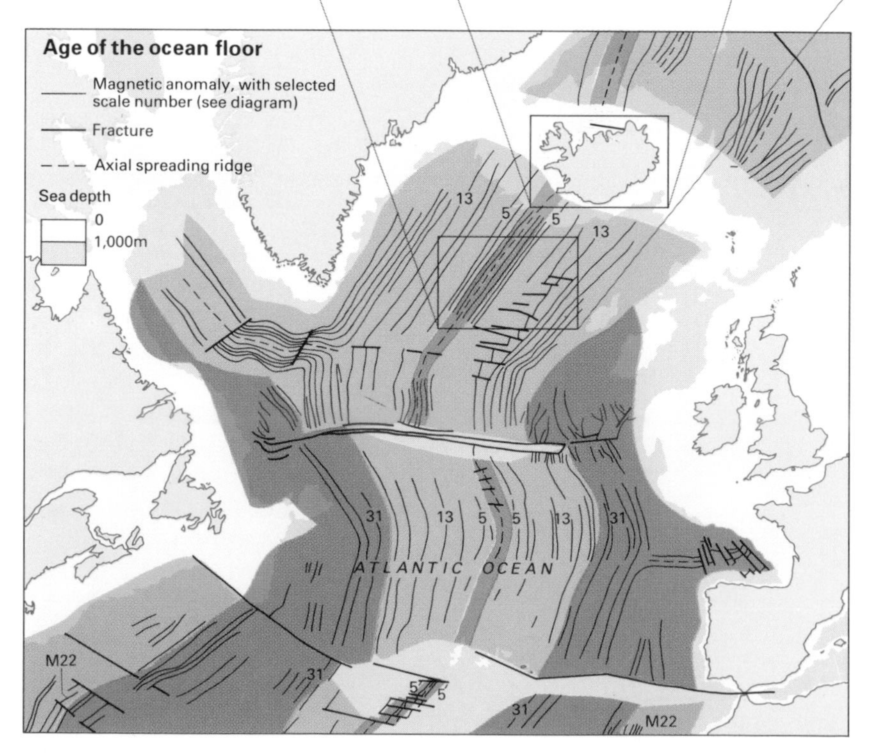

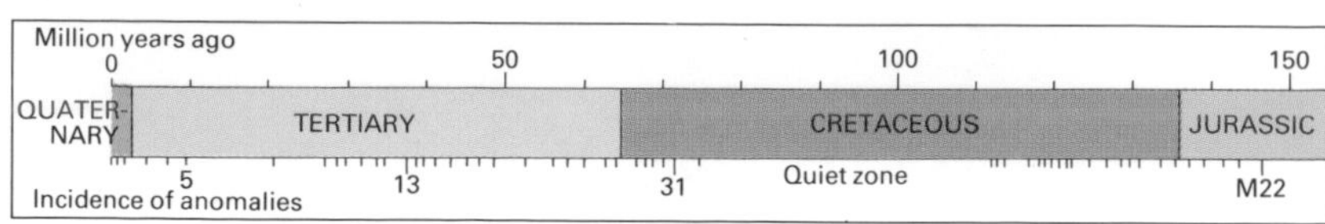

► Further proof comes with the dating of the rocks at each side of the mid-Atlantic ridge. The youngest rocks are found on the ridge axis itself. Farther and farther away from the axis there are older and older rocks. The oldest rocks are about 190 million years old, indicating the time that the Atlantic Ocean was born as a rift valley.

It is not just the continents that are moving – the ocean floor is moving as well. In fact the ocean floor is the prime mover, pushing and carrying the continents along. The surface layer involved is called the "lithosphere", a rigid outer shell about 75km thick comprising the crust and the uppermost layer of the mantle beneath it (➧ page 33). Through all the major oceans is a huge system of rugged mountain ranges more-or-less linked, about 80,000km long, and standing up to 3km above the adjacent ocean basins. The system has no single name. In the Central and South Atlantic it is the mid-Atlantic ridge, in the North Atlantic to the southwest of Iceland it is the Reykjanes ridge, in the Pacific it is the East Pacific rise, and so on. But whatever the various sections are called, they have one thing in common. In a narrow band along their axes, molten material from the asthenosphere (the mantle layer immediately below the lithosphere) is rising to form new lithosphere. When the rising magma gets closer to the surface it cools, solidifies, and moves sideways in both directions to make way for even newer material behind it. The ocean floors are thus in a state of continuous creation and spreading. The ocean ridges exist as mountains partly because of the general upward pressure of material from below but largely because of the expansion of the rocks due to the heat below. As new oceanic lithosphere moves away from a ridge it cools and contracts. As a result, its surface drops with respect to sea level, and the depth of water above it increases. Out to a distance from the ridge equivalent to a lithospheric age of about 70 million years, the depth to the ocean floor follows a remarkably simple law depending entirely on the contraction. The water depth is proportional to the square root of the lithosphere's age; so the age of the oceanic lithosphere near the ridge can be found simply by measuring water depth.

Assuming the Earth to be neither contracting nor expanding, oceanic lithosphere must be "destroyed" at precisely the same rate as it is being created. This destruction occurs along narrow regions known as subduction zones where the spreading lithosphere bends downwards, usually at an angle of something like 45°, reenters the Earth's interior, slowly melts, and becomes reassimilated. The surface expression of a subduction zone is often a long narrow trench in which water depths can exceed 10km, which compares to the 3-5km of the average ocean basin.

Which slab – or "plate" – of lithosphere is pulled down and which remains on the surface depends on circumstances. Oceanic lithosphere spreading eastwards from the East Pacific rise, for example, meets the west coast of South America and plunges down into the Earth's interior beneath that continent. The rule is that, because continental lithosphere is lighter than oceanic lithosphere, continents will always stay at the surface. On the other hand, there are some areas in which two regions of spreading ocean floors meet, in which case subduction occurs to the less vigorous of the two. Finally, if an area of oceanic lithosphere being consumed beneath a continental edge pushes or carries a landmass into the subduction region, there will be a continental collision. As neither landmass can be subducted, one tends to override the other and mountains are formed from both. When India collided with Asia, for example, the Himalayas began to form with India tending to underride the mainland. As the spreading ocean floor is still forcing India northwards, this process is continuing, thickening a continental lithosphere that is already about the thickest on Earth.

A direct test of spreading

Most Earth scientists were convinced by the magnetic evidence for ocean floor spreading, but they also wanted more direct proof. There was an obvious test. If the oceanic lithosphere is spreading, it will get older with increasing distance from a ridge. The opportunity to verify this came in 1969 when the United States initiated the Deep Sea Drilling Project, a long-term program to drill cores from the oceanic crust. From the data obtained it was clear that the age of the deepest sediment at any site is close to the age of the ridge-produced igneous rock immediately beneath it. The survey found that the ages increased with distance from the ridge. A survey in the Atlantic showed that during the past 80 million years, the South Atlantic has been spreading from each side of the mid-Atlantic ridge at a rate of about 2cm a year. This movement is still continuing today.

The buildup of the sands of time

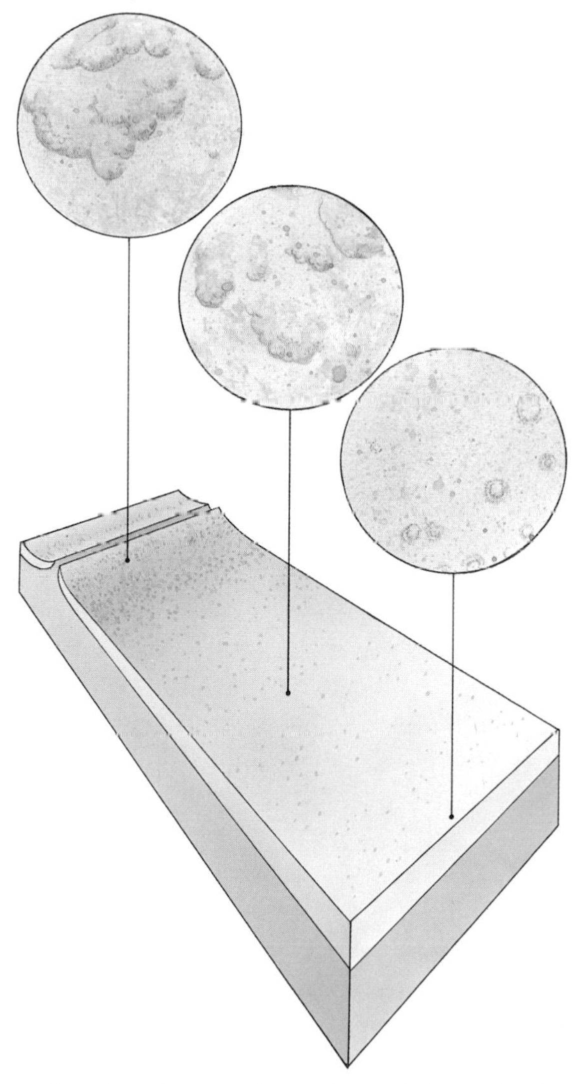

▲ *The aging of the ocean floor can be seen by the sediment upon it. Submarine photographs taken on the East Pacific rise show clean, sediment-free volcanic structures on the ridge axis. Five kilometers away the sediment is gathering in the hollows. At ten kilometers the structures are completely covered by a long-accumulating blanket.*

The skin of the Earth

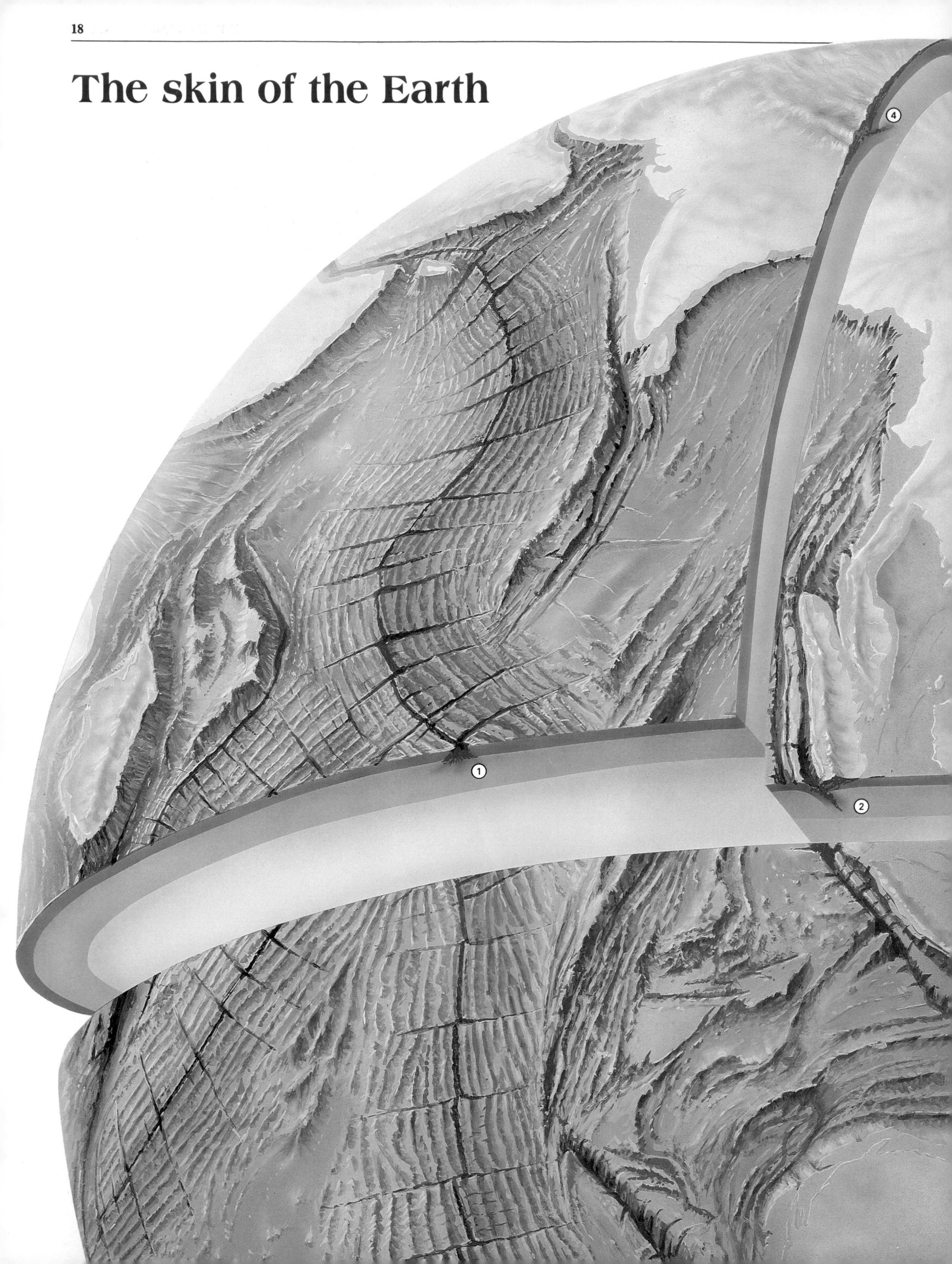

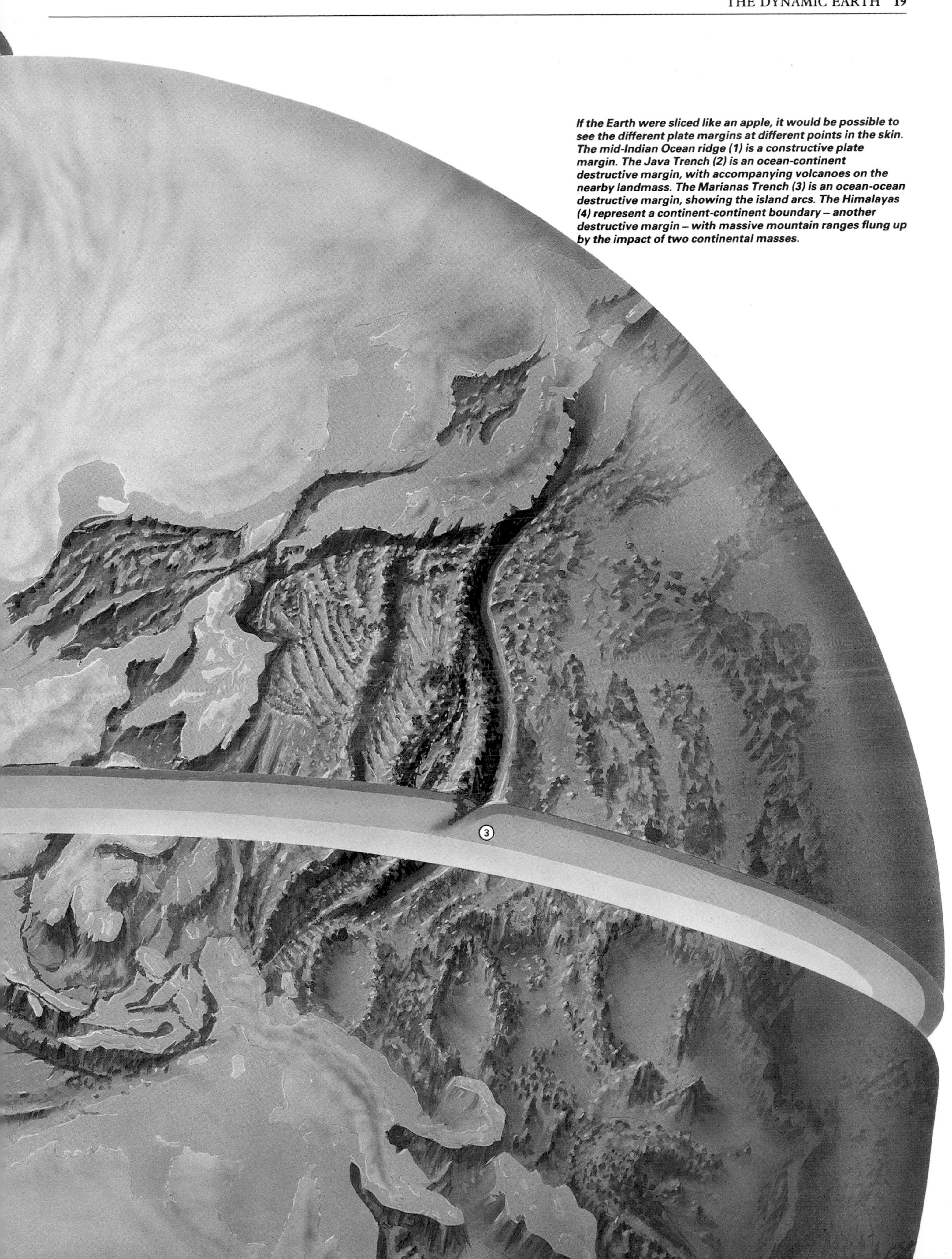

If the Earth were sliced like an apple, it would be possible to see the different plate margins at different points in the skin. The mid-Indian Ocean ridge (1) is a constructive plate margin. The Java Trench (2) is an ocean-continent destructive margin, with accompanying volcanoes on the nearby landmass. The Marianas Trench (3) is an ocean-ocean destructive margin, showing the island arcs. The Himalayas (4) represent a continent-continent boundary – another destructive margin – with massive mountain ranges flung up by the impact of two continental masses.

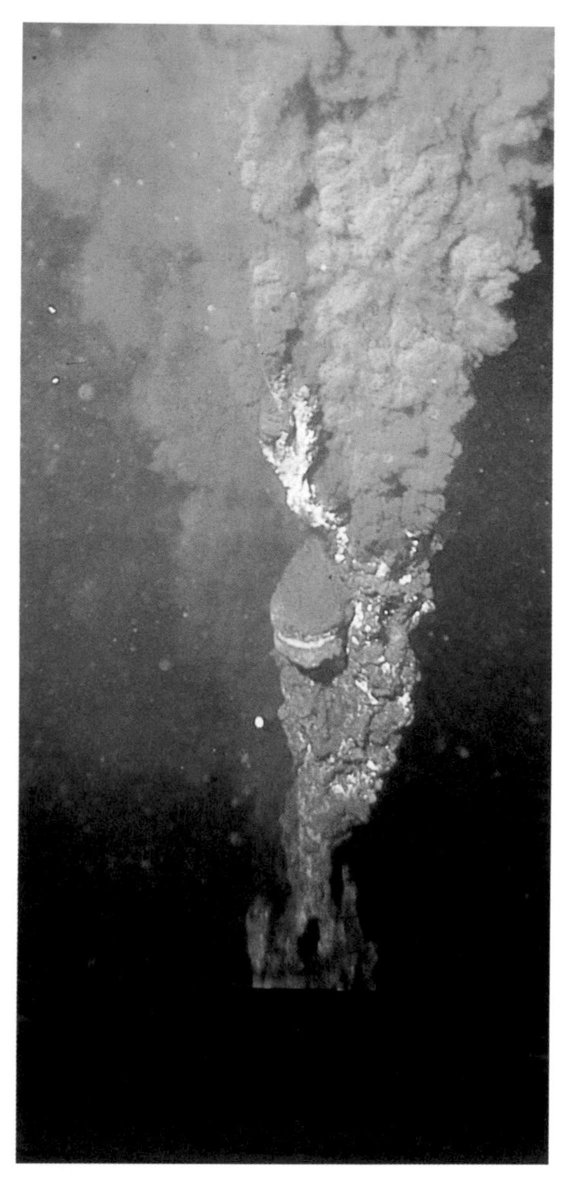

▲ ***The intense volcanic activity at constructive plate margins was unknown until a few decades ago. Only at isolated spots, such as Iceland and Tristan da Cunha in the Atlantic Ocean, does ridge material actually appear above the surface of the sea. Now, with the exploration of the ocean depths, the undersea volcanoes that encircle the world along the oceanic ridges can be observed directly. One manifestation of this activity is the presence of black smokers – hot springs of mineral-rich water that belch out of the ocean floor in places where the ridge is particularly active.***

Ocean floor spreading – the implications

Areas of continental collision are extremely complicated, as are oceanic subduction-trench zones. Geologists understand neither very well. Subduction zones are probably regions in which material from the descending oceanic lithosphere is metamorphosed and added to the edges and undersides of continents (➧ pages 101-8). They are certainly regions where, because of the intense temperatures and pressures generated by the downgoing slab, extensive volcanic activity is induced along the continental edge. Sometimes, however, the volcanic activity has moved (migrated or jumped) some distance away from the land, forming a volcanic island arc out in the ocean. In the marginal, or back-arc, basins then left between island arc and coastline there is often a mini-system of ocean floor spreading, apparently independent of the main spreading on the seaward side of the arc. But among the many uncertainties about ocean floor spreading and its consequences one thing is perfectly clear. None of the present oceans is floored by lithosphere more than 200 million years or so old. So all of today's ocean basins were formed since the breakup of Pangea, which leaves Earth scientists wondering what lay beneath Panthalassa.

Continental drift and ocean floor spreading have provided a new perspective on the Earth's basic surface divisions. Before the reality of these phenomena was accepted, the natural boundaries at the Earth's surface were seen to be the continental margins, some of which are tectonically active (for example, the west coast of South America) and some of which are passive (the east coast of South America). The discovery of ocean floor spreading, the new appreciation of the nature of oceanic ridges and the revised perception of continents as passive bodies in the global context, have made it clear that the more fundamental boundaries are the tectonic zones, irrespective of their positions in relation to the continents.

According to the theory of plate tectonics, the narrow tectonic bands at the Earth's surface are the boundaries between blocks, or plates, of the lithosphere. There are 15 major lithospheric plates of various sizes, although there are also a number of more minor ones (microplates). A few of the plates (for example, the Pacific) are almost entirely oceanic, but most incorporate significant proportions

Visiting the ridges

Most exploration of the ocean floor and the layers beneath it, including drilling of the crust, is carried out from ships at the surface. But in the years from 1971 onwards a team of French and American scientists used submersibles to make a series of personal visits to the rift valley that runs along the axis of the mid-Atlantic ridge, specifically to a section of the rift about 650km southwest of the Azores (Project FAMOUS). They found the floor of the narrow rift zone to be covered with fresh lava flows in the form of pillow lavas, the globular shapes that lava often takes when rapidly cooled under water (➧ pages 64-5). Moreover, the valley floor was riddled with fissures up to 10m wide and oriented parallel to the ridge axis as if the bed of the sea were being pulled apart by enormous stresses. The scientists also discovered the heat flow through the floor to be very high, implying the presence of hot magma close to the suface. All this is consistent with the rift zone being a site where magma upwells regularly and the lithosphere is being pulled apart.

The discovery of "smokers"

Since 1977 the Americans have been using the submersible Alvin to make similar visits to sites along the East Pacific rise. Because the East Pacific rise is spreading three times faster than the mid-Atlantic ridge, it is less rugged and the rift is less well developed. The lava flows there are also smoother, although pillow lavas do exist. The most spectacular discovery made by Alvin, however, was the presence of strong jets of hot water emerging from vents in the ocean floor in very active areas. These jets have come to be called "smokers" because they appear as dense black or white clouds as a result of a heavy load of sulfides of such metals as iron, zinc, manganese and copper. Water penetrates the hot rocks in the rift zone, dissolves out the minerals, and redeposits some of them in columns, or "chimneys", around the vents.

The high chemical content of the water allows vast quantities of chemical-feeding bacteria to exist. These, in turn, give rise to all sorts of strange life forms that feed on them (➧ pages 121-8).

of both oceanic and continental lithosphere. No plate is solely continental. An important property of plates is that they have a high torsional rigidity, which means that they are difficult to distort in a horizontal direction, but a low flexural rigidity, which means they can easily be distorted vertically. A reasonable analogy would be a piece of paper floating on water, which exhibits precisely those properties. The analogy is even more exact in that the asthenosphere is mobile, probably because it is partially molten. The lithosphere can therefore be envisaged as floating on the asthenosphere. In greater detail, each plate can be envisaged floating on the asthenosphere, jostling against, and interacting with, its neighbors, which are doing likewise.

▲ *Some plates seem to be bounded entirely by constructive margins. Africa rests on one such. Only where it meets Europe is there a margin that can be thought of as destructive with its accompanying volcanic, earthquake and mountain-building activity. This means that all other plates, and all continents carried by them, must be moving away from Africa.*

Plate boundaries – the major types

There are three types of plate boundary. First, there are constructive boundaries along which new oceanic lithosphere is created. These are the oceanic ridges. Second, there are destructive boundaries along which old oceanic lithosphere is consumed. These are the subduction zones. The third type are the conservative boundaries, along which lithosphere is neither created nor destroyed. These are the "transform" faults – the zigzags along which the oceanic ridges are offset. Between the offset sections of ridge, the two adjacent plates are moving past each other, but no new material is being produced here. All three types of plate boundary generate earthquakes, but only constructive and destructive boundaries are associated with volcanic activity (➧ pages 25-32).

Any given plate is defined by a combination of at least two of the three types of boundary, but the particular combination may change with time. Because oceanic lithosphere is always being created and destroyed, the size and shape of a plate are continuously changing. The plate's interactions with its neighbors will therefore be perpetually modified, and this will inevitably mean that at some time or other the nature of its boundaries will change. Incidentally, because of the impermanence of the oceanic parts of plates, Earth scientists are going to find it very difficult, if not impossible, to work out the global tectonic history of the Earth farther back in time than the past 200 million years or so.

Faulting across ocean ridges

The ocean ridge system, although continuous on the large scale, actually comprises many short sections offset from each other by a special type of fault known as a transform fault. The section of each transform fault that lies between the adjacent sections of ridge is active. Here two plates move against each other and earthquakes occur. Beyond the ridge segments the fault is passive, since the material at each side of the fault belongs to the same plate and is moving in the same direction. Here there is no earthquake activity, although there may be some minor movement and fracturing as the two sides of the fault settle down to a common surface level. Transform faults arise because the geometry of the spreading process makes it impossible for an oceanic ridge to trace out a curved path on the Earth's surface. The curve is taken up by the fault's offsets.

The zig-zag plate margin

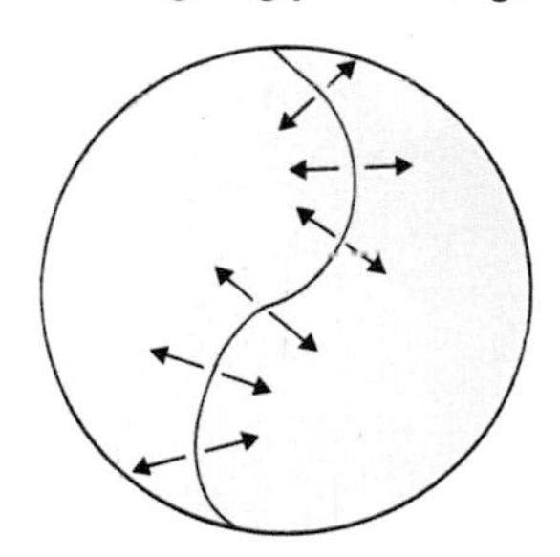

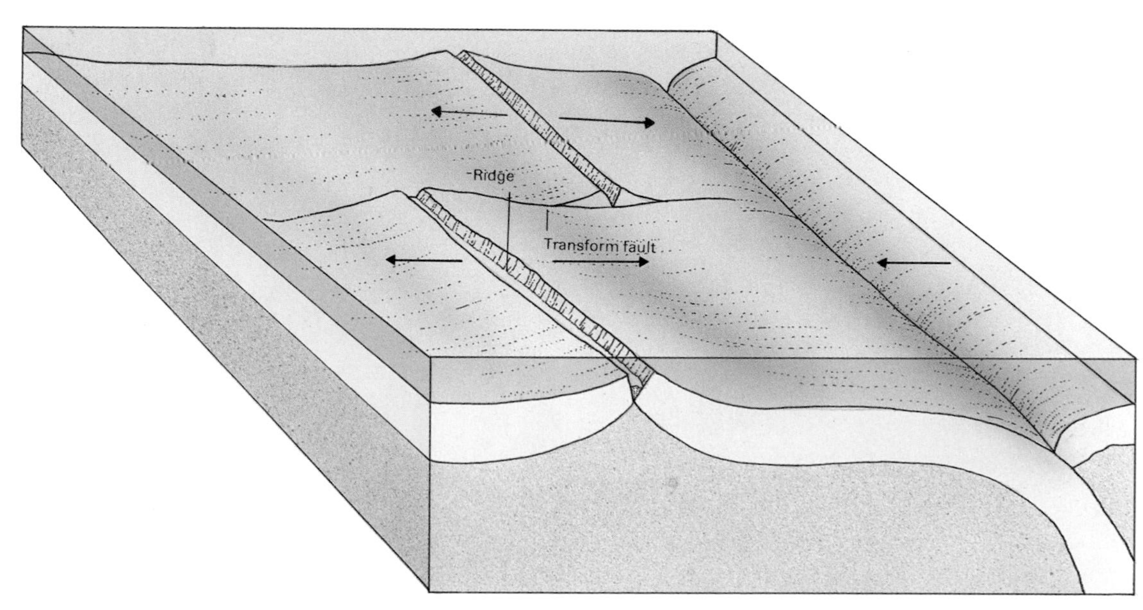

▲ ► *An oceanic ridge cannot be curved. Growing plates move directly away from their constructive margins at right angles to them. Growth from a curved ridge would involve too much stretching and crumpling. To form a curve, the shape is taken up by short straight stretches offset by transform faults.*

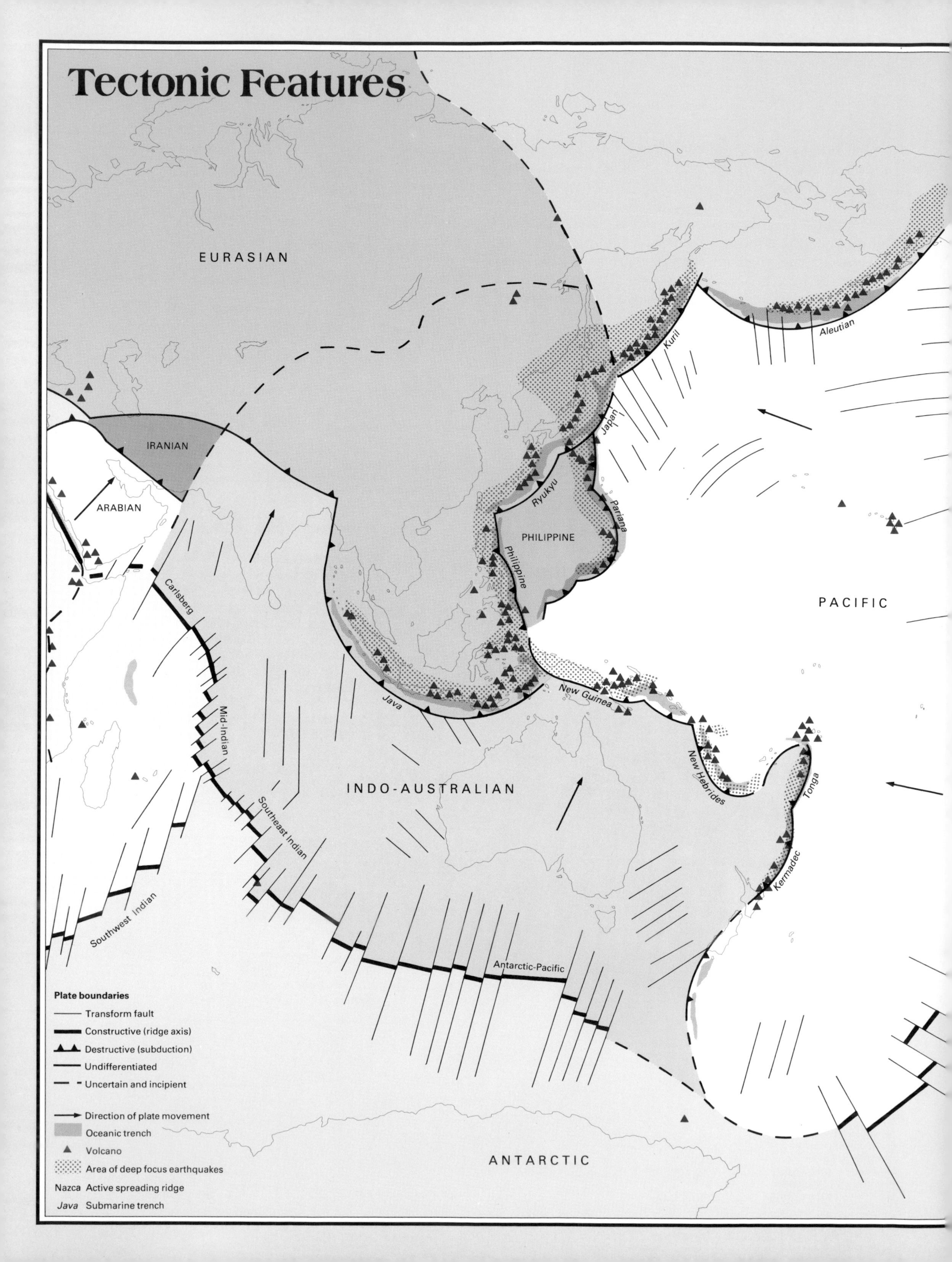
Tectonic Features
EURASIAN
IRANIAN
ARABIAN
PHILIPPINE
PACIFIC
INDO-AUSTRALIAN
ANTARCTIC
Kuril
Aleutian
Japan
Ryukyu
Mariana
Philippine
Carlsberg
Mid-Indian
Java
New Guinea
New Hebrides
Tonga
Kermadec
Southeast Indian
Southwest Indian
Antarctic-Pacific
Plate boundaries
Transform fault
Constructive (ridge axis)
Destructive (subduction)
Undifferentiated
Uncertain and incipient
Direction of plate movement
Oceanic trench
Volcano
Area of deep focus earthquakes
Nazca Active spreading ridge
Java Submarine trench

NORTH AMERICAN
EURASIAN
Reykjanes
JUAN DE FUCA
HELLENIC
Puerto Rico
CARIBBEAN
COCOS
Cocos
Mid-Atlantic
AFRICAN
NAZCA
SOUTH AMERICAN
Peru-Chile
Nazca
East Pacific rise
South Sandwich
SCOTIA
ANTARCTIC

See also
Earthquakes and Volcanoes 25-32
Internal Structure 33-44
Dating the Rocks 73-80
The Earth Through Time 81-100

The mobility of the asthenosphere gives the overlying lithospheric plates the freedom to move horizontally. But the existence of that freedom does not ensure that the plates will actually move. A driving force is required. One of the earliest suggestions was that the release of heat by the decay of radioactive elements in the mantle, might set up thermal convection currents in the asthenosphere, or perhaps even throughout the whole mantle, and that these currents would then exert a drag on the base of the lithosphere, pulling it along. This remains a possibility, although Earth scientists now believe that there are bigger forces available in the oceanic lithosphere itself.

First, because the oceanic ridges stand high and the lithosphere slopes downwards away from them, there is a gravitational sliding force acting in the direction of spreading. This may be regarded as a push away from the ridges and is therefore called the ridge-push force. Second, when the oceanic lithosphere begins to descend into the Earth's interior at a subduction zone, it has become cold and dense – certainly colder and denser than the asthenosphere into which it is descending. As a cold lithospheric slab can move down into the Earth through up to 700km before it finally melts and becomes assimilated, it can exert a considerable pull on the lithosphere still on the surface. This is the slab-pull force. Third, there is a suction force tending also to pull the plate into the asthenosphere.

Earth scientists have tentatively concluded that the slab-pull force is the largest by far. This does not mean that the other forces play no part at all but only that, if they do act, their effect is small in comparison. As well as the driving forces, however, the plates are also subject to resistive forces. For example, assuming that the plates are not being driven by convective currents in the asthenosphere, when a plate is pulled over the asthenosphere by some other force the asthenosphere will exert a viscous drag on the base of the lithosphere. This will tend to slow the plate down. The downgoing slab will also meet resistances, there will be frictional resistance along transform faults, and there will be obvious resistance when two continents collide. Plate motions are thus governed by a complicated combination of both positive and negative forces, and the particular combination will vary from plate to plate and from time to time.

A plate tectonic lacuna

Although a high proportion of Earth scientists, including those in China, accept the basic tenets of plate tectonics, there is one country in which this is conspicuously not the case – the USSR. The reasons for this have to do with politics, human nature and the organization of science in that country.

During the 1930s, the Soviet geologist M.A. Usov, following Josef Stalin, made it clear that ideas originating in capitalist countries were not to be tolerated and that scientific theories must be seen to conform to the Marxist theory of history, to dialectical materialism. In geological terms that came to mean dedication to theories involving the domination of vertical movements of the Earth's crust, in which there is "a struggle between two conflicting factors...compression and expansion" and in which "periods of suppressed struggle...are followed by a revolutionary phase".

A product of that time was Vladimir Beloussov who, during the 1960s and 1970s, became the most influential geologist in the USSR, holding a number of major positions. Aided by the monolithic nature of the Soviet scientific establishment and isolationism (Soviet scientists were seldom free to attend western conferences or to see western scientific periodicals), Beloussov managed to suppress most discussion of plate tectonics and maintain support for his own theory of "oceanization" in which the ocean basins are formed by a "foundering" of continental crust and its subsequent flooding by oceanic magmas.

During the late 1970s there were vague signs of change. Plate tectonic ideas began to make a tentative appearance in Soviet publications, although all reference to western plate tectonic terminology was eschewed. Geologists from the USSR were for a few years participating members of the Deep Sea Drilling Project. But by and large Soviet academic geologists continue to disregard the discoveries that have revolutionized Earth science everywhere else.

Forces on a plate

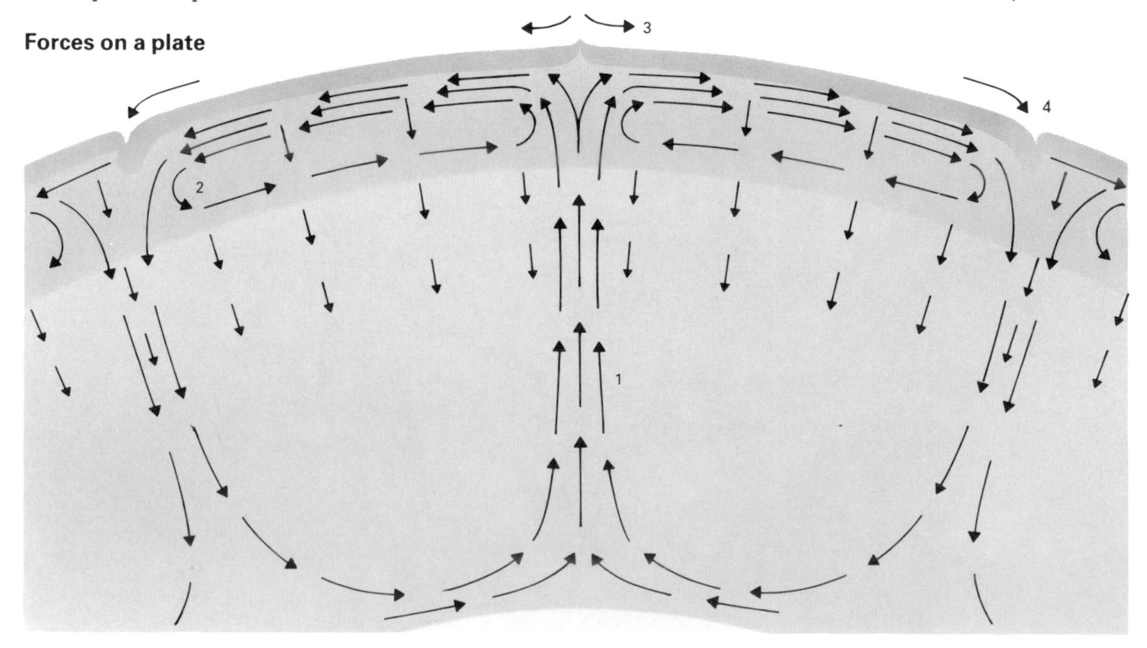

◄ The plates move, we know that. What geologists do not know is the force that moves them. It has always been believed that convection currents in the mantle would do the job, just like the movement of scum on the surface of boiling jam in a pan. There are two types of possible convection currents. One circulates throughout the whole mantle (1). The other is resricted to the asthenosphere (2). There are also important surface processes. A ridge-push force (3) is produced by the weight of plate material bearing down from an oceanic ridge. A slab-pull force (4) results from the cold edge of a plate sinking into the asthenosphere at a destructive margin, pulling the rest of the plate along behind it.

Earthquakes and Volcanoes

2

Plate movements – the origin of earthquakes... Depths of earthquake foci – shallow, intermediate and deep...Measuring earthquakes – intensity and magnitude...Volcano types...Volcanic emissions – gases dust and bombs...PERSPECTIVE... Tsunamis...Benioff zones...Seismometer principles...Historic earthquakes...Volcanic hot spots...Historic eruptions

An earthquake is a sudden release of energy that has accumulated slowly in a local region of the Earth's crust or upper mantle (➧ page 34). The point (in reality, a volume of perhaps a few cubic kilometers) at which the energy is released is called the focus, or hypocenter, and the point on the Earth's surface directly above the focus is known as the epicenter. When an earthquake takes place it emits vibrations, or waves, which can cause considerable disruption in the vicinity of the epicenter, the extent of the damage depending on the size of the shock, its depth, and what man has built in the epicentral region. For example, if an earthquake focus lies just beneath a city, it is probable that buildings and civil engineering structures will be damaged or even destroyed, and that in the process people will be injured or killed. Irrespective of whether or not there are people and artefacts present, however, an earthquake is likely to produce fissures in the ground, cause uplift and subsidence, and trigger landslides and avalanches.

Tsunamis

When an earthquake occurs at the bottom of the sea the disturbance of material may have wide-reaching effects. The displacement of water will produce a wave, or a series of waves, that will spread out from the area. In the open sea these waves will travel at several hundreds of kilometers per hour but have an amplitude of only a meter or so. The distance between successive wave crests may be 150km. Hence ships, even quite close to the site of the original disturbance, may not notice their passage.

However, once the wave reaches shallower water it slows down and builds up to a towering wall of water, up to 30m high. Such an occurrence is sometimes called a tidal wave, but the correct term is the Japanese word "tsunami". Its effect can be particularly devastating to coastal communities.

▸ Any seabed disturbance caused by an earthquake or a volcano may produce a tsunami – a wave of water that may build up and sweep inland causing terrible damage.

▾ Earthquakes and volcanoes are usually found in definite zones on the Earth's surface. These zones coincide with the active plate margins, both constructive and destructive, and the transform faults. This is no coincidence, since it is the movements of plates at these margins, as they pull apart or grind over one another, that give rise to the phenomena of earth tremors and volcanic eruptions.

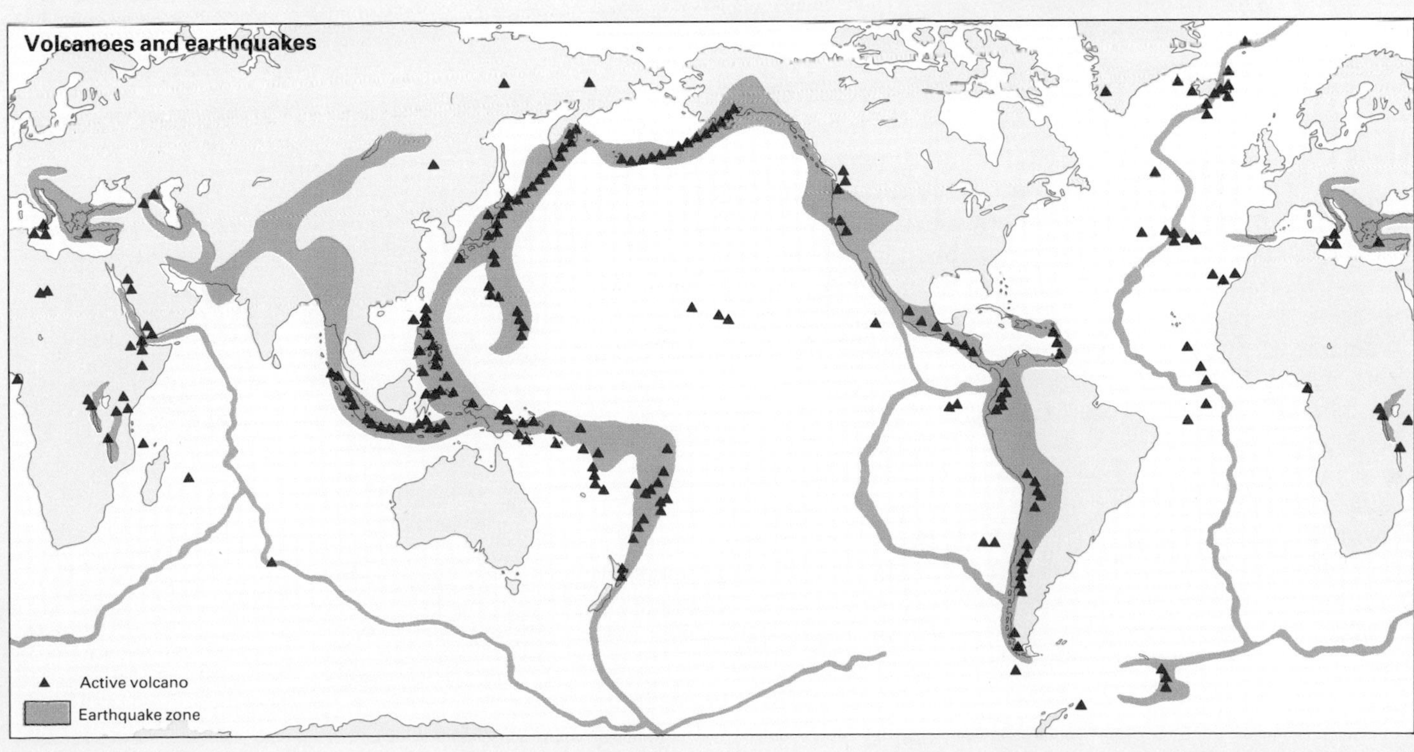

The largest recorded earthquake was in 1960 in Chile – it had a magnitude of 8·9 – as powerful as a 100 megaton atomic explosion

Benioff zones

When the oceanic lithosphere descends into the Earth's interior at subduction zones, strains are set up which, when released, generate earthquakes. Because the lithosphere moves downwards at an angle of about 45°, the earthquake foci lie along a band at the same angle. This characteristic earthquake pattern, which is one of the most important pieces of evidence in favor of subduction, is very well illustrated by the Tonga region of the South Pacific. Away from the Tonga trench in a roughly NW direction, the earthquake foci become deeper and deeper as the lithosphere reaches greater and greater depths. This phenomenon was discovered by Hugo Benioff, the Californian seismologist, which is why subduction zones are now often called Benioff zones.

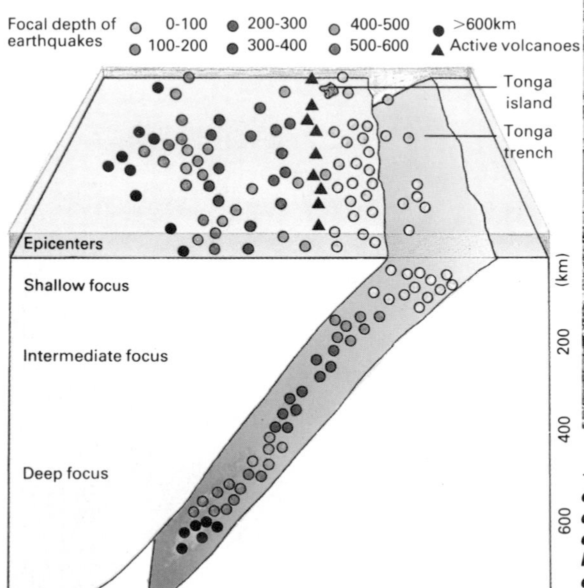

◄ Measuring the depths of earthquake foci at destructive plate margins demonstrates that they lie on a sloping plane corresponding to the plane of the descending plate.

▼ The San Andreas fault is the source of many serious earthquakes in California. It can be an obvious landscape feature, as here where it crosses residential San Bernardino.

▲ When earthquakes occur in populated areas destruction and loss of life can be great. An earthquake of intensity 10 on the Mercalli scale produced this damage in Sicily.

Most earthquakes derive their destructive power from the motions of the Earth's lithospheric plates. As the plates interact with each other along their margins, strain is built up until some of the rocks involved can take it no more. They then suddenly rupture, generating seismic activity. Most of the world's earthquake epicenters therefore lie along narrow bands that coincide with plate boundaries – the oceanic ridges, transform faults and subduction zones. On the other hand, some earthquakes, and not always small ones, occur within plates, far from plate edges. In North America, for example, the most damaging earthquakes in historic times have taken place not in California, through which runs the San Andreas fault, but in the eastern part of the country, in the Mississippi Valley and South Carolina. Geologists have only the haziest ideas of why and how such mid-plate shocks occur, suspecting that they are due to deeply-buried and not yet fully-deactivated faults left over from an earlier phase of plate tectonics.

In the vertical direction, earthquakes are classified as shallow if their foci lie at depths of 0-70km, intermediate in the depth range 70-300km, and deep if they occur below 300km. The deepest shock ever recorded occurred at a depth of 720km beneath the Flores Sea in the East Indies in 1934. Earthquakes do not occur at *all* depths down to 700km, nor are they distributed uniformly. Beneath Japan they occur

▼ *An earthquake of a particular magnitude will produce effects of differing intensities in different places. This will be due to such factors as distance from the epicenter and the stability of the soil and rocks. In this Californian earthquake the greatest intensities were in a zone parallel to the San Andreas, where the movement was greatest. The contours describe five points based on the Modified Mercalli Scale.*

Modified Mercalli Scale

1 Not felt except by few.
2 Felt by few at rest. Delicately suspended objects swing.
3 Felt noticeably indoors. Standing cars may rock. Duration estimated.
4 Felt generally indoors. People awakened. Cars rocked. Windows, etc. rattled.
5 Felt generally. Some plaster falls. Dishes, windows broken. Pendulum clocks stop.
6 Felt by all. Many frightened. Chimneys, plaster damaged. Furniture moved. Objects upset.
7 Everyone runs outdoors. Felt in moving cars. Moderate damage to structures.
8 General alarm. Very destructive and general damage to weak structures. Little damage to well-built structures. Walls, monuments down. Furniture overturned. Sand and mud ejected. Changes in well-water levels.
9 Panic. Total destruction of weak structures. Considerable damage to well-built structures. Foundations damaged. Underground pipes broken. Ground fissured and cracked.
10 Masonry and frame structures commonly destroyed. Only best buildings survive. Foundations ruined. Ground badly cracked. Rails bent. Water slopped over banks.
11 Few buildings survive. Broad fissures. Fault scarps. Underground pipes out of service.
12 Total destruction. Acceleration exceeds gravity. Waves seen in ground. Lines of sight and level distorted. Objects thrown in air.

How earthquakes are measured

Until the 1930s, the only way in which seismologists could measure an earthquake was to observe its effects at the surface. They would assess the degree to which shaking was perceptible, estimate the amount of damage done, and measure the visible deformation of the Earth itself. They could then express the power of the earthquake at various distances from the epicenter on an arbitrary scale known as an intensity scale.

Earthquake intensity scales have been in use since 1811, but they have been continually improved. That used today in most countries is the Modified Mercalli, which has 12 points (I-XII) covering the whole range of earthquakes experienced. Seismologists plot the intensities of an earthquake on a map and join the points of equal intensity by lines known as isoseismals. If the ground in the area were uniform, the isoseismals would be circles about the epicenter, but this is never the case in practice. There are irregularities depending upon crustal conditions. For example, an earthquake will usually have a more destructive effect (and hence the intensity will be higher) where the ground is soft than where there is hard rock.

at many depths, but there are far more deep earthquakes in the 300-400km range than in any other 100km interval. By contrast, in the San Andreas fault zone of California almost all seismic events occur in the upper 20km, and beneath South America there are hardly any shocks at all between 300km and 550km. Around the world as a whole, there are three times as many intermediate earthquakes as there are deep ones, and about ten times as many shallow events.

Nor are earthquake epicenters distributed uniformly along the plate boundary regions. Almost all deep earthquakes, 90 percent of intermediate ones and 75 percent of shallow ones occur around the margin of the Pacific Ocean, the circum-Pacific belt. Most of the remaining big earthquakes take place in the Alpine-Himalayan belt. The oceanic ridges and transform faults, on the other hand, are generally associated with the smaller and shallower events. In fact, the world's earthquake activity is heavily biased in favor of shallow shocks; they release about 75 percent of seismic energy compared to only about 3 percent from deep events. Moreover, it is the shallow earthquakes that cause most damage. The deepest earthquake known to have caused damage and death (about 1,000 people) was that in Romania in 1940, which had a recorded depth of about 160km, but this was very exceptional.

The earthquake most destructive of human life killed 830,000 people.

Seismic risk maps
In earthquake-prone countries, it is useful to have available maps showing how the degree of seismic danger varies from place to place, before planning civil engineering projects. Such seismic risk maps can take various forms depending on the scale. A local risk map covering a small area within a known earthquake region would try to indicate the potential damage from such an event on a zone-by-zone basis. For this purpose intensity studies of recent earthquakes are essential, for they take into account varying ground conditions and other factors that could have an influence on the amount of damage done by an earthquake. On the other hand, a seismic risk map of a continent might seek simply to show the probability of an earthquake occurring on a region-to-region basis. In the seismic risk map of the United States (right), the numbers indicate the maximum ground acceleration (as fractions of the acceleration due to gravity) to be expected from earthquakes over a 50-year period. Some of the danger zones are far from the San Andreas fault.

Earthquake risk

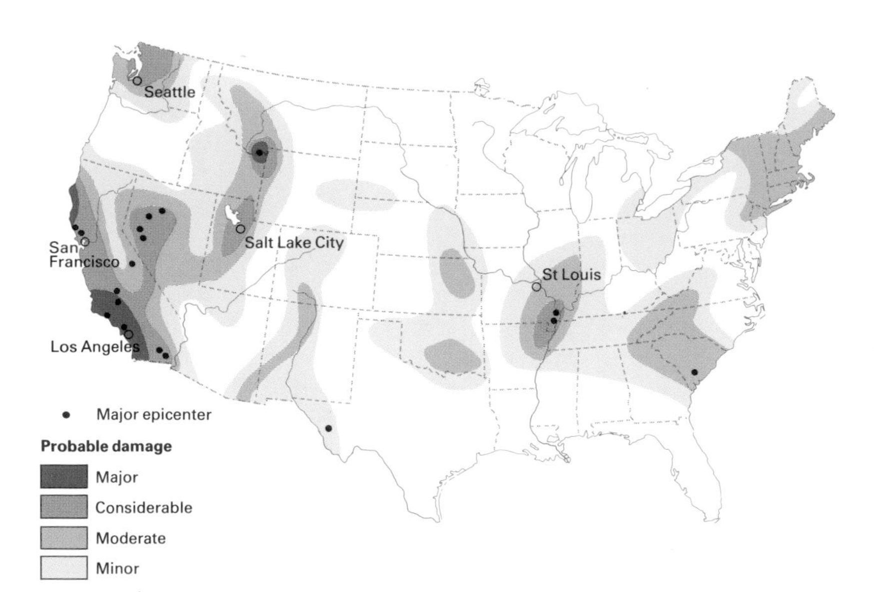

The seismometer principle
Earthquake waves are detected and recorded using seismometers, or seismographs, the principle of which is extremely simple (below). A rotating chart recorder is fixed to a rigid plate which is attached to the Earth. A heavy mass with a pen attached is then suspended from the frame by a spring or wire, with a pivot restricting motion to one direction. When the Earth and frame are jolted by a seismic vibration, the frame moves but the mass tends to remain stationary because of its inertia. There is thus relative movement between the pen and the rotating chart.

A seismometer station would normally have three instruments, two measuring horizontal motions in two directions and one measuring vertical motions. Modern instruments are still based on the first seismometer which used a weight, developed in the 1880s. The main difference is that the mass is magnetic and is suspended in an electric coil which is free to move. Relative movement between the mass and the coil sets up electric signals in the coil which are then amplified and recorded. Networks of seismometer stations exist across the world, and these can pinpoint the focus of any earthquake that occurs.

Magnitude & range of earthquakes

Magnitude	Annual average
8·0–8·9	<1
7·0–7·9	10–20
6·0–6·9	100–200
5·0–5·9	800–1,000
4·0–4·9	6,000–10,000
3·0–3·9	50,000–100,000
2·0–2·9	>300,000
1·0–1·9	>700,000

Recording earthquakes

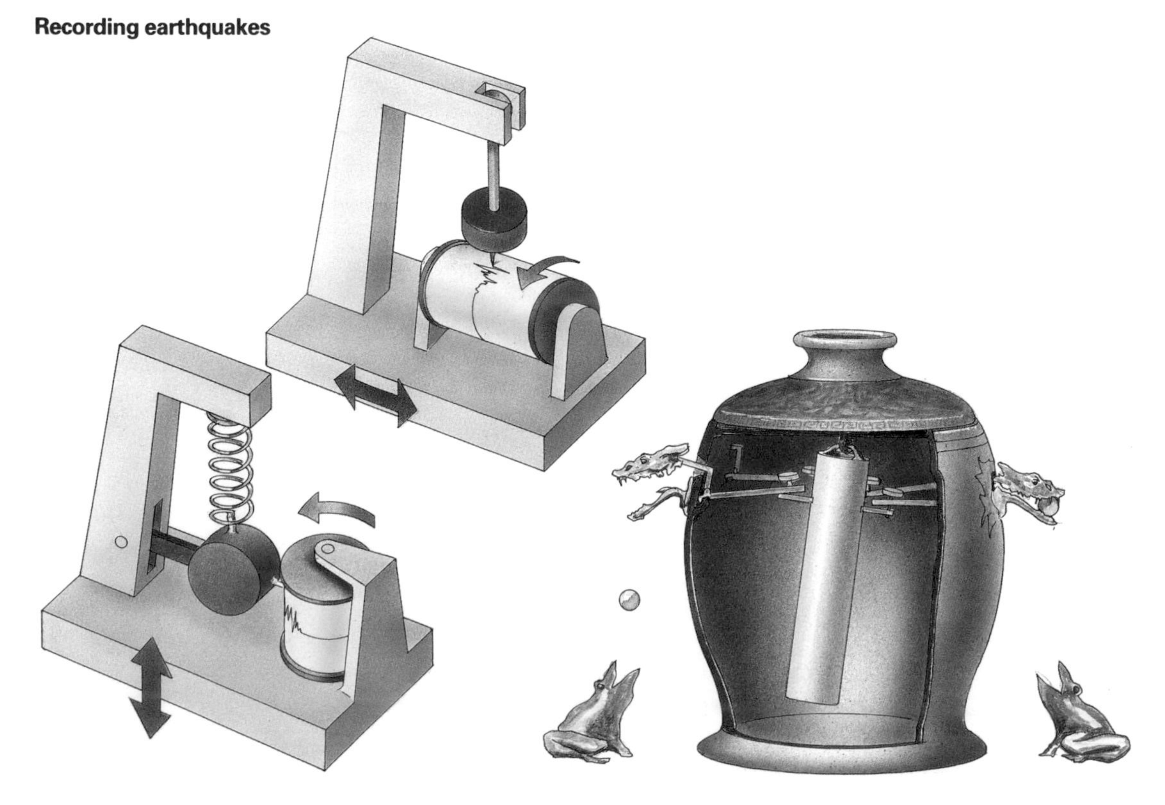

◄◄ The principle of the seismometer is very simple. A suspended weight that is heavy enough will remain stationary while everything else shakes around it during an earthquake. A pen attached to this stationary weight will then mark out a trace on a moving surface. The surface is usually a rotating drum that will produce a continuous trace. Seismographs are mounted in threes so that both vertical and horizontal movements are recorded.

◄ In this ancient Chinese earthquake sensor a swinging weight inside the dome knocked a ball out of a dragon's mouth. The position of the dragon indicated the direction of the earthquake.

► One of the most famous earthquakes of historical times was that which devastated San Francisco in 1906. Over 700 people died and thousands were made homeless by the event and by the fire that followed.

When seismologists refer to "the intensity of an earthquake" as a single figure, they mean the maximum intensity at the epicenter. Because intensity depends on much more than the energy released, however, it is not a very good measure of the absolute strength of an earthquake. A much more satisfactory concept from this point of view is that of magnitude, although intensity studies are still carried out because they make useful contributions to seismic risk maps.

Earthquake magnitude is a much more precise measure of the intrinsic size of an earthquake, for it ignores the destruction and just concentrates on the energy released. It is calculated from the amplitude of the waves emitted as recorded by seismometers, and is expressed on a scale ranging from zero upwards. Each successive point on the scale represents a factor of 10 increase in wave amplitude; so that, for example, a magnitude 5 earthquake emits waves 10 times that of a magnitude 4 event and 100 times that of a magnitude 3 event, and so on. In other words, the magnitude scale is logarithmic. Each point on this scale represents a factor of 30 increase of the energy released. The magnitude scale is sometimes called the Richter scale after its inventor, Californian seismologist Charles F. Richter.

The very biggest earthquakes ever recorded have magnitudes in the range 8·5-9·5. Earthquakes with magnitudes greater than 7·5 are called "great", those in the range 6·5-7·5 are "major", those between 5·5 and 6·5 are termed "large", and those of magnitudes 4·5-5·5 are "moderate". The rest are "small". Big earthquakes are less frequent than little ones. Each year more than a million earthquakes happen.

Some earthquakes, especially shallow ones, are clearly associated with the slipping of faults, some of which are visible at the Earth's surface and some of which may be traced underground. However, at depths of only a few tens of kilometers the pressures become so high that, in theory, no fault should be able to slip; the frictional forces are too great to permit it.

Historic earthquakes

The magnitude 8·3 San Francisco earthquake of 1906 is probably the most famous earthquake in history, not because it was the biggest or because it killed the most people but because it was the first great earthquake to occur in a western industrial nation after the birth of seismology in the late nineteenth century. It was thus the first seismic event to be studied in detail using scientific techniques. In fact, only about 700 people died, and most of those perished in the subsequent fire which also caused most of the $400 million worth of damage. The earthquake itself lasted less than a minute, although aftershocks smaller than the main shock rumbled on for months afterwards. The San Francisco earthquake was important geologically because it produced visible effects that clearly related it to activity along the San Andreas fault. Horizontal displacement, up to a maximum of 7m, took place along the 300-400km of the fault.

The earthquake most destructive of human life was the Shansi Province, China, event of 1556 (magnitude unknown) which killed 830,000 people, although China came close to that again when 650,000 perished in the magnitude 7·6 Tangshan Province event of 1976. Of the nine earthquakes known to have killed more than 100,000 people no fewer than six occurred in China, two in Japan and one in India. More people have died in China than elsewhere, but this is because China is a large country with a high population density. In terms of the number of people killed per unit of seismic energy, the most dangerous place is the Mediterranean region, followed in order by Iran-Pakistan-Afghanistan, Central Asia, South America, Japan-Formosa, and India. North America comes eighth on the list, after New Zealand.

The lava of Mount St Helens was so silica-rich that entrepreneurs were able to manufacture commemorative glass ashtrays from it

Volcanic activity

Volcanism occurs when molten rock, or magma, from the Earth's interior is able to exploit a weakness in the lithosphere and thus rise towards the surface. It has two basic styles, although the most conspicuous is by no means the most important. That involving the greater volume is fissure volcanism, in which magma rises and some is extruded through linear fissures or fractures. Eighty-one percent of all magma reaches the surface, or at least the upper crust, in this way; most of it does so at oceanic ridges where new oceanic lithosphere is being generated. By far the greater part of the Earth's volcanic activity thus occurs out of sight below the ocean surface.

The other 19 percent of magma rises at particular points rather than along linear fissures, some of it going to form the more familiar central-vent volcanoes. Some 12 percent is generated at destructive plate boundaries, but less than half of that comes out at the surface. Another 6 percent gives rise to central-vent volcanoes on the ocean floor, so the bulk of that is out of sight too. Only about 0.6 percent of magma goes to form the conventional sort of continental volcano. Geologists are under no illusions that as far as the Earth's overall workings are concerned, central-vent volcanoes are of very little importance. Most volcanism takes place at plate boundaries and, like earthquakes, is a direct result of plate interactions.

Yet it is continental central-vent volcanism that has been studied in the most detail, for the obvious reason that it is the most accessible. The products of such volcanism can be liquid, solid or gaseous even though they all began as liquid magma. The liquid form is known as lava. Lava usually emerges from volcanic vents at temperatures of 800-1200°C and, depending on its composition and quantity and the local topography, flows for distances of up to several tens of kilometers. As it does so it cools from the outside, slows down and finally solidifies in one of three basic forms. Highly mobile lava cooling in air usually solidifies into rope-like folds known by the Hawaiian term *pahoehoe*. The slower-moving variety is likely to solidify into a fragmented mass with either a conspicuously jagged (*aa* lava) or a lumpy (block lava) surface. Lava cooling beneath water is quenched much more rapidly, tending to break up into sack-like blocks (pillow lavas).

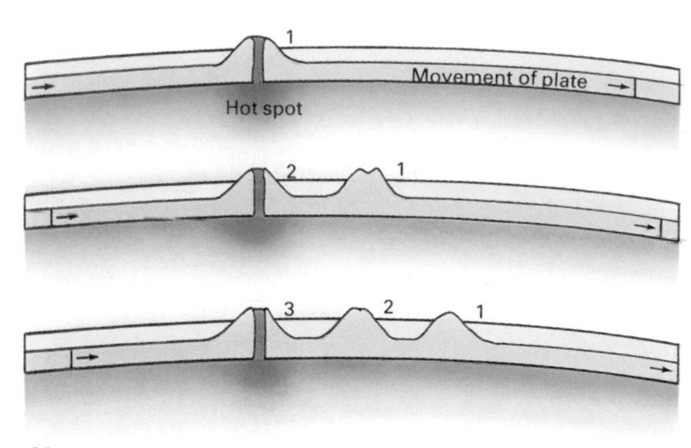

Hot spots

Central-vent volcanoes within plates (those away from plate boundaries) occur on both continents and the ocean floors. They are particularly prevalent in the Pacific where there are more than 10,000, some having risen above sea level to form volcanic islands but most having remained submerged (seamounts). Moreover, they are not all distributed randomly. Many are clearly members of linear chains of oceanic volcanoes that increase in age from one end to the other. The classic example is the Hawaiian-Emperor chain of 107 islands and seamounts that stretches for about 6,000km across the Pacific. The members of this chain get younger right up to the currently-active volcanic island at one end, the island of Hawaii itself.

One way of looking at active volcanoes is to regard them as "hot spots" – points at which magma from the asthenosphere is "burning" its way up through the lithosphere. But many geologists argue that hot spots may not be produced by material from a region as shallow as the asthenosphere. Instead, they envisage narrow (about 100km diameter) plumes of magma rising from the lower mantle. Furthermore, some have suggested that these mantle plumes are fixed in position in the mantle, thereby offering a neat explanation of the linear volcanic chains (above). If the mantle plumes are fixed, the volcanic activity they produce at the surface will appear to progress as the overlying lithospheric plates move over them. They will generate precisely the sort of volcanic island chains actually observed. Geologists have not yet been able to prove that mantle plumes exist. They do not even agree on how many there might be, estimates ranging from 16 to more than 100.

▼ Volcanoes at constructive plate margins tend to have quiet eruptions, with great outpourings of runny lava. This is largely because of the composition of the lava which is low in silica. Iceland has many gentle-sided volcanoes produced by this process. One such is the island volcano Surtsey off the south coast.

▼ Hot-spot volcanoes have the same type of lava as those at constructive plate margins. Hawaii thus has similar volcanoes to those in Iceland.

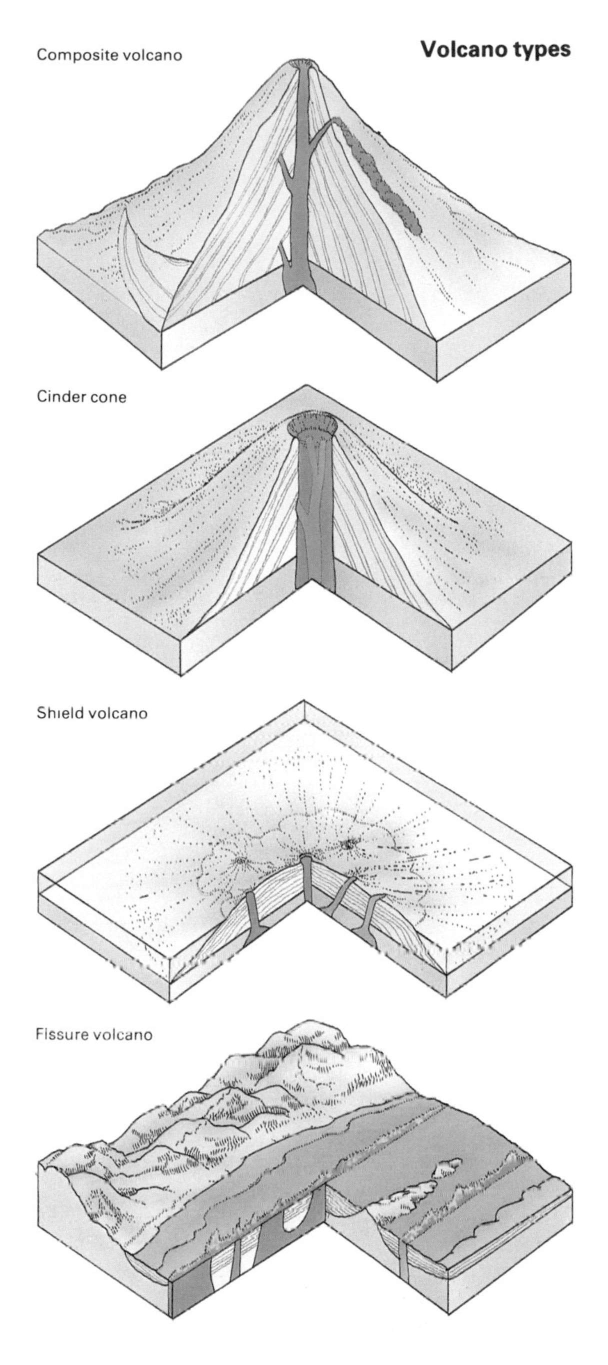

The volcanic cone

Beneath the typical volcano (above) lies a magma chamber, a reservoir of molten rock that can be regarded as a staging post for magma on its way from the asthenosphere to the surface. The magma chamber connects to the central vent, or conduit, at the top of which is a crater. This last feature can be the result of lava sinking back into the vent or of debris having been blasted away, but in either case it is likely to have been enlarged by erosion. When a volcano is dormant, some or all of the magma in the underlying chamber may drain away, in which case the top of the cone may collapse to form a large basin-shaped depression known as a caldera. If the central vent of a volcano becomes blocked during an eruption, lava may be diverted sideways to form secondary vents and cones on the flank of the main cone.

Volcanoes that eject ash form the familiar "cinder cones" comprising layers of cinder and dust from successive eruptions. Less violent eruptions that eject lava can also form cones, although if the lava is plentiful a broad structure much larger than the typical cinder cone results. Known as a shield volcano, this structure may be tens of kilometers wide and several kilometers high, being built up of hundreds or thousands of individual lava flows. More commonly a volcano will produce lava and cinders at different times, leaeing to alternauing layers of the two types of material. Stiff, slow-moving lava will produce tall volcanoes. Fissure eruptions do not form volcanoes but release flows of very fluid lava that can cover areas up to 500 square kilometers.

▲ Volcanoes at destructive plate margins have a lava that is rich in silica. Such a lava is very viscous and does not flow easily. Consequently the eruptions here can be very violent as pressures build up inside the volcanic vents and blast away any solidified lava that is blocking them. Ash and dust can form cauliflower-shaped clouds above the volcano, and floods of hot ash can flow destructively down the steep slopes. These events were seen during the well-publicized eruption of Mount St Helens in Washington State in 1980.

See also
The Dynamic Earth 9-24
Predicting Natural Disasters 241-4
Manmade Earthquakes 245-8

Volcanoes emit gases, mainly carbon dioxide and steam. In magma at great depth, such volatile substances remain dissolved in the fluid. As the magma rises the pressure drops and the gases appear as bubbles. Once at the surface, much of the gas escapes, often with explosive violence, although some bubbles are left forming porous rock such as pumice. When gases explode from a volcano some of the magma is shot into the air and fragmented, falling back to the ground as particles of various sizes. There can be dust (particles less than 0·35mm in diameter), ash (0·35-4·0mm), lapilli (4-32mm) and bombs (greater than 32mm). Bombs can be very large indeed, up to 100 tonnes in particularly violent eruptions. These fragments form pyroclastic rocks (pyroclastics) or, collectively, tephra. They are generally blown upwards, although on occasions they can travel along the ground in combination with hot gases forming a deadly flowing mixture known as a ***nuée ardente***. Volcanic activity ranges from the violent ejection of pyroclastics at one extreme to the quiet extrusion of lava at the other.

◄ Krakatau, seen here six months before its explosion in 1883, was an extensive island. Now there is a vast caldera below the sea level of the Sunda Strait.

▲ The image of a dog killed in Pompeii was one of the many casts of the city's inhabitants produced by pouring plaster into a hollow in the solid ash.

Some historic eruptions

Of the many thousands of volcanic eruptions that have taken place in historic times, the few that have become famous are remembered not because they were the most violent but generally through the benefit of good communications.

One of the best known eruptions is that of Vesuvius (Italy) which in AD79 overwhelmed the city of Pompeii, asphyxiating most of its 20,000 inhabitants. Rain subsequently gave rise to a huge mudflow that buried the nearby city of Herculaneum. Terrible though this disaster was, it would probably be little known today if Pliny the Younger had not provided a vivid eyewitness account of it, if archeologists had not begun to excavate Pompeii in 1748, and if Lord Lytton had not later written a best-selling novel about the episode (The Last Days of Pompeii).

An even more famous eruption was that of Krakatau (Indonesia) which in 1883 killed 36,000 people and produced abnormal air waves, sea waves and distinctive red sunsets around the world. Although this was a larger than average explosive eruption, violently ejecting about 18 cubic kilometers of material, it is by no means the largest known. Only 68 years before, for example, an eruption of Tambora had ejected 100 cubic kilometers of rock. The chief reason for Krakatau's fame is that the eruption was the first of any great size to occur after the spread of the electric telegraph. News of the explosion was tapped around the world within hours, enabling its diverse worldwide effects to be related to a single cause.

The most famous eruption of very recent times was that of Mount St Helens (USA) in 1980. It ejected less than 1 cubic kilometer of rock, which makes it a smaller than average eruption. Mount St Helens is known to almost everyone chiefly because it received the full force of the modern media in a country in which communications are technologically the most advanced in the world.

Internal Structure

3

The body of the Earth – the inaccessible vastness...Earth's zone of weakness – where the plates slide about...Ocean crust – the young section of the skin...Continental crust – the old... Mantle structure – the greatest volume of Earth material...The core...PERSPECTIVE...Seismic waves...Their behavior and their uses... The Mohole fiasco

Most of the Earth's interior is inaccessible to Man. The deepest he has been able to get in person is to the bottom of the deepest mine, which is less than 4km. In addition, he has been able to sample the crust to somewhat greater depth by drilling. The most ambitious drilling program has been carried out by the Russians, who by 1984 had reached 11km on the way to their target depth of 15km. But the Earth's average radius is 6,370km. So even if the Russian plan is successful, it will only involve less than one quarter of one percent of the distance from the Earth's surface to its center.

The vast bulk of the Earth's interior must therefore be investigated by indirect methods. The most important of these is the use of seismic waves. These are the vibrations emitted by earthquakes and explosions and their behavior provides information on the Earth's internal physical structure. Other conclusions about the Earth's internal properties may be drawn from study of the Earth's gravitational, magnetic and thermal fields.

Seismic data show that there are three major discontinuities, or boundaries in the Earth. These separate the planet's interior into four distinct zones – the crust, the mantle, the outer core and the inner core.

The crust is the Earth's outermost layer. In comparison with the Earth's radius it is very thin indeed, although its thickness varies widely. Beneath the oceans the crust has an average thickness of about 7km and most of it lies within the range 5-11km. The continental crust, by contrast, has an average thickness of about 35-40km, which conceals an even greater variation. Beneath the Central Valley of California, for example, the crust is only about 20km thick; but beneath parts of the Himalayas, where India is being thrust under the main Asian landmass, it is up to 90km thick. Altogether the crust accounts for about 0·6 percent of the Earth's volume and 0·4 percent of its mass.

Seismic waves: the seismologist's probes

Earthquakes emit vibrations, or waves, in all directions. Some of the waves travel through the Earth's interior and are thus known as body waves (because they pass through the body of the Earth). There are two types of body wave – P waves (primary) and S waves (secondary). Other waves are restricted in their travel to the vicinity of the Earth's surface and are hence called surface waves. There are also two types of surface wave – Love waves and Rayleigh waves.

When a seismic wave passes through a material, it sets particles of the material vibrating. The chief differences between P, S, Love and Rayleigh waves, apart from where they travel, are in the ways that the particles vibrate. In the case of P waves the particles vibrate backwards and forwards in the direction along which the wave is traveling. The medium through which the wave travels is therefore alternately compressed and extended (dilated). In S waves, on the other hand, the particles vibrate in a direction at right angles to the direction in which the wave is traveling. It is thus a shaking motion and the medium is distorted as the wave passes through – up and down, from side to side, or a combination of the two, depending on the precise nature of the earthquake.

In Love waves the particles also travel from side to side at right angles to the direction in which the wave is traveling. In this respect they are similar to one of the forms of S wave. They differ from S waves, however, in that the amplitude of the side-to-side vibration decreases rapidly with depth, whereas that of the S wave remains constant. In Rayleigh waves the particles move in vertical ellipses that lie in the direction of wave travel; they thus undergo a rolling motion much as near-surface water particles when a sea wave passes. The amplitude of Rayleigh waves also decreases rapidly with depth.

The magnitude of an earthquake may be calculated from either body waves or surface waves, or both, depending on which is most convenient. Body wave magnitudes often differ from their corresponding surface wave magnitudes, but not usually by much. The body waves are the ones chiefly used to determine the Earth's internal structure, however.

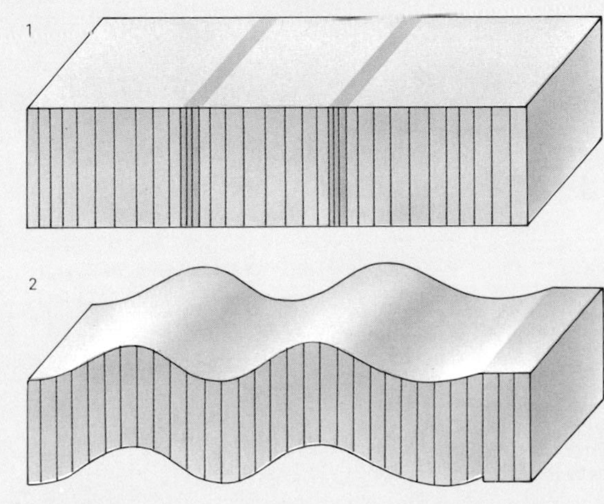

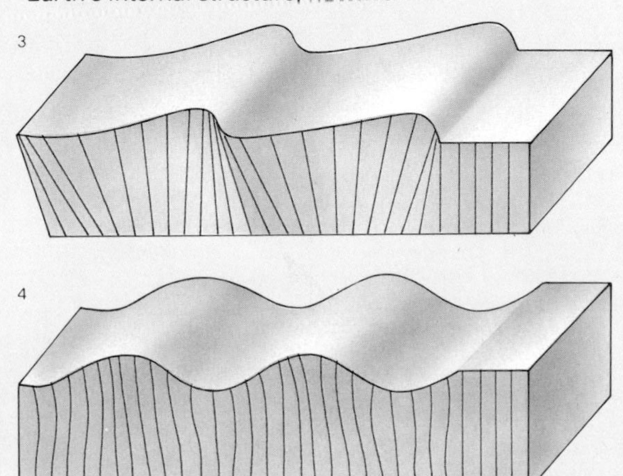

▸ The different types of seismic wave can be shown schematically. The P wave (1) is a compressional pulse, alternately compressing and extending the rocks through which it passes. The S wave (2) produces a shaking action. Of the surface waves the Love wave (3) sets up a kind of swinging motion. While the Rayleigh wave (4) is very much like the a sea wave, dying out at depth. These waves do not usually come singly, but are superimposed on one another.

Apart from the ocean and the various magma chambers, the outer core is the only section of the Earth which is liquid

A section through the Earth

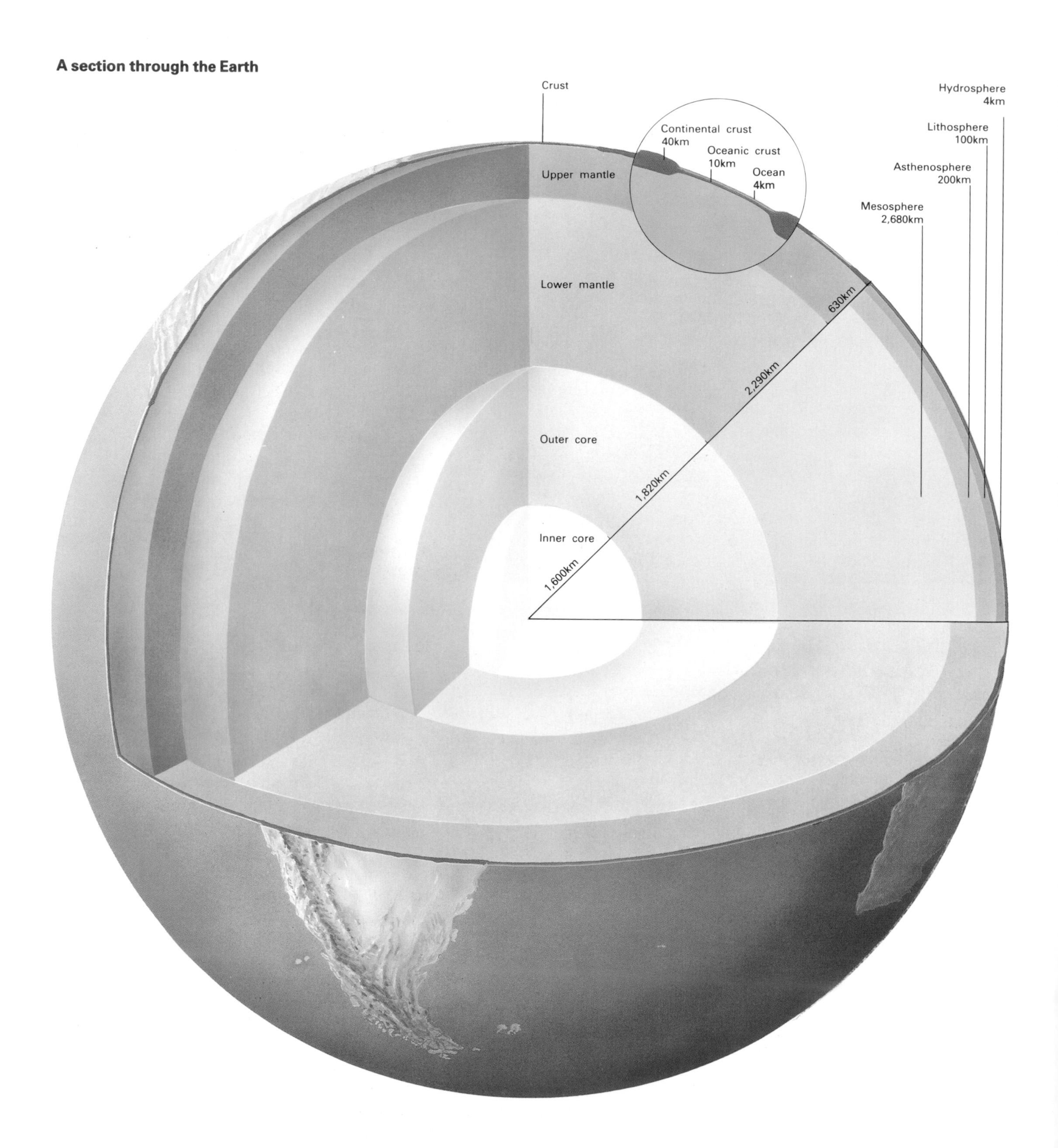

▲ The Earth's crust is the minutest skin on the surface of the globe. The bulk of the Earth consists of the stony mantle, which can be divided into the upper mantle and the lower mantle because of its physical properties. Below this is the core which, again, can be divided into two sections – an inner and an outer core. The inner core is solid, while the outer core is thought to be liquid.

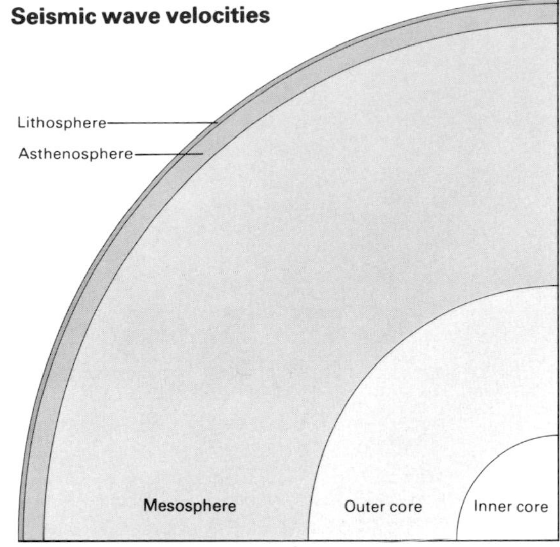

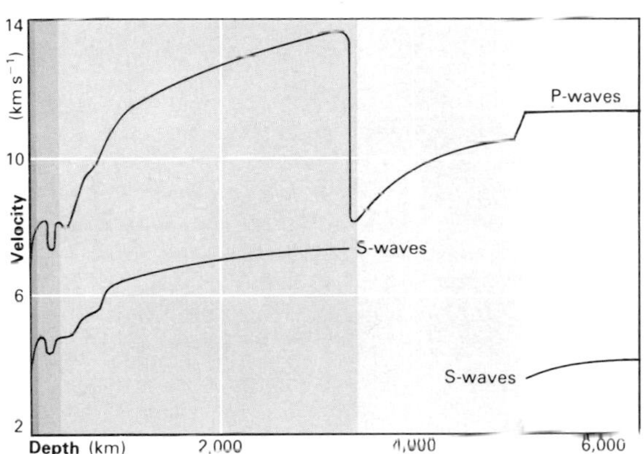

Seismic wave velocities

The velocity of seismic waves gradually increases with depth in the Earth down to the core-mantle boundary. In the upper crust, velocities are highly variable, reflecting the wide variety of rocks present. Thus in the upper continental crust, P wave velocities generally lie in the range 5·9-6·2km s^{-1}, whereas in ocean-bottom sediments they usually vary from 1·5 to 2·5km s^{-1} and in the igneous rocks immediately below the sediment they tend to be in the range 4·5-5·5km s^{-1}. At the base of the crust, whether continental or oceanic, P wave velocities are generally in the vicinity of 7·0km s^{-1}.

There is then a sudden jump to 7·9-8·1km s^{-1} across the Moho, and thereafter the velocity increases to about 13·5km s^{-1} at the base of the mantle. Across the core-mantle boundary there is another sudden jump in the velocity, but this time it is a decrease – to about 8·1km s^{-1}. But the increase then starts again so that the velocity has risen to about 10·2km s^{-1} at the base of the core. There is then a sudden jump to about 11·1km s^{-1} across the the outer-inner core boundary followed by by an increase to about 11·3km s^{-1} at the center.

Although seismic velocity increases generally with depth throughout the mantle, there is one small but very important exception to this rule. The asthenosphere – the region in the topmost part of the mantle on which the plates move – has slightly lower seismic velocities than expected.

The next layer down, below the crust, is the mantle, which accounts for about 82 percent of the Earth's volume and 67 percent of its mass. The sharp discontinuity between the crust and the mantle, the crust-mantle boundary, is known as the Mohorovičić discontinuity (or Moho for short) after the Yugoslav seismologist who discovered it in 1909 (▶ page 37). The base of the mantle is marked by the core-mantle boundary which lies at a depth of about 2,885km. Below the core-mantle boundary – or the Gutenberg discontinuity as it is sometimes called, after the German-American seismologist Beno Gutenberg – lies the core. The outer core, down to a depth of about 5,155km, is liquid and is the only layer of the Earth in this state. Another distinct boundary separates the outer from the inner core. The inner core is solid.

Although the three boundaries are sharp (which means that they are almost certainly less than a few kilometers thick), their positions can only be determined to within a few percent. Moreover, slightly different figures are obtained from different methods of analysing the data. This explains in part why different geologists quote slightly different depths for the boundaries. The other reason is that it is not clear that the boundaries lie at precisely their same respective depths all round the Earth. This is certainly true for the Moho and may also be true for the deeper discontinuities. Some seismologists believe, for example, that the surface of the core-mantle boundary undulates with peak-to-crest differences of at least several kilometers.

The Earth's zone of weakness

In the upper mantle, at depths of about 75 to 250km, there is a region in which the seismic velocity is a few percent lower than in the regions immediately above and below. This is known as the low-velocity zone. It cannot be completely fluid, because S waves travel through it. Nor does it appear to be chemically distinct – it is no different in chemical composition from the the material that overlies and underlies it. It is almost certainly a region in which, because the temperature is very close to the melting point of the stony material of which the mantle is made, the mantle is partially molten. Though essentially solid, the material in the low-velocity zone can flow very slowly. It forms a zone of weakness in the Earth.

The crust-mantle and core-mantle boundaries almost certainly represent chemical discontinuities; above and below each, the chemical composition is different. The existence of the low-velocity zone provides an alternative way of looking at the Earth's interior based not on chemical boundaries but on the variation of physical properties.

The low-velocity zone is known as the asthenosphere. The whole of the region above it – that is the whole of the crust together with that part of the upper mantle between the base of the crust and the top of the asthenosphere – is called the lithosphere. The lithosphere is the rigid outer shell of the Earth from which the tectonic plates are formed. In other words, the base of the lithosphere, or the top of the asthenosphere, marks the bottom of the plates. (The plates are therefore not crustal plates, but lithospheric plates.) The tectonic plates "float" on the asthenosphere, and this is the zone of weakness that gives them the freedom to move. If there were no asthenosphere the surface of the Earth would be rigid – there would be no plate tectonics, no seafloor spreading and no continental drift.

By far the largest part of the mantle lies below the asthenosphere and is known as the mesosphere.

A uniform Earth?

1

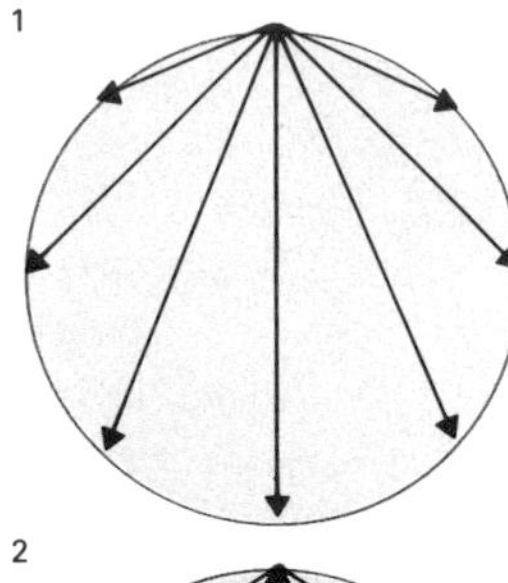

2

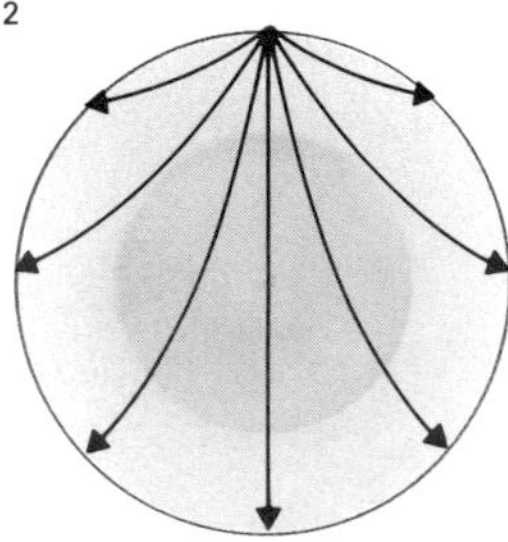

▲ *Even if the Earth were not divided into layers, the seismic waves would still not pass through it in straight lines. The waves would speed up as they passed through denser material compressed at depth, and would be refracted, like light waves in water.*

Wave paths through the Earth

The velocity of a seismic body wave depends on the density and the elastic properties of the material through which the wave is passing. Moreover, since P and S waves are not influenced by the same elastic characteristics of a material, they travel at different velocities even within the same medium. In fact, P waves usually travel at about 1·7 times the velocity of S waves. A seismometer detecting an earthquake will therefore pick up first the P waves and then, a little later, both P waves and S waves together.

By noting the times taken for body waves to travel from earthquakes to seismometers, seismologists have been able to trace wave paths through the Earth and determine how the wave velocity varies from place to place. With a few exceptions, within any given layer in the Earth both P wave and S wave velocities increase with depth. As a result, body waves do not travel in straight lines (apart from the one traveling directly along a radius) but are continuously bent, or refracted, back towards the surface. Furthermore, where there is a sharp boundary in the Earth (for example, between one type of material and another) there will be a sudden change in the wave's velocity and hence its direction. By studying such sudden refractions, seismologists have been able to determine the positions of the crust-mantle and core-mantle boundaries.

The Moho – the boundary between the crust and the mantle – was discovered in 1909 when Andrija Mohorovičić noted that two sets of P waves arrived at a seismograph from the same earthquake. The S waves also arrived in two sets. Mohorovičić reasoned, correctly, that the faster set must have passed down through an upper layer of the Earth, traveled for some time in a second layer where the velocity was higher, and then moved up through the upper layer again to the seismograph – all in less time that it took for the other set of waves to travel directly from the earthquake to the seismograph entirely in the upper layer. He calcualted the P wave velocity in the upper layer at 5·6km s^{-1} whereas that in the lower layer was 7·7km s^{-1}, and that the boundary between the layers lay at a depth of 45–54km.

Body waves also provide information about the physical state of the core. At angles of arc more than 103° away from an earthquake, no direct S waves are received at all. There is an S wave "shadow zone". This is because S waves cannot pass through fluids. The existence of the S wave shadow zone thus indicates that at least the outer core is liquid.

P waves do pass through fluids. However, the refractive and geometric properties of the core and mantle are such that P waves are redirected so that they cannot emerge at all at angles of 103° and 143°. From the positions of the two shadow zones, seismologists calculate the radius of the Earth's core to be about 3,485km. The radius of the core is thus more than half the radius of the whole Earth.

Primary wave paths

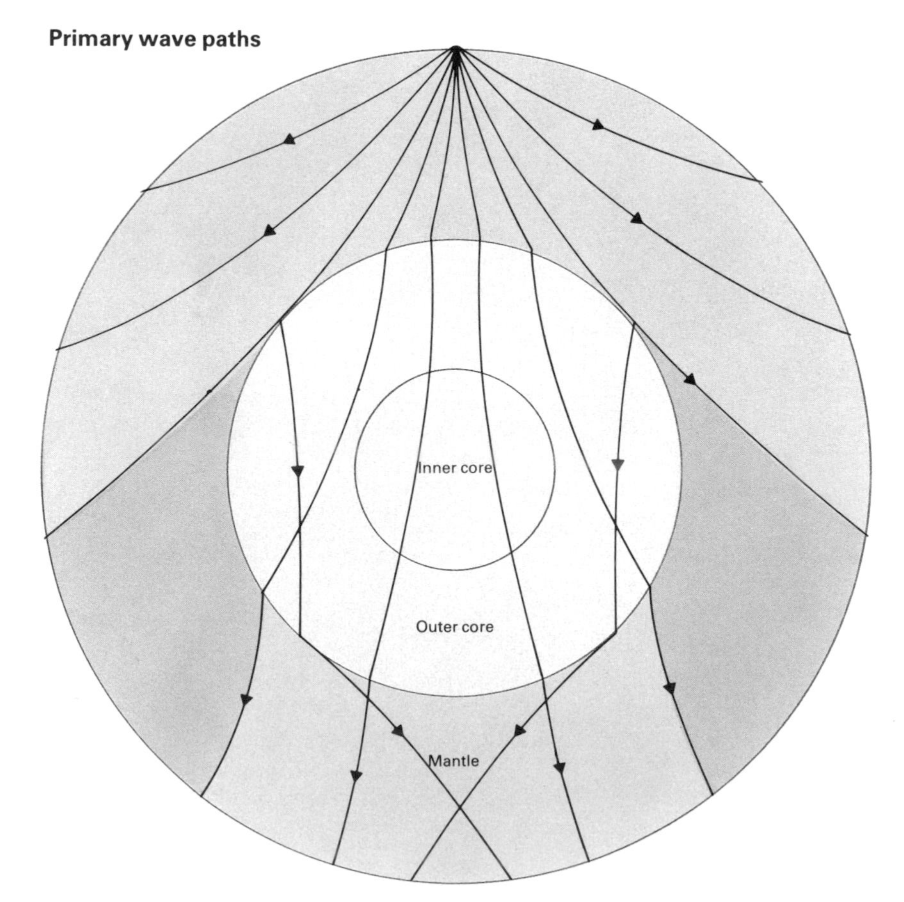

◄ *P waves are refracted gradually by the increasing density of the layers. They are also refracted abruptly at the boundaries between the layers.*

▼ *Because of the refraction there are some areas of the Earth's surface not reached by P waves from a particular earthquake. A "shadow zone" results.*

► *The P wave shadow zone is incomplete. Some waves that pass into the inner core are refracted out again and may reach the surface feebly in the shadow zone.*

► *The S wave shadow zone is more extensive. Shearing waves cannot through liquid and since none pass through the outer core this must be a liquid layer.*

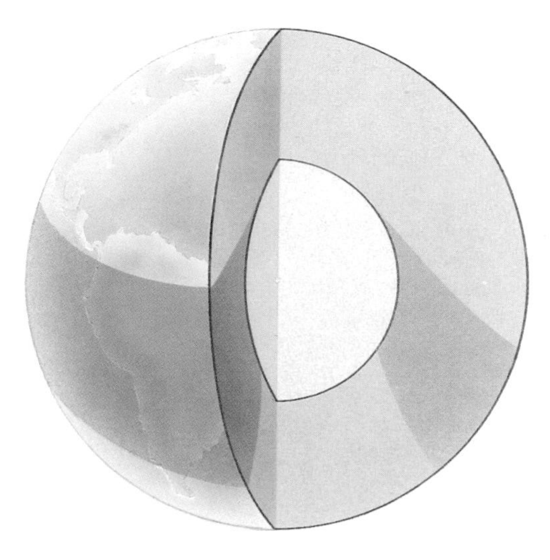

P waves in the shadow zone

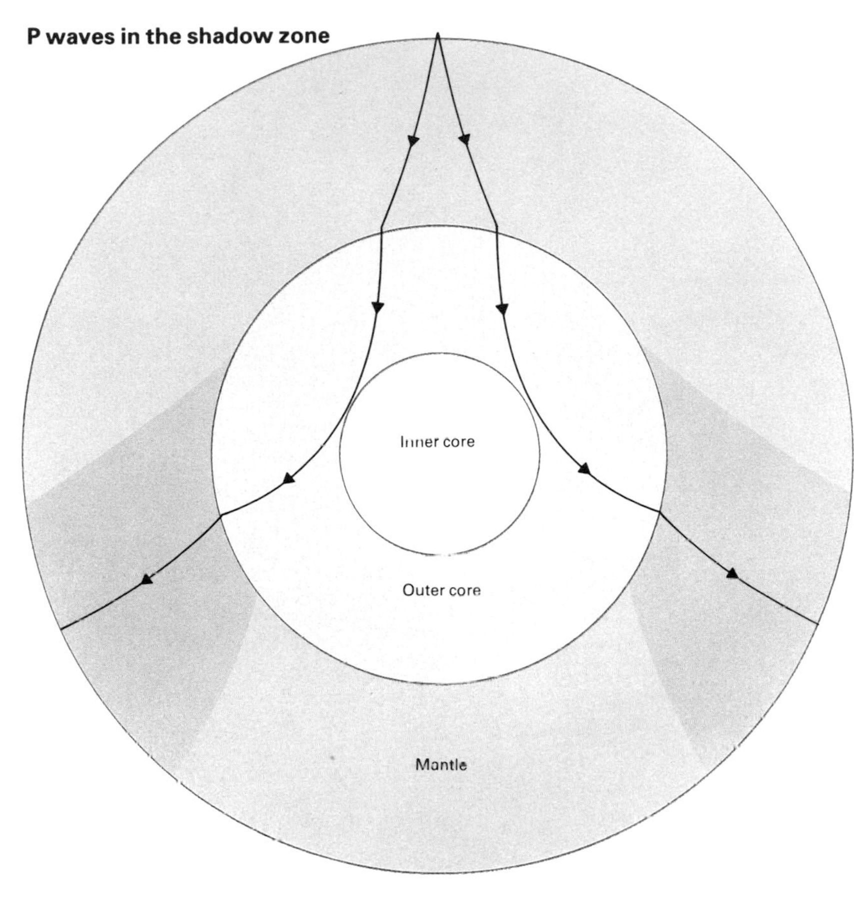

Secondary wave paths

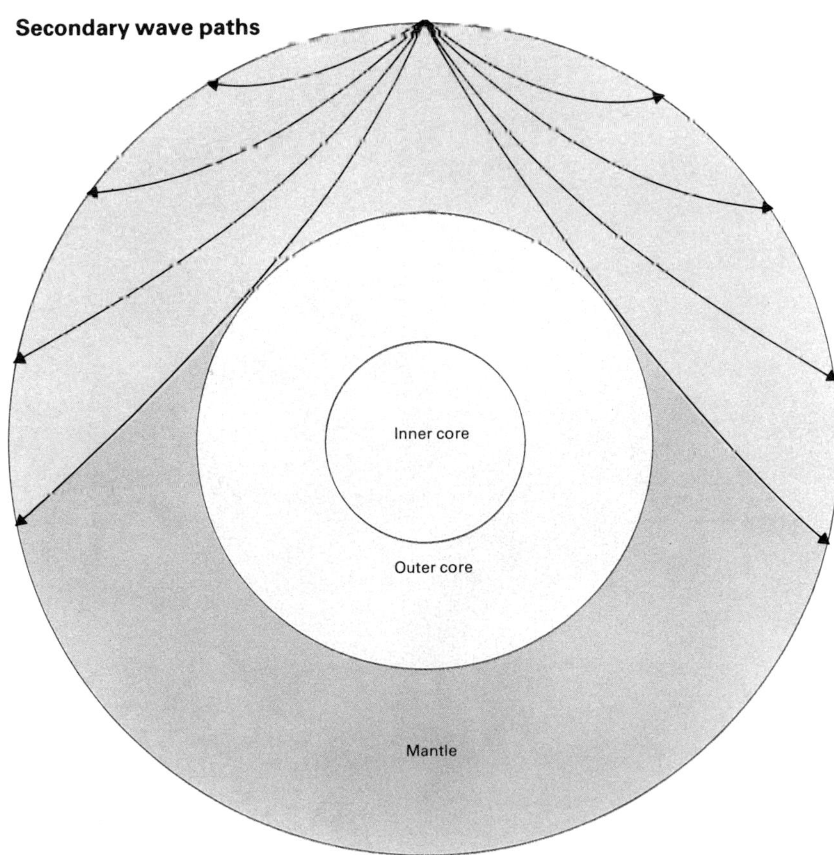

▲ *Mohorovičić, the discoverer of the Moho.*

Mohorovičić and the crust-mantle boundary

Andrija Mohorovičić was born on 23 January 1857 in the seaport of Volosko, then part of the Austro-Hungarian Empire but now in Yugoslavia. He studied mathematics and physics at Carl IV University in Prague under Ernst Mach and then became a secondary-school master for 11 years. During that time he set up a meteorological station (1887), began to publish papers on meteorology, and thus came to the attention of the director of the Meteorological Observatory in Zagreb. The director persuaded Mohorovičić to join the staff, and when he (the director) retired in 1891 Mohorovičić was given the position and hold it until his own retirement in 1922. He died on 18 December 1936.

Though (like Alfred Wegener) a meteorologist, Mohorovičić became interested in seismology because of local earthquakes. Zagreb had been struck by a destructive earthquake in 1880, as a result of which the Observatory had acquired a seismograph, albeit a poor one. When another severe earthquake struck in 1901, Mohorovičić suggested that a proper seismograph station be set up. This was done in 1906. The first earthquake to be recorded properly by the new station was that of 8 October 1909 (about magnitude 6). It was to become the subject of Mohorovičić's first seismological paper in 1910 (at the age of 53) and the chief source of his later fame.

Mohorovičić analysed the seismic data from the 1909 earthquake and its aftershocks, and discovered the boundary between the crust and the mantle that now bears his name.

In Mohorovičić's words, "there must be an abrupt change of material, which composes the interior of the Earth, because there must occur an abrupt change of velocity of the earthquake waves." The upper layer was the crust and the lower layer was the uppermost mantle.

Mohorovičić continued with seismological studies. By 1936 when he died, only eight of his 21 published papers had been about his original subject of meteorology.

The structure of the oceanic crust

For many years following the discovery of the Moho, the structure of the crust and mantle was investigated using seismic waves from the earthquakes. Since the 1950s, however, most such studies have made use of waves from explosions (which are also called seismic waves). The advantages of explosions are that they can be generated at will and can be set off at accurately known times and locations. The main disadvantage is that (with the exception of nuclear detonations) explosions can only be made powerful enough to penetrate the crust and upper mantle. They cannot therefore be used to determine the Earth's gross structure, especially at very great depth, but they are ideal for delineating the structure of the Earth's upper layers.

Explosion studies show that the oceanic crust comprises three distinct layers. Layer 1, at the top, is the poorly consolidated sediment lying on the ocean floor. Its thickness is very variable, ranging from zero on the crest of active oceanic ridges to 3km near some passive continental margins. Over most of the ocean floor, however, it is usually no more than a few hundred meters thick. The next layer down, layer 2, is harder rock and has an average thickness of 1·5km. Cores obtained by the Deep Sea Drilling Project demonstrate that layer 2 is largely basalt, although some small parts of it could be consolidated (hardened) sediment. Layer 3 is about 5km thick. It has not been penetrated by drilling and so its nature is unknown. It is probably either basalt or its coarse-grained chemical equivalent, gabbro. Below layer 3 lies the mantle.

Structure of the continental crust

The structure of the oceanic crust is simplicity itself compared to that of the continental crust. Oceanic crust is comparatively thin, not too variable in thickness and, above all, young. None of it is more than about 200 million years old. The continents, by contrast, contain rocks up to almost 3·8 billion years old which are both much thicker and more variable. They have also undergone erosion, distortion, fragmentation, suturing, uplift, depression and much else besides. This longer and more complex history is inevitably reflected in the structure of the continents as they appear today.

Not that the full complexity of the continental crust is easy to detect. The accompanying diagram (right) shows sections through the continent in six quite different types of environment of widely different ages. The first is a stable shield area of Precambrian age (older than 570 million years) in Wisconsin; the second is the Basin and Range province of the United States, a region of modern continental rifting within Precambrian rocks; the third is a 400 million year old continent-continent collision zone in northern Scotland; the fourth is a 100 million year old ocean-continent destructive plate boundary in southern California; the fifth is a modern ocean-continent destructive plate boundary in the central Andes; and the sixth is a modern continent-continent collision-zone in the central Alps. Each section shows the structure as determined by the seismic refraction method and gives the P wave velocities in km s^{-1}.

The lowest layer in each case is the mantle, where the P wave velocity is 7·9km s^{-1} or greater. Above the mantle lies the crust, which varies from less than 40km to almost 70km in thickness and which may contain two, three or four distinct layers. Unfortunately, these sections do not indicate how complex the continental crust really is. To examine the finer detail, a different seismic technique is needed.

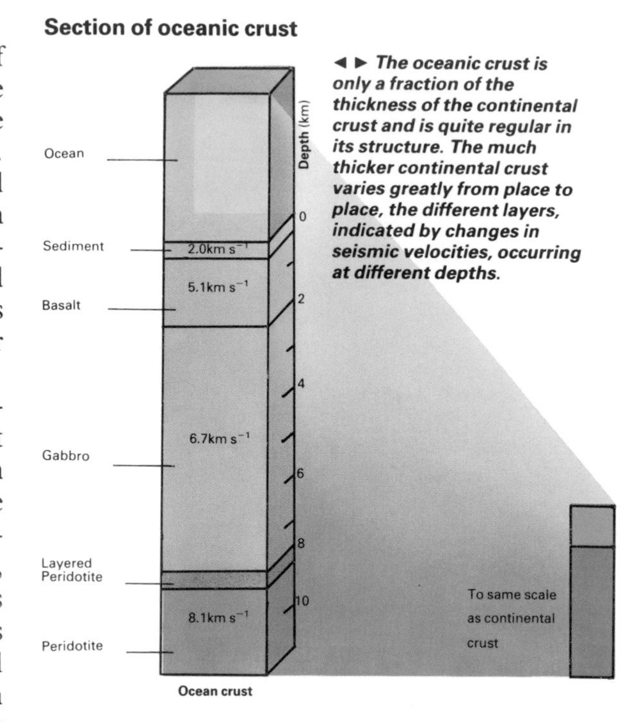

◄ ► The oceanic crust is only a fraction of the thickness of the continental crust and is quite regular in its structure. The much thicker continental crust varies greatly from place to place, the different layers, indicated by changes in seismic velocities, occurring at different depths.

Section of continental crust

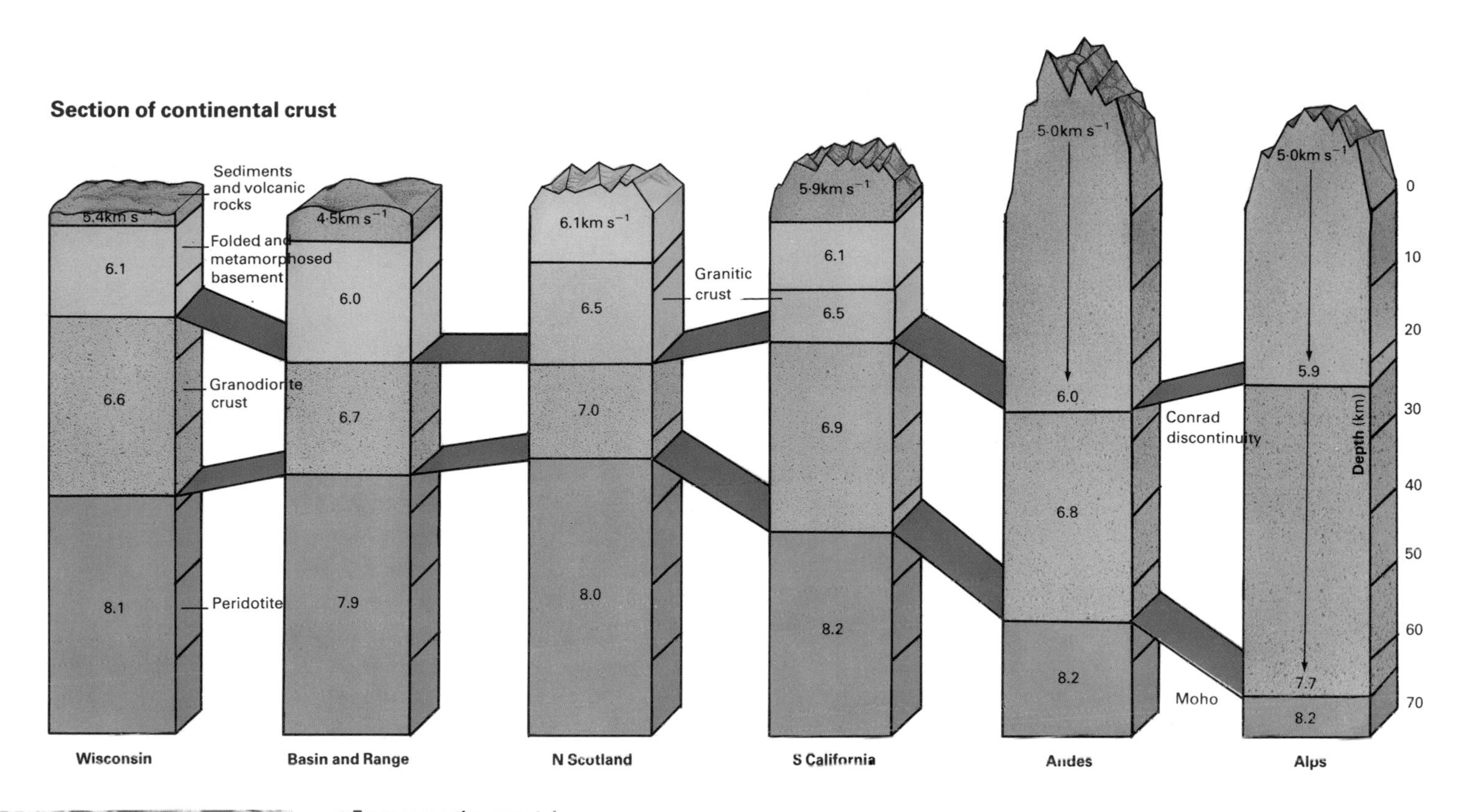

◄ *From space the mountain ranges of the Earth look very high and rugged. In fact the height is not great compared with the vast area of land seen at one time. Their relief is usually emphasized by a covering of snow on the highest peaks and ridges, as in this view of the western Alps.*

► ▼ *A section through the Alps highlights the extreme complexity of the continental crust. The mountain range was formed by the movement of the African Plate against the Eurasian Plate, twisting the landscape into folds and thrusting each fold into another, producing the tangle of rock structures seen today.*

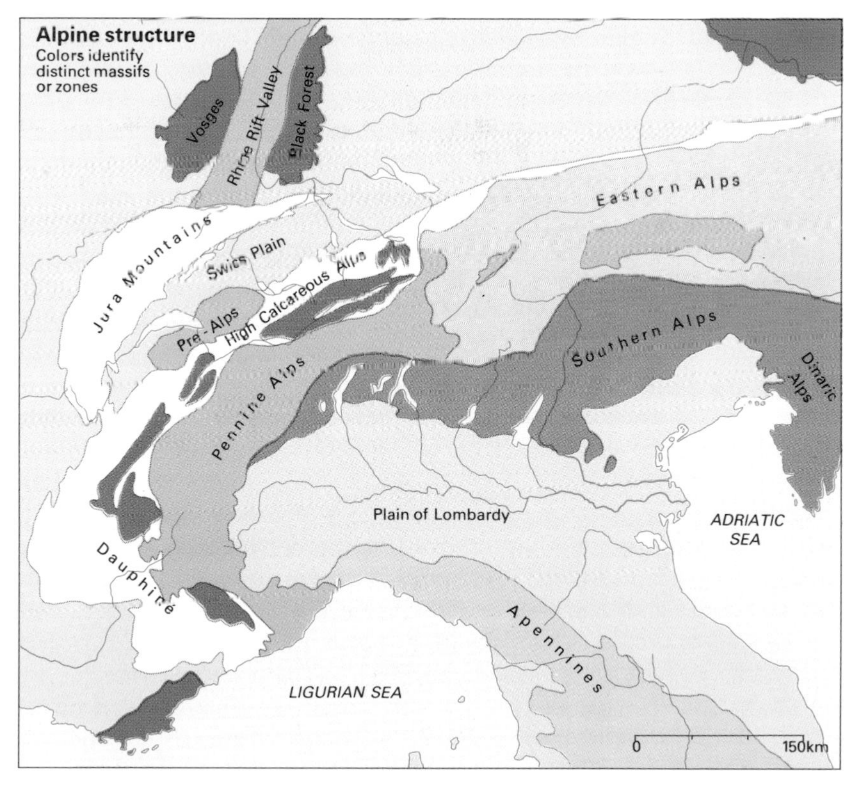

Tectonic section across Western Alps

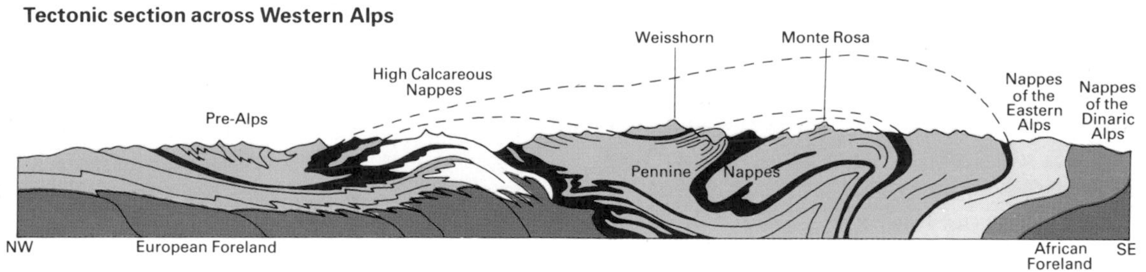

Vibrations through the Earth reveal complex structures unimagined from the surface

Seismic refraction

Explosions, like earthquakes, emit waves in all directions. (Unlike earthquakes, however, the waves are predominantly P waves; S waves are very weak or non-existent.) If an explosion is produced at the surface, above a region where the Earth is layered, one of the many emitted waves will strike the first boundary at just the correct angle to be refracted horizontally into the second layer. It will then travel just inside the second layer, giving off at all points waves that will travel back up to the surface to reach whatever sensors or seismometers have already been placed there. (This only works as long as the velocity of seismic waves in the second layer is higher than that in the layer above, which is usually the case.) In addition, each seismometer will receive from the explosion a direct wave that has traveled entirely in the upper layer.

The simplest procedure is to arrange the explosion "shot point" and a series of seismometers in a straight line. By recording the arrival times at the seismometers of both the direct and the refracted waves, seismologists can then calculate the P wave velocity in the upper layer, the P wave velocity in the second layer, and thickness of the upper layer.

Other waves striking the first boundary will, instead of being refracted to the horizontal, be refracted down towards the second boundary. One of these will strike the second boundary at just the correct angle to be refracted horizontally into the third layer, and waves from this will also ultimately reach the seismometers. Thus it becomes possible to determine the thickness of the second layer and the P wave velocity in the third layer.

This is the most important experimental technique in the whole of geophysics. It can be extended in principle to any number of layers and boundaries (but the more there are, the more complicated the mathematics becomes); all that is necessary is that the explosion be powerful enough to send waves down to the appropriate depths. The technique can also be used where the boundaries are not horizontal (in which case the mathematics becomes even more complicated). Seismic refraction can be carried out both on land and from a ship at sea.

Measuring the bending of the waves

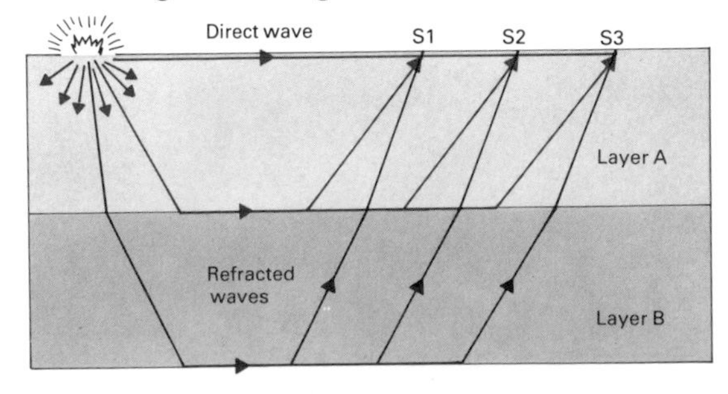

▲ ***In seismic refraction surveys the shock waves are refracted when they meet a different material. Sensors on the surface can measure by how much the waves are refracted and can work out the subsurface structures.***

Seismic reflection profiling has been used to investigate the detailed structure of the deep continent in a number of parts of the world, but nowhere are the results more spectacular than in Kansas. In northeastern Kansas the surface rocks are mostly flat-lying, little-deformed sediments with few outcrops of the basement beneath. Sampling of the outcrops, shallow drilling and seismic refraction studies have long suggested that the underlying geology is probably as bland as the landscape. One seismic refraction experiment, for example, seemed to indicate that, beneath the thin veneer of sediment, there was a single crustal layer with little variation in composition.

Seismic reflection profiling, by contrast, reveals a remarkable picture of criss-crossing reflecting interfaces that indicate very great structural complexity. The precise geological interpretation of the complexity is uncertain, although the presence of so many physical boundaries would seem to imply considerable diversity in chemical composition. It certainly suggests extensive faulting, folding, intrusion and metamorphism, at least in the middle part of the crust. In the lower part of the crust and in the upper part, immediately beneath the sediment cover, there are fewer reflectors, which could mean that the rocks there are more uniform. But it could also mean that even seismic reflection profiling has too little resolving power, in which case the continental crust could be even more complex than anyone has envisaged.

The complexity of the Kansas crust is particularly startling because the simple surface geology misled geologists into expecting comparable simplicity at depth. However, similar crustal complexity is observed in many other areas investigated by seismic reflection profiling. In some cases the subsurface structure can be seen as the downward extension of the diverse geological features observed at the surface, but in other cases (such as in Kansas) the geology at depth shows a complexity not hinted at by surface features.

Seismic reflection profiling

The seismic refraction method is a fairly blunt instrument. It is excellent for determining the gross structure of the Earth as a whole and the broad pattern of layering in the upper part, but on finer and finer scales it becomes less and less effective. It only provides the average seismic properties throughout each of the regions investigated and it can only detect strong boundaries. The method is incapable of resolving very fine structural characteristics and is thus likely to make the subsurface structure look simpler than it really is.

To determine fine structure, a rather different seismic technique must be used – seismic reflection. In this method a wave from a seismic source at the surface is transmitted almost vertically downwards and is reflected directly back to the surface from the various rock interfaces below. The experiment is repeated at many points along a line that can be anything from a few kilometers to hundreds of kilometers long. In this way a profile is built up of all the reflecting interfaces beneath the line.

At sea, seismic reflection profiling is carried out from a continuously moving ship; the ship carries the wave source and the wave detectors are towed behind the ship in a line known as a "streamer". In this case the source is usually an airgun and the detectors are hydrophones – pressure-sensitive devices that convert pressure waves in the water into electrical signals. On land the source is usually either a chemical explosion or an electromechanical device that produces waves by vibrating the ground. The detectors are geophones, which are very similar to the seismometers used in earthquake detection.

Seismic reflection has been used for many years by the petroleum industry in the search for oil traps, but academic geologists have made extensive use of it only since the 1970s. In 1975 the Americans set up the Consortium for Continental Reflection Profiling (COCORP), based at Cornell University, whose object is to run long profiles (typically 50–100km long) across areas of geological interest. Similar work is carried out in Europe, Canada, Australia, China and the USSR.

What COCORP and similar schemes are showing is that the deep geology of the crust is no less complex and variable than the geology at the surface. This means that a nation can no longer consider its geological knowledge complete until it has thoroughly investigated the structure down to the Moho throughout the whole country. As long as the money is forthcoming (for seismic reflection profiling is expensive), the search for basic geological information in the late 20th century could well match that of the late 19th.

▲ The landscape of Kansas is a flat and featureless plain. It was always assumed that this meant that the subsurface geology was likewise bland and regular.

► Modern seismic reflection equipment consists of road vehicles that send seismic pulses directly into the ground without disturbing the area with explosions.

► Seismic reflection reveals the distribution and shape of underground features by measuring the time taken for seismic waves to echo back from whatever lies below the surface. The technique revealed that the structure of Kansas had a complexity that was not even hinted at by the topography. The lines on the seismic reflection trace indicate all kinds of folds, faults and boundaries in the top 60 kilometers of continental crust.

The continental crust is normally the only part of the Earth's structure that can be sampled directly

▲ *Inge Lehmann, who postulated a solid inner core.*

The solid inner core

It is not strictly true to say that no P waves emerge between 103° and 143°. No strong direct P waves emerge, but some very weak ones do. During the 1920s and early 1930s the possible origins of these faint P waves were the source of much discussion, a number of hypotheses were tried. Then in 1936 the Danish seismologist Inge Lehmann, one of the first women seismologists, showed that they could be explained easily on the assumption that the Earth's core is not completely fluid. If there is a solid inner core, some P waves will undergo refraction within it and at the inner core-outer core boundary, and hence be redirected so as to emerge as weak signals in the P wave shadow zone. Lehmann's suggestion was readily endorsed by the top seismologists of the time and has been widely accepted ever since, although there is still no definite proof that the solid inner core exists.

The structure of the upper mantle

In the absence of information to the contrary, geologists always assume that things are simple. They are usually proved wrong in the end. They have always known from observation that the geology near the surface is complicated, but long after seismic techniques were invented they maintained that the lower crust was more or less uniform. Indeed, they did not really begin fully to appreciate the complexity of the lower crust until the late 1970s, and there may be many surprises to come even yet. It was likewise long assumed that the upper mantle is uniform. But this has also proved false.

The most conspicuous example of variation in the upper mantle is the asthenosphere, but there are also more subtle differences. For example, although seismic velocity increases gradually with depth in the mantle (except in the asthenosphere) the increase is not always very smooth. At a depth of about 400km and again at about 650km there are particularly large increases in seismic velocity that amount to minor jumps. In other words, at those depths there are seismic discontinuities, albeit not as strong as the ones at the Moho and the core-mantle boundary. As a result, that part of the mantle between 400 and 650km is sometimes known as the mantle transition zone.

The significance of these two discontinuities is still being debated. The consensus appears to be that they are not major chemical boundaries, such as the crust-mantle and core-mantle boundaries are thought to be, but that they probably represent phase changes (changes of physical state without change in overall chemical composition). Some geologists also believe that there may be some significance in the fact that the most conspicuous of the discontinuities – namely, that at 650km – is close to the depth below which no further earthquakes occur. Be that as it may, the very existence of the two discontinuities shows at the very least the upper mantle is not very uniform in a vertical direction.

Nor does it seem to be laterally uniform either; in other words, at any given depth in the mantle the properties are not necessarily constant around the Earth, or even over a region. For example, beneath the United States the seismic P wave velocity in the uppermost mantle just beneath the Moho varies at least over the range 7·8-8·3km s^{-1}.

◄ ***The only time oceanic crust is directly visible is in rare rock sequences called ophiolites thrust up at destructive margins. Ophiolites consist of silica-poor igneous rocks mixed up with oceanic sediments. This outcrop is in Cyprus.***

► ***The mantle appears to have some internal structure. In the upper mantle beneath the United States the P wave velocities vary from place to place, indicating that something in it is changing over the area. Geologists do not yet know what it can be.***

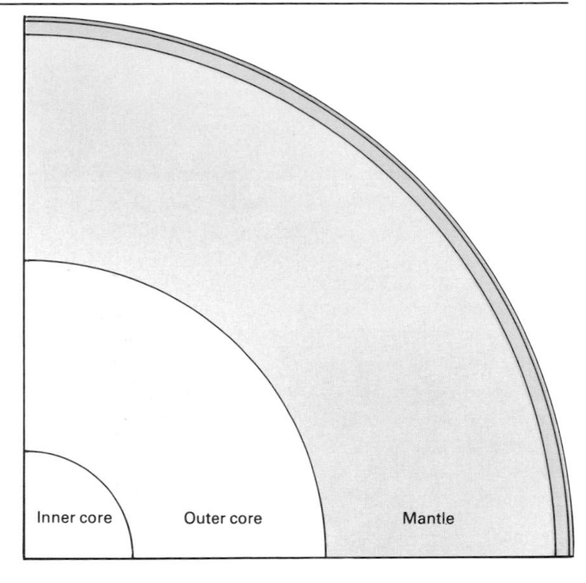

Since seismic velocity depends upon density and elasticity, the figures suggest that these parameters also vary laterally.

For now, most geologists are content to believe that the lower mantle below the transition zone really is uniform. Time will tell whether or not that, too, is an illusion created by ignorance.

Density, pressure and gravity

The average density of the Earth, which is calculated from the planet's mass and volume, is 5,520kg m^{-3}. The measured densities of most rocks near the Earth's surface lie in the range 2,500-3,000kg m^{-3}. The Earth cannot therefore be a body of uniform density, for there must be heavier-than-average material at depth to compensate for the lighter-than-average rocks at the surface. In short, the Earth's density must vary with depth. Unfortunately, the determination of that variation is a difficult problem. In principle it could be obtained from the seismic velocity, whose variation throughout the Earth is fairly well known. But seismic velocity depends not only on density but also on the Earth's elastic properties, which are also unknown.

There are several different approaches to the problem of determining the density variation in the Earth, most of which are highly involved. One makes use of the blunderbuss strategy, and was developed by the American Frank Press. Press used a computer to generate five million random models of the density variation. He then discarded all those that were unsuitable on the grounds that they were inconsistent with the Earth's mass, its rotational characteristics, the seismic velocity variation, and so on. In the end he was left with only three plausible models, all of which were similar and not very different from models obtained by other methods. The three were then averaged, in which form they are probably very close to the truth.

Once the density variation is known it is easy to calculate the variation of elastic properties (from the seismic velocities) and also the variation of pressure and gravity. Not surprisingly, pressure increases steadily towards the Earth's center. However, gravity remains almost constant throughout the mantle and only drops off in the core, where a large proportion of the Earth's mass (about 33 percent) is concentrated, but only about 16 percent of its volume.

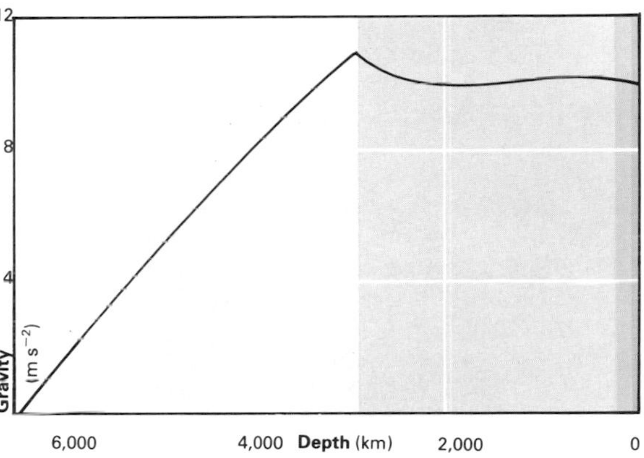

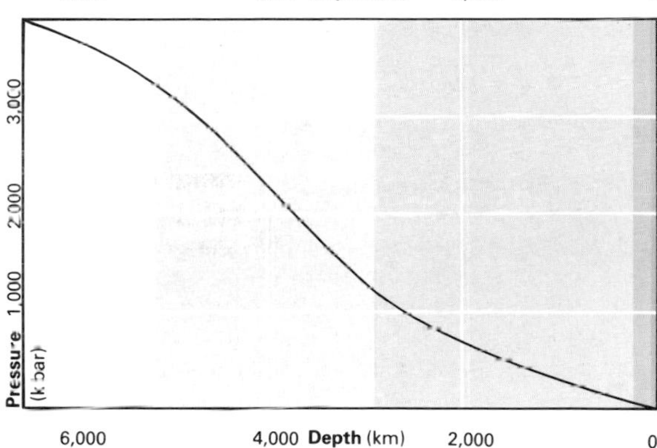

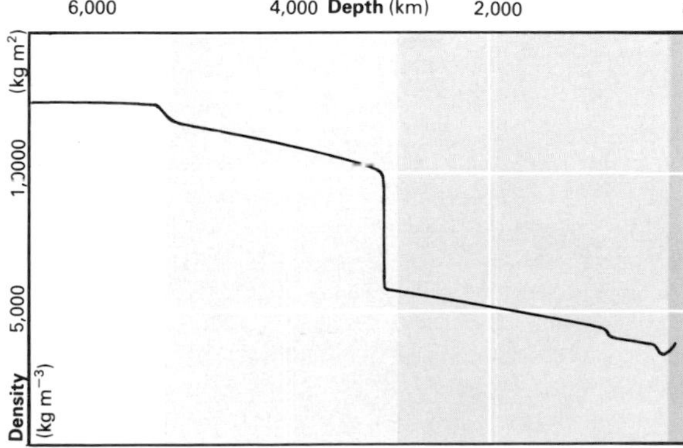

▲ ***Beneath the Earth's crust physical properties change with increasing depth. Gravity remains relatively stable through the mantle, but then drops off rapidly in the core. This is because after a certain point there is more material above than below exerting a gravitational attraction. Pressure increases with depth in a fairly gentle curve. Density varies with the different kinds of materials that exist in these lower regions. The core material is about twice as dense as the material of the mantle.***

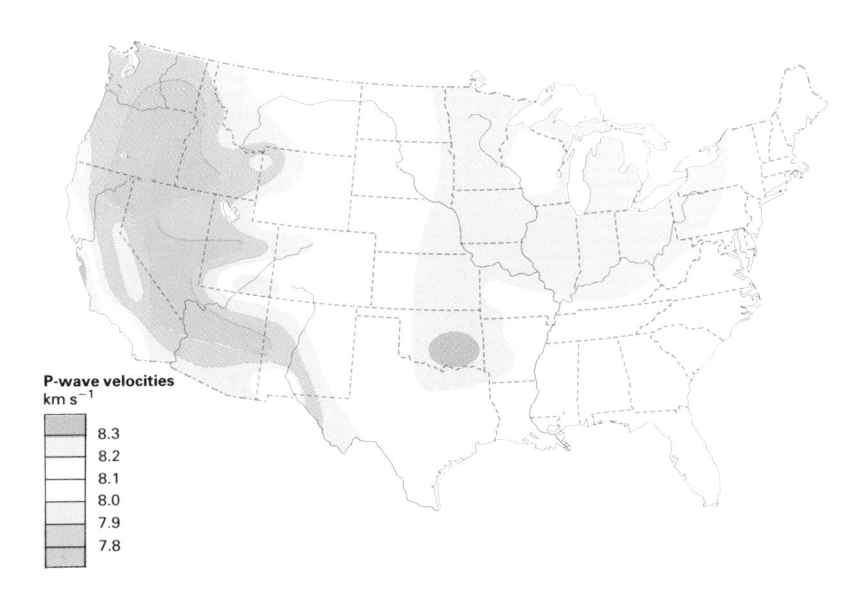

See also
Magnetism, Gravity and Heat 45-56
Internal Composition 57-68
The Growth of Continents 101-8

The Mohole fiasco

In 1957 a small group of distinguished Earth scientists met in Washington DC and conceived a daring scheme – to drill a hole right through the Earth's crust and into the mantle. The scientists were "members" of the American Miscellaneous Society (AMSOC), an "organization" that held no formal meetings and had no membership list, bye-laws, officers or publications. In fact, AMSOC did not really exist at all; it was a joke organization postulated as a harmless caricature of conventional, stuffy scientific societies. Its only manifestation was an occasional get-together over drinks in a Washington club.

The idea of drilling to the Moho – or Project Mohole as it came to be called – may not have been entirely serious either. The scientists themselves were never entirely clear what it was meant to achieve for science, and the technology of the scheme was quite clearly beyond the state of the drilling art. But the 1950s was the time when the US government had begun to support science in a big way. Science in general was awash with money, producing a state of euphoria that made it possible to feel that government money would be available for almost anything scientific. Besides, Earth scientists were beginning to feel resentful that the big money was going into physics and space.

But whatever the motive for suggesting it, the project soon gathered a momentum of its own. AMSOC became affiliated to the prestigious National Academy of Sciences, planning funds were obtained from the National Science Foundation (the body set up in 1950 to give research grants to universities), an oceanographic engineer was appointed to supervise the project, and a small test hole was drilled off the coast of California. This Phase 1 was a brilliant success, receiving public acclaim from President Kennedy. It set a world record for the time for drilling 197m into the ocean floor beneath almost 3km of water.

But thereafter, the scheme began to fall apart. None of the scientific participants was experienced in the financial planning of technological projects. One month the cost of Project Mohole was said to be $5 million and the next it was put at $14 million. Some said it would be better to drill an intermediate-depth hole before the final hole to the mantle; others wanted to go straight for the main goal. Moreover, by this time (1961-2) the whole affair was taking up so much time that most of the original scientists wanted to leave. So did the National Academy of Sciences, which was not set up to run major engineering projects.

In the end, the planning was put out to tender by the National Science Foundation. The winning bid was not the Foundation's first choice but was from an engineering firm in the constituency of the chairman of the Congressional committee that was responsible for allocating money to the National Science Foundation itself. Then after a long delay the firm estimated the cost of the project at $68 million, although the Foundation conceded that it might rise to $125 million. Congress baulked at the expense and the political controversy surrounding the choice of firm. In August 1966 it cut off all funding. Project Mohole was dead.

▲ *CUSS 1 was the drilling ship designed and built for Phase 1 of the Project Mohole. The technology of the ship was the only success of the whole enterprise. Sideways-pointing thruster motors enabled her to stay at a particular spot despite tides and currents. The disturbance near the bow is one of the thrusters in operation. Her stern deck is stacked with 18-meter lengths of drill pipe.*

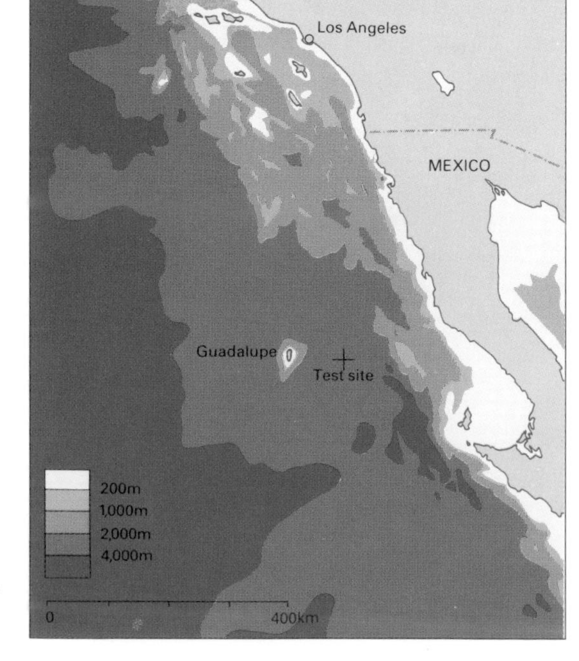

► *The site of Phase 1 was off Baja California. This phase, successful though it was, was nothing compared with the ultimate goal of drilling through the whole crust.*

Magnetism, Gravity and Heat 4

Magnetic field – components and origin...Gravity – effects of mountains and ice caps...Anomalies – where the heavy rocks lie...Temperature variations – the changes in temperature with depth...The origin of the Earth's heat...PERSPECTIVE...An ancient compass...The Magnetic field in space...Magnetism in the past...Uplift of Scandinavia...Bouguer anomaly...Temperature measurements...Mantle convection

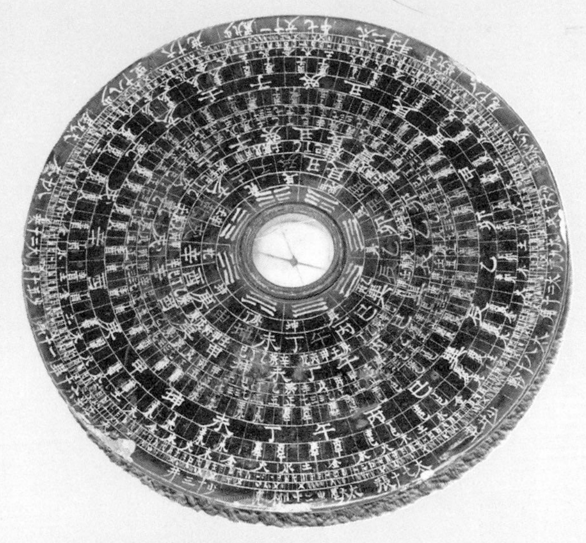

▲ **The ancients appreciated the magnetic field of the Earth. By the first century BC the Chinese were building compasses, using them for fortune-telling.**

Like some other planets, the Earth has a magnetic field (the geomagnetic field). Although it looks immensely complicated on a map, 95 percent of the field is produced by a simple dipole and is thus identical in shape to that produced by a humble bar magnet. Two things make the field appear complicated. One is that the dipole at the center of the Earth lies not along the Earth's axis of rotation but at 11° to it. This means that the simple field pattern produced by the dipole is not symmetrical about lines of latitude or longitude. Secondly, superimposed upon the dipole is a small irregular field known as the non-dipole field.

Because of the 11° inclination of the dipole and the presence of the non-dipole field, a compass needle will not generally point towards true (geographic) north but at an angle to it. This angle is called the magnetic declination and is sometimes to the east of north and sometimes to the west. A map of declination, known as an isogonal chart, shows that the declination can be much bigger than most people probably realize. Even within middle latitudes (between 45°N and 45°S) the declination can be more than 40° (to the southwest of Australia), and if all latitudes are considered it rises to much higher values than that. On the other hand, there are places where the declination is zero and compass needles do point to true north

A normal compass needle is mounted on a vertical pivot and is thus constrained to rotate in the horizontal plane. However, the geomagnetic field has a vertical component as well as a horizontal one and this can be measured using a dip needle – a compass needle mounted on a horizontal pivot so that it rotates in a vertical plane. If there were no non-dipole field, the dip needle would point vertically downwards at two points known as the geomagnetic poles. These are antipodal – at 79°N, 70°W and 79°S, 110°E. The geomagnetic poles are 11° of latitude away from the geographic poles because the dipole is inclined at 11° to the rotational axis. But because of the non-dipole field, the two magnetic dip poles are actually some distance away from the geomagnetic poles. Moreover, because the non-dipole field is irregular, these two points are not antipodal. They lie at about 70°N, 101°W (north magnetic dip pole) and 67°S, 143°E (south).

The Chinese spoon

The existence of the Earth's magnetic field as such was not discovered until AD 1600, although people had been making use of magnetism long before that. The ancient Greeks were familiar with lodestone, a strongly magnetic natural rock, but they never realized it had direction-seeking characteristics. That discovery fell to the Chinese, who by at least the first century BC had constructed a compass in the form of a lodestone spoon balanced upon a smooth board. This remarkable instrument was used not for navigation but for geomancy, the art of prophesy.

The Earth's field in space

The magnetic field extends many Earth-radii into space but its shape is distorted by the solar wind. The charged particles streaming out from the Sun compress the field at the forward edge and greatly elongate it on the opposite side. The field is thus confined to a zone called the magnetosphere, the edge of which is the magnetopause. On the solar side of the Earth, the magnetopause lies at about 10 Earth-radii, although its precise position at any time depends on the variable intensity of the solar wind. Away from the Sun, the magnetopause extends to beyond 60 Earth-radii. The velocity of the solar wind particles is so high (about 1,000km s^{-1}) that a shock wave forms at the forward edge, producing a complex interaction zone – the magnetosheath.

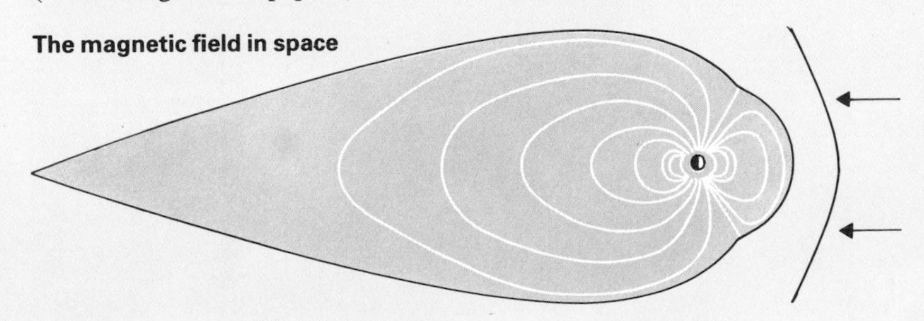

◄ ► **A theoretical magnet with a single north magnetic pole, would be forced away from one of the Earth's poles, swing out into space and be attracted back down to the other pole. The possible paths that it could take define the Earth's magnetic field (right). Radiation from the Sun distorts the field into a teardrop shape (left).**

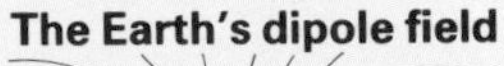

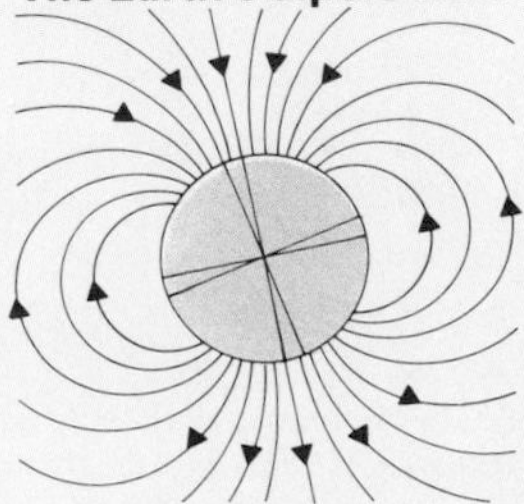

The Earth's magnetic field is not constant – it varies over area, and through time

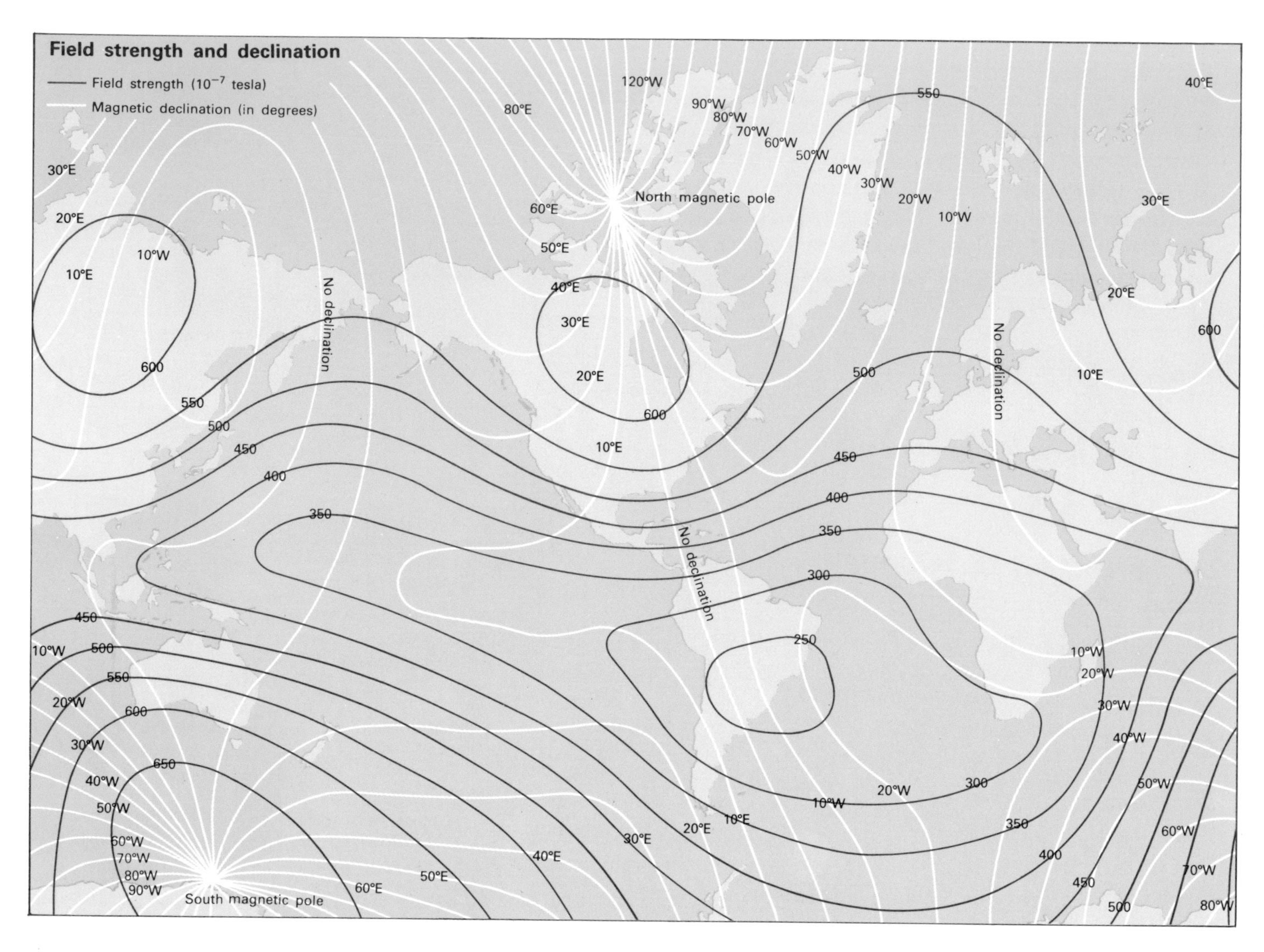

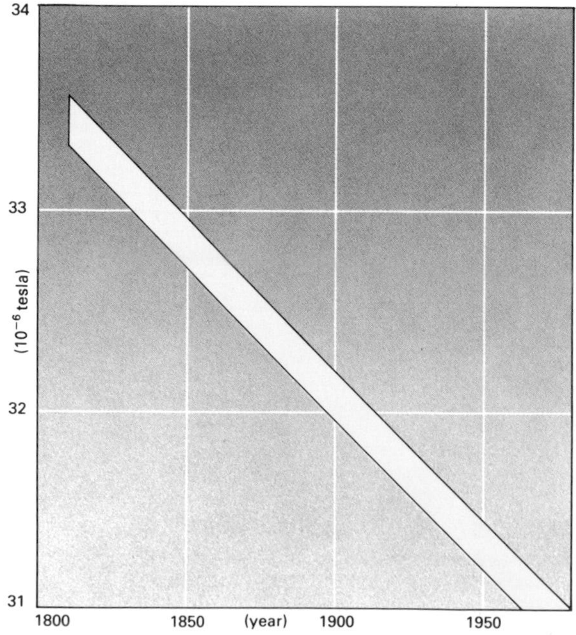

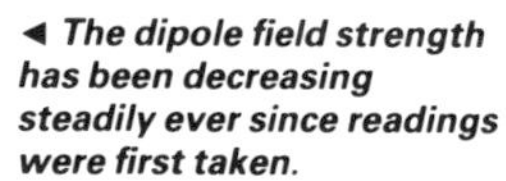

◄ *The dipole field strength has been decreasing steadily ever since readings were first taken.*

▲ *The strength of the dipole element of the Earth's magnetic field varies from place to place. The map shows the variation of field strength in 1975, along with the magnetic declination at various parts of the world.*

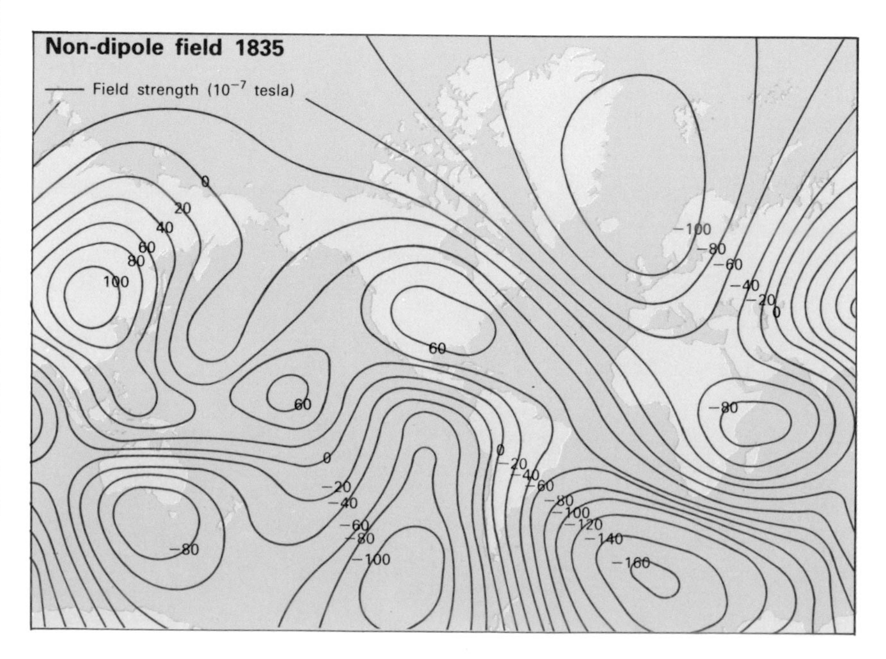

►► *The non-dipole element of the magnetic field is much weaker, but it also changes from time to time. The maps show the intensity of this component of the field in 1835 (right) and 1965 (far right).*

The geomagnetic field is further complicated by the fact that both its dipole and non-dipole elements change with time. The dipole changes its position only very slowly. Over the past 150 years or so, the geomagnetic poles have stayed at the same latitude (79°), and they have changed longitude at a rate of only 0·042° a year. On the other hand, the dipole has decreased in strength over the same period by more than 7 percent. If it continues to decrease at that rate, the dipole field will have disappeared altogether by about AD 4000. That could happen, but it is equally possible that the dipole strength could start to increase again at any time.

The non-dipole field changes much more rapidly than the dipole. The nature of the changes can easily be seen by comparing charts of, say, the vertical component of the non-dipole field for 1835 and 1965. Though the charts are broadly similar, in the 130 years between the two the curves and loops have changed their shapes, positions and magnitudes. A detailed mathematical analysis shows that while undergoing changes the non-dipole field is drifting generally westward at a rate of about 0·2° longitude a year. This rate of change is about an order of magnitude greater than that of the dipole field. Indeed, changes in the non-dipole field are so rapid that they can be observed on a year-to-year basis. This process inevitably gives rise to problems in navigation, which is why mariners' charts must be updated every few years.

As accurate measurements of the Earth's magnetic field have only been made since the early 1800s, longer term changes in, and characteristics of, the field have not been observed directly. However, some information about the field over longer time scales may be obtained from paleomagnetic studies. These show, among other things, that the Earth has had a magnetic field for at least 2·6 billion years and probably longer, that throughout the history of the field the dipole component is likely to have been as dominant as it is now, that the strength of the dipole component has decreased by about 50 percent during the last 2,000 years or so, that from time to time the direction of the dipole field has reversed completely, and that when the field reverses it does so remarkably rapidly by the standards of geological time – in less than 10,000 years.

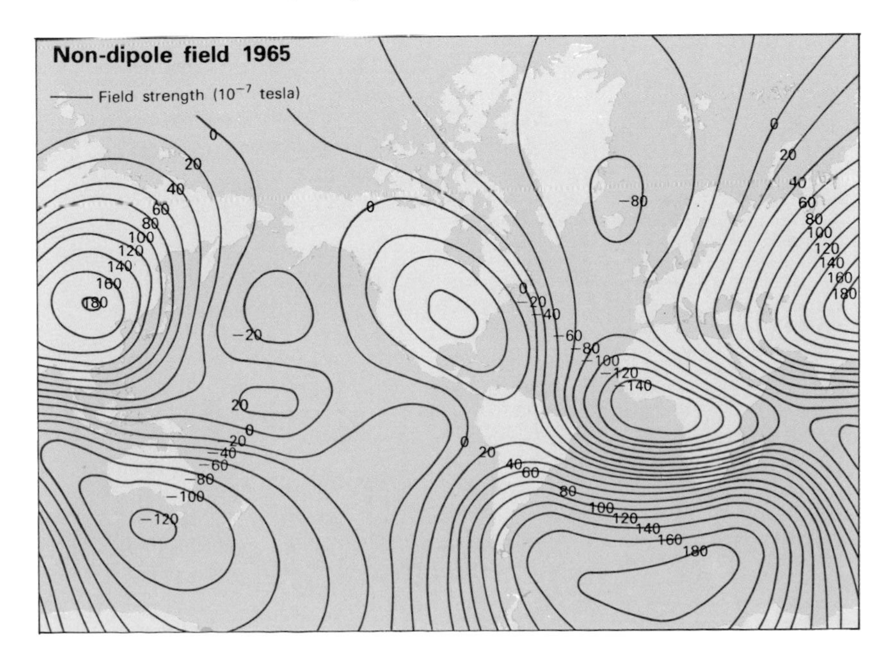

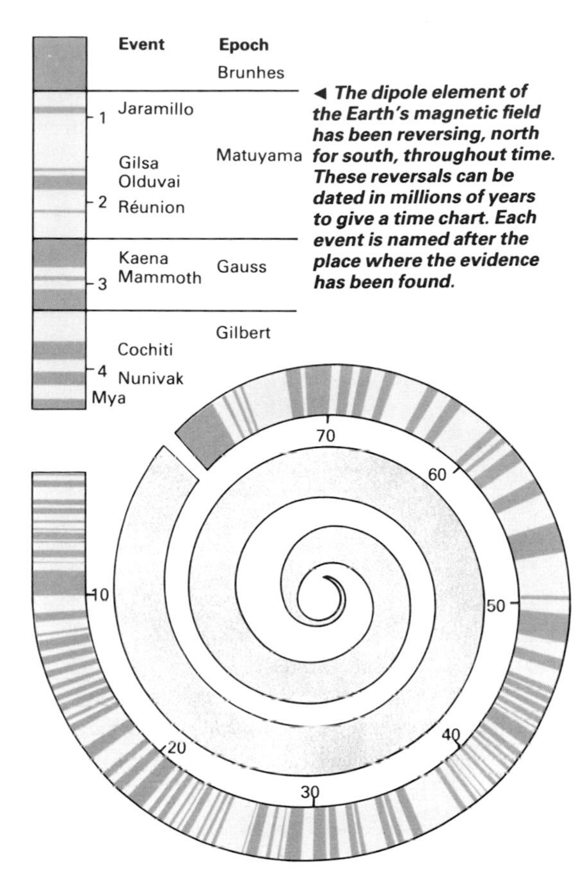

◄ ***The dipole element of the Earth's magnetic field has been reversing, north for south, throughout time. These reversals can be dated in millions of years to give a time chart. Each event is named after the place where the evidence has been found.***

Paleomagnetism

Paleomagnetism is the study of magnetism in rocks. Most rocks contain very small particles of magnetic material, usually iron-titanium oxides. When a rock is formed, these particles acquire a magnetization parallel to the direction of the Earth's magnetic field at that time and place. In many rocks this magnetization, though weak, is very stable, maintaining its integrity for sometimes billions of years. Such rocks are therefore "fossil magnets" with details of the Earth's magnetic field locked into them.

Paleomagnetic studies were begun in the 1950s, when sensitive magnetometers were developed, with the aim of exploring the size and shape of the Earth's magnetic field in the past. The data provided valuable information about the drifting of continents, and so the emphasis changed.

It was soon discovered that the field is continually reversing. At various times in the past, therefore, a rock that would now become magnetized in one direction (normal) would then have become magnetized in precisely the opposite direction (reversed). By careful dating, Earth scientists could plot the pattern of reversals over the past few million years. This pattern was then used as proof of ocean floor spreading.

Unfortunately, the accurate polarity-time scale based on continental rocks cannot be extended back more than about 4-5 million years because it is not possible to date the rocks accurately enough. However, a less accurate scale going back many tens of millions of years has been constructed from ocean floor anomalies. Some very old continental rocks are reversely magnetized, so the Earth's magnetic field has obviously been reversing for at least several hundred million years.

The fact that part of the geomagnetic field changes rapidly provides an important clue about the field's origin. The simplest possible explanation of the field is that the Earth is a permanent magnet. But if parts of the solid Earth were to move in such a way as to produce field changes of the rapidity observed, the planet would long ago have disintegrated! The field must therefore originate in a part of the Earth in which rapid movement can take place without violent disruption, and that means in the fluid outer core.

The core is thought to consist largely of iron, which is a good conductor of electricity. Conductors that move in magnetic fields are known to generate electric currents, and electric currents in conductors are known to produce magnetic fields. The only viable explanation of the geomagnetic field, therefore, is that it arises in some way from motions (probably thermal convection) in the iron of the outer core. These motions are complex, and the exact way in which they generate the field will probably never be fully understood.

The attraction of gravity

The second of the Earth's fields is gravity. During the mid-19th century, a team of surveyors led by Sir George Everest carried out a triangulation survey of northern India. In the course of their work they determined the precise distance between two towns in two different ways – by measurement across the land and from astronomical observations. Much to their surprise they found that the two measured distances differed by about 150m in approximately 600km. As this was much too much to be explained by experimental error, even by the standards of those times, Sir George and his colleagues concluded that the plumb line used in their astronomical measurement could not have been pointing towards the center of the Earth as expected, but must have been deflected towards the Himalayas. The mountains must have been exerting a gravitational attraction on the plumb bob and pulling the line away from the vertical.

To prove the point, Sir George estimated the mass of the mountains and calculated what effect they should have on a plumb bob. He then had another surprise. The mountains were deflecting the plumb line by far less than they should have been. If they had been exerting their full attraction, they would have produced a discrepancy in the distance between the two towns not of 150m but of 450m. A similar discovery was made about a century before in the Andes by Pierre Bouguer, but, in the more developed scientific climate of the 19th century, it was Everest's "Indian puzzle" that attracted the attention.

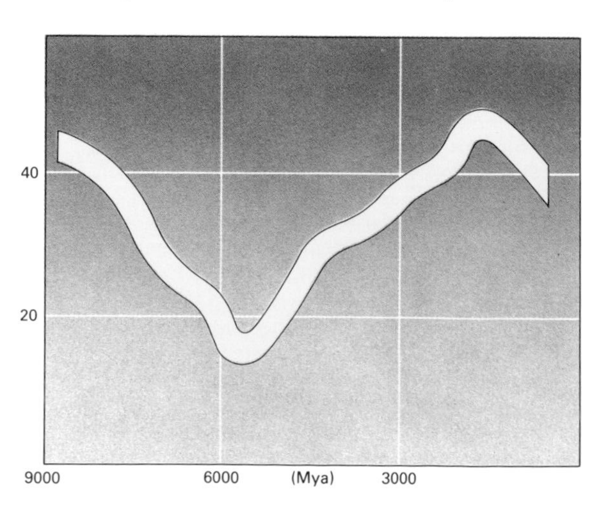

◄ When a brick is baked, the magnetic particles in the clay minerals align themselves with the Earth's magnetic field. This is then sealed. Archeologists can analyze ancient bricks and determine the strength of the field at the time the bricks were baked. From this it has been discovered that the dipole field was decreasing until about 6,000 years ago, then increasing until about 1,500 years ago, since when it has been decreasing again. Further archeological evidence suggests that climatic disruption accompanied the weak magnetic field.

The self-exciting dynamo

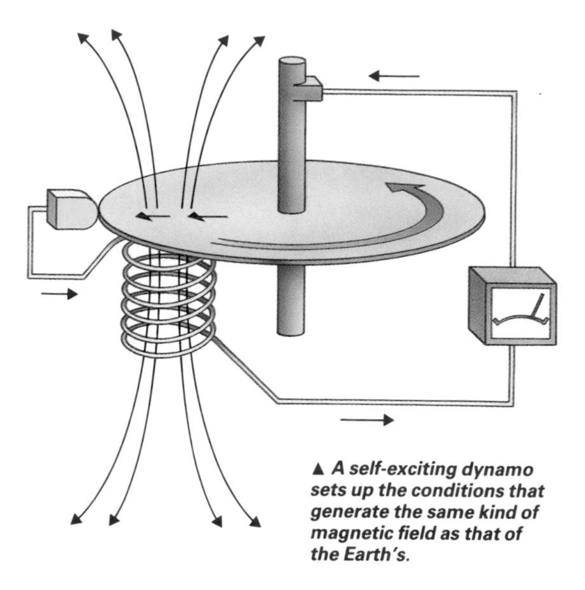

▲ A self-exciting dynamo sets up the conditions that generate the same kind of magnetic field as that of the Earth's.

The self-exciting dynamo

The real magnetic field-producing mechanism in the Earth's outer core can hardly be envisaged. The principle involved may well be that of the self-exciting dynamo, which can easily be understood in terms of a simple electro-mechanical analogue. The apparatus consists of a solenoid (electric coil) located beneath a copper disk mounted on an axle which is kept rotating. A wire then goes from the edge of the disk to the axle via the solenoid and a current meter. The connections with the disk and axle are made by rubbing contacts (brushes).

All that is needed to set the process going is the application of an initial magnetic field for a short time. If a conductor moves in a magnetic field a current is induced, and so the initial magnetic field will induce a current in the spinning disk. This will flow round the circuit and, in particular, through the solenoid. But a current flowing through a coil produces a magnetic field, and so the current already generated will produce a magnetic field in the solenoid, which will generate further current in the disk, and so on. The apparatus therefore goes on generating a magnetic field for ever. (It is not a perpetual motion machine, however, as energy has always to be used to keep the disc rotating.)

By connecting two pieces of equipment together it is possible not only to generate a perpetual magnetic field but also to generate one that from time to time reverses spontaneously. In fact, a slightly more complicated version of the system using cylinders instead of disks was actually built at the University of Newcastle-upon-Tyne in England during the 1960s – and it worked.

In the real Earth, the iron of the outer core is the equivalent of the conducting material of the disk and the source of energy that keeps the conductor moving is probably heat from the decay of radioactive isotopes. The initial magnetic field that set the field generation process going could well have been a stray field from the Sun.

The discovery of mountain roots

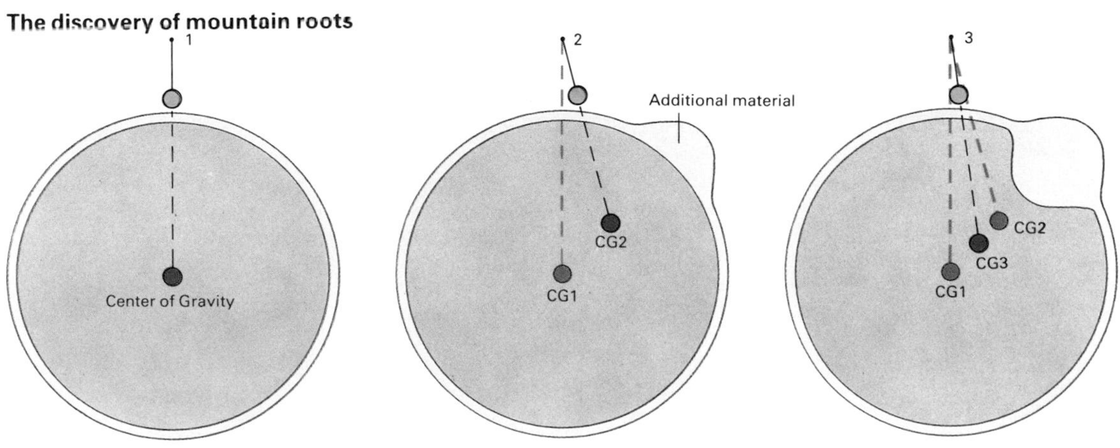

◄ ▲ *Theoretically a plumb bob should point to the center of the Earth (1) – the Earth's center of gravity (CG1). Near a massive mountain it would be expected to deflect to one side by the gravitational force of the mountain's mass, and indicate CG2 (2). In fact the deflection is smaller than expected, indicating CG3 (3). The reason is that there is light mountain material deep below the Earth's surface, rather than the denser material that makes up the bulk of the Earth. Studies in the Himalayas showed this effect.*

During the 1850s Sir George Airy, the Astronomer Royal, saw that the gravitational attraction exerted by the extra mass of the mountains must be partly counteracted by a deficiency of mass beneath. In other words, mountains must have hidden "roots" made of rock whose density is lower than that of its surroundings. This explained the discrepancy in Everest's findings.

The principle involved is that of isostasy, the geological version of Archimedes' principle of buoyancy. Because the continents are composed of comparatively light material, they float on the denser asthenosphere beneath, much as a piece of wood floats on water. If an extra load is placed on part of a continent, that region will sink until it reaches a new position of floating balance. The partially molten material in the zone of asthenosphere immediately below will slowly flow aside as the base of the lithosphere, and with it the base of the crust, moves downwards. In practice, however, the base of the lithosphere probably does not move down very much at all in any real sense, for that part of the lower lithosphere moving down will become partially molten itself – that is, it will become part of the asthenosphere. If the load is removed, the continental region involved will then rise again, the new asthenosphere will be converted back into lower-temperature lithosphere, and old asthenospheric material will flow back in to take its place.

The most spectacular example of a continental load is a mountain. The extra mass of the mountain depresses the base of the crust, forming a root of crustal rock that is lighter than the upper mantle surrounding it. The crust is therefore particularly thick in mountainous regions, not just because the mountains rise above the surrounding plains but because the base of the crust is bent downwards. As the mountain slowly erodes away the crust will gradually get thinner. When the continental load is ice, the crustal thickness does not change but the continent still falls and rises.

All of the Earth's lithosphere aspires to isostatic balance, and most of it has achieved the state in broad terms. Some regions are not quite in equilibrium, however. Fennoscandia is a case in point.

The bouncing land surface

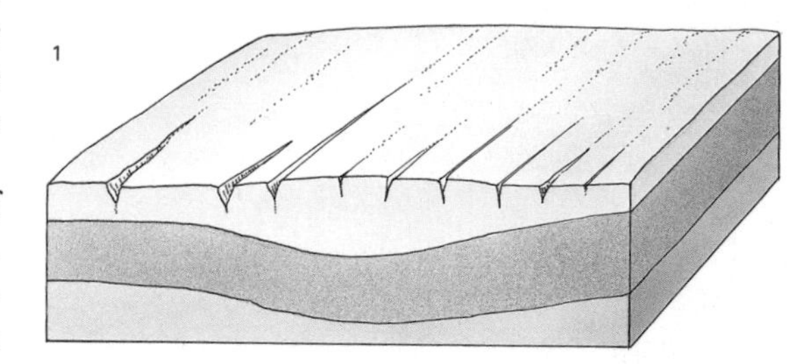

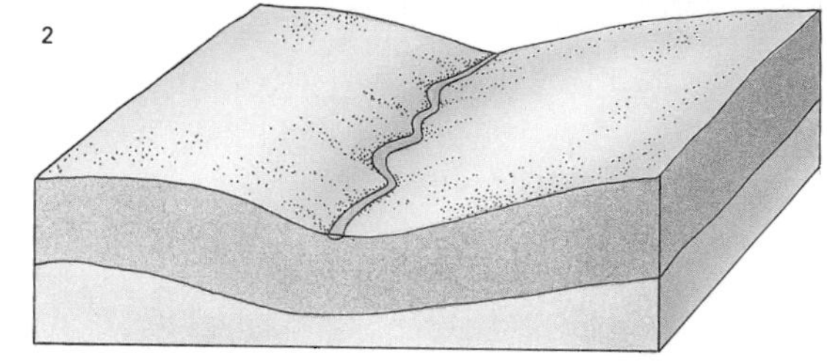

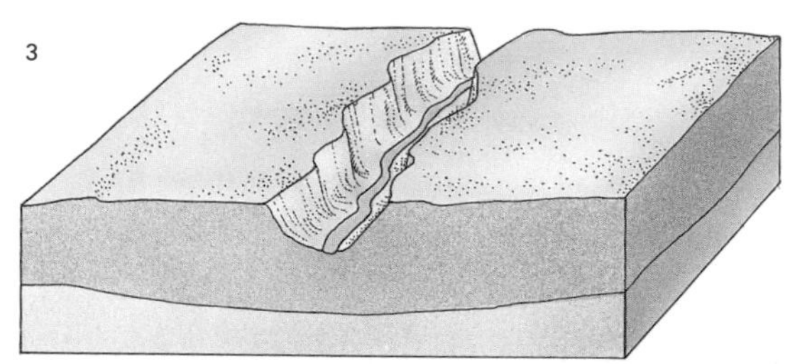

▲ During an ice age the weight of the glaciers and ice caps depresses the crust immediately beneath, down into the mantle. After they melt, it takes some time for the crust to spring up again. When it does spring up, the rivers cut down to their former levels.

► Lake Bonnevile in Utah was much bigger during the Ice Age. Since then much of it has evaporated. Now with the loss of the great weight of water the underlying crust is rebounding and former shorelines lie well above the new water level.

Mountain roots

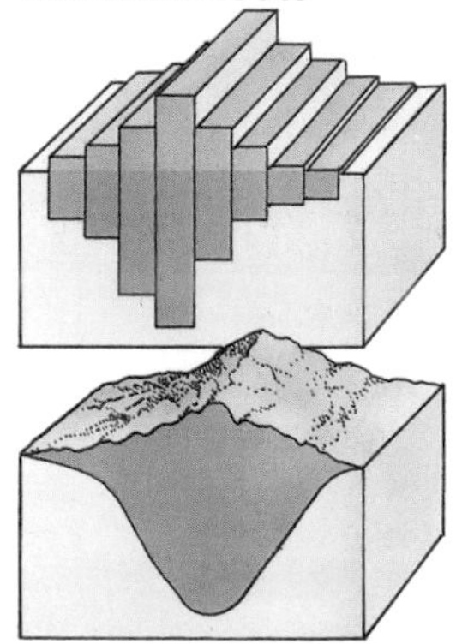

▲ Sir George Airy had the right idea in 1855 when he suggested that mountains were like blocks of wood floating on water. The blocks that protruded highest above the surface also extended to the greatest depths below. This led to the understanding of mountain roots.

The Fennoscandian uplift

Mountains are created and depress the continental lithosphere so slowly that the phenomenon is hardly measurable. But nature has also provided a grand experiment that illustrates the principle of isostasy at work at an observable rate. One of the areas covered by ice during the last Ice Age was Fennoscandia – the countries of Norway, Sweden, Finland and Denmark. Under the growing mass of ice the Fennoscandian lithosphere was gradually pushed downwards. When the ice retreated about 10,000 years ago the region began slowly to rise, and is still rising.

Beaches form at sea level, but around the edges of Fennoscandia there are numerous beaches at various heights above sea level. These must have formed when Fennoscandia was lower in the water than it is now. By dating and measuring the altitudes of these raised beaches, geologists have been able to construct a "contour" map showing by how much the area has risen over the past 5,000 years. The greatest uplift has occurred over western Sweden where the ice was presumably thickest. Here the land has been rising at an average rate of 2cm a year for at least 5,000 years, which suggests that it has probably risen something like 200m since the ice retreated.

The progress of the Fennoscandian uplift has been plotted in detail. As the ice began to melt about 100,000 years ago and the land began to rise, the Baltic area was covered by a fresh-water lake – the Baltic Ice Lake. Further melting of the ice raised the sea level round about and flooded the fresh-water lake, making it a salt-water inlet called the Yoldia Sea. Then, as the land continued to rise, the area was cut off from the sea once more and another fresh-water lake – the Ancyllus Lake – was formed. Finally the uplift slowed, and the rising sea level flooded the lake to make another salt-water body that eventually became the Baltic as we know it today. The various lakes and seas were named after fossils that were widely distributed in them at the time.

There is still a large negative gravity anomaly over Fennoscandia, showing that isostatic adjustment is by no means complete. In other words, the region has not yet returned to isostatic equilibrium. From the size of the anomaly, geologists estimate that there is still another 200m of uplift to come.

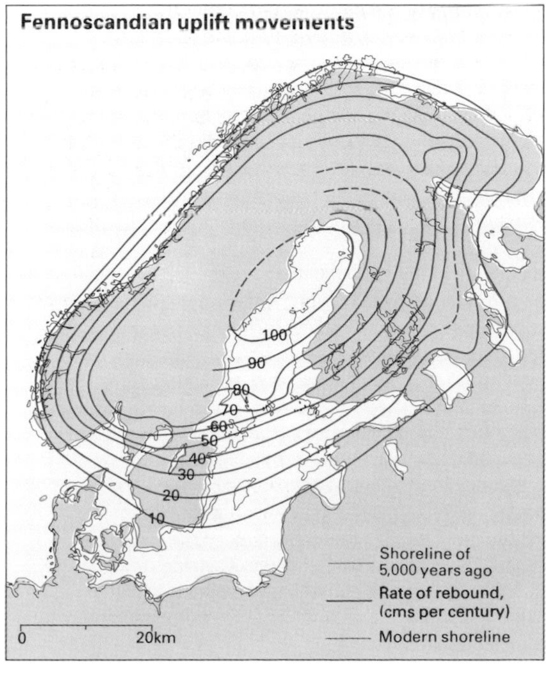

◄ The Scandinavian region was compressed under a great weight of ice during the Ice Age. Now that the ice has melted it is rising up again at a measurable rate. The rate of rebound is greater at the head of the Gulf of Bothnia (a meter per century) than elsewhere in the region.

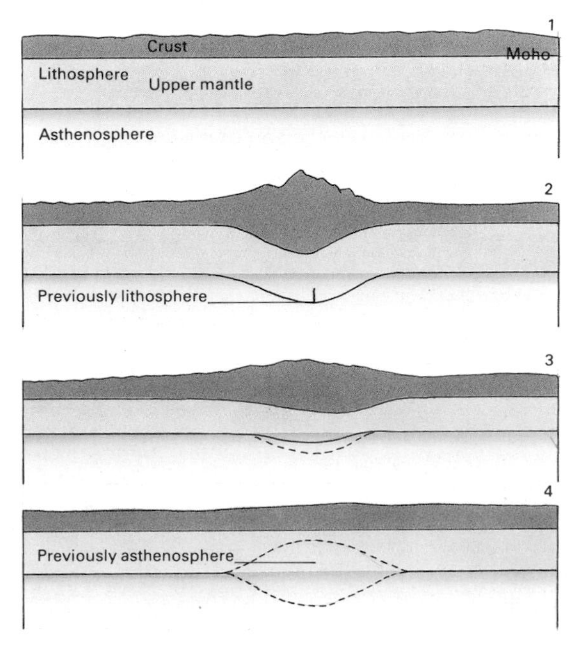

► When a growing mountain sinks, its roots may simply press the lower boundary of the lithosphere into the asthenosphere, displacing the soft asthenosphere material. However, it seems more likely that, as the lower boundary of the lithosphere is depressed, it melts to become part of the asthenosphere, so the boundary remains constant (1,2). As the mountain erodes the reverse happens, with asthenosphere rising and solidifying (3).

Gravity anomalies

The discovery that mountains have low-density roots, and that the roots produce a measurable effect on the local gravity field, soon led to the idea that gravity observations could be used to provide information on subsurface rock bodies. This is the principle of gravity surveying or, more accurately, gravity anomaly surveying. The idea here is that a subsurface rock body with a density different from that of the surrounding rock will produce at the Earth's surface a value of gravity different from what it would have been if the subsurface body had not been there. In other words, the body produces a gravity anomaly. In principle, then, it should be possible to determine the size and shape of an anomaly by measuring the gravity at various points above the subsurface body and comparing the readings with one taken at a reference site, at sea level, below which there is no such body.

In practice, gravity anomaly surveying is a little more complicated than that because the difference between the gravity value above a subsurface body and that at a sea-level reference site can be due to a number of effects other than that of the body itself. To obtain the true anomaly from the subsurface body it is therefore necessary to eliminate such effects. This can be done, with care, and the resulting anomaly is called the Bouguer anomaly.

Bouguer anomalies occur on various scales, reflecting the sizes of the bodies producing them. Academic geologists, for example, may be interested in large-scale anomalies produced in the vicinity of large-scale regional geological features such as mountains or trenches, or they may wish to investigate smaller bodies such as subsurface granite masses with horizontal dimensions of tens of kilometers. Economic geologists are more likely to have an interest in searching out valuable ore bodies or oil traps. And at a very much smaller scale, archeologists sometimes use gravity anomalies in their search for buried artefacts.

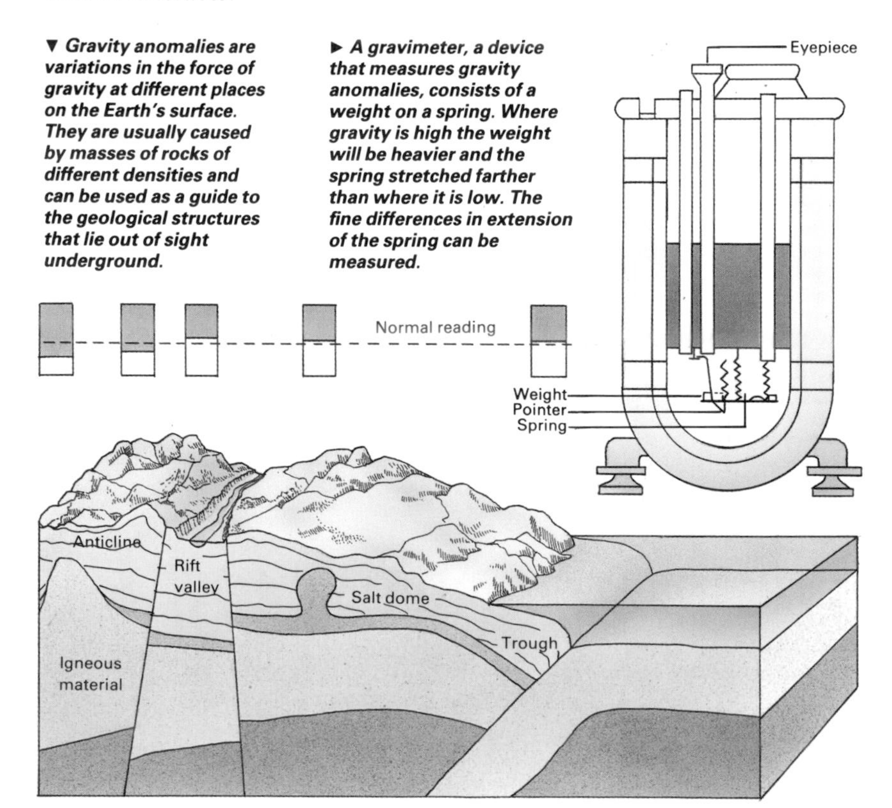

▼ Gravity anomalies are variations in the force of gravity at different places on the Earth's surface. They are usually caused by masses of rocks of different densities and can be used as a guide to the geological structures that lie out of sight underground.

► A gravimeter, a device that measures gravity anomalies, consists of a weight on a spring. Where gravity is high the weight will be heavier and the spring stretched farther than where it is low. The fine differences in extension of the spring can be measured.

The Bouguer gravity anomaly

To measure the gravity anomaly produced by a subsurface body, it is not sufficient simply to compare the gravity at a site (B) above the body with that at a site (A) at sea level. The comparisons must be made under the same conditions. The surface of the Earth is so irregular that some important corrections must be applied to the measured gravity in order to obtain the true value of the anomaly.

First, it is necessary to remove the effect of differences in latitude between the two points. Because of the rotation of the Earth, it is not perfectly spherical. It has an equatorial bulge and a polar flattening, making the radius at the equator about 21km greater than at the poles. Other things being equal, the gravity at the equator will therefore be lower than at the poles because a point on the equator is farther away from the Earth's center. Moreover, the gravity at the equator will be even smaller because unlike at the poles, there is a centrifugal force tending to make things fly outwards. The variation of gravity with latitude between the equator and the poles at sea level is expressed by a standard formula known as the International Gravity Formula (IGF). If sites A and B are at different latitudes, the IGF must be used to make the "latitude correction", reducing both sites to the same latitude for comparison.

Second, it is necessary to remove the effect of altitude differences. Since gravity gets smaller with increasing distance from the Earth's center, if A and B are at different heights there will be a difference in gravity between them for that reason alone. The correction to reduce both sites to the same altitude for comparison, the "free-air correction", is easy to make because gravity is simply inversely proportional to the square of the distance from the Earth's center.

Third, it is necessary to remove the effects of attraction of the material between the level of site B and that of site A. If the site is on land, for example, it is necessary to calculate what effect the rock between B and sea level will have on the measured gravity at B and remove it from the reading. This is known as the "Bouguer correction".

Finally, if there are hills or valleys, or both, in the vicinity of site B these, too, will have an effect on the gravity measured there, just as the Himalayas affected Sir George Everest's plumb line. The effect of such topographic variations must be calculated and allowed for. This is the "topographic correction".

Once all these corrections have been made, the way is clear to make a meaningful comparison of the gravities at sites A and B. Any remaining difference in gravity is known as the "Bouguer gravity anomaly", and must be due to a subsurface body beneath B. Having obtained the Bouguer anomaly, it is then possible to begin to think about the shape and nature of the body producing it. If the gravity at B is greater than A – a positive anomaly – there must be an excess of mass (higher-density rocks) beneath B. On the other hand, if the gravity at B is smaller than at A – a negative anomaly – there must be a deficiency of mass (lower-density rocks) beneath B.

▼ *The worldwide variation in gravity can be measured using a satellite such as the European Space Agency's Starlette. It is small but extremely heavy and its orbit can be calculated precisely. Any variation in this orbit is due to variations in the gravitational attraction of the Earth and can be measured by the reflection of Earth-based lasers from its mirrors.*

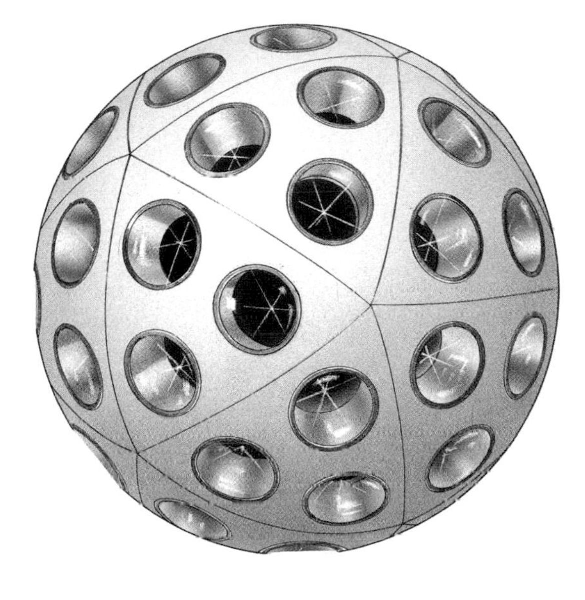

▼ *Satellite measurement of the worldwide gravity anomalies shows that the Earth is not a perfect sphere. In some places the sea level surface differs from the theoretical sea level surface by up to 100m. This is due to variations in density below the crust, and many of these anomalies seem to be associated with active plate boundaries.*

The need for an International Gravity Formula

There is one major problem that all users of gravity data have to face, and that is that a gravity anomaly alone can never provide a unique interpretation of the subsurface body that produces it. For any anomaly that is measured there is always an infinite number of possible combinations of size, shape, depth and density that can explain it. Thus although the existence of a gravity anomaly may be sufficient to demonstrate that a subsurface body of anomalous rocks exists, the complete specification of that body requires not just gravity data but information from such sources as seismic surveys and boreholes.

The idea of an International Gravity Formula is that it should express the value of gravity all over the Earth's surface at sea level and at the equivalent level in the continents. The formula therefore gives gravity at various latitudes taking into account the variations introduced by the Earth's shape and rotation. Geophysicists have found, however, that the IGF only gives the value of gravity at sea level to a very good approximation. Very accurate observations of gravity, both at the Earth's surface and from satellites, show that the surface over which the IGF actually applies is not precisely the sea-level surface but one undulating gently around it, being sometimes above and sometimes below. This surface is called the geoid and it differs from sea level by up to about 100m.

The very large-area deviations of the geoid from the sea-level surface are, like other gravitational anomalies, the result of density variations in the Earth. But in this case they are unlikely to be due to variations in the crust or lithosphere, for they do not correlate at all with the distribution of continents and ocean basins. The variations involved must therefore be in the mantle. They are very poorly understood, but they are almost certainly one manifestation of large-scale convection currents in the mantle. And those, in turn, result from temperature variations in the Earth.

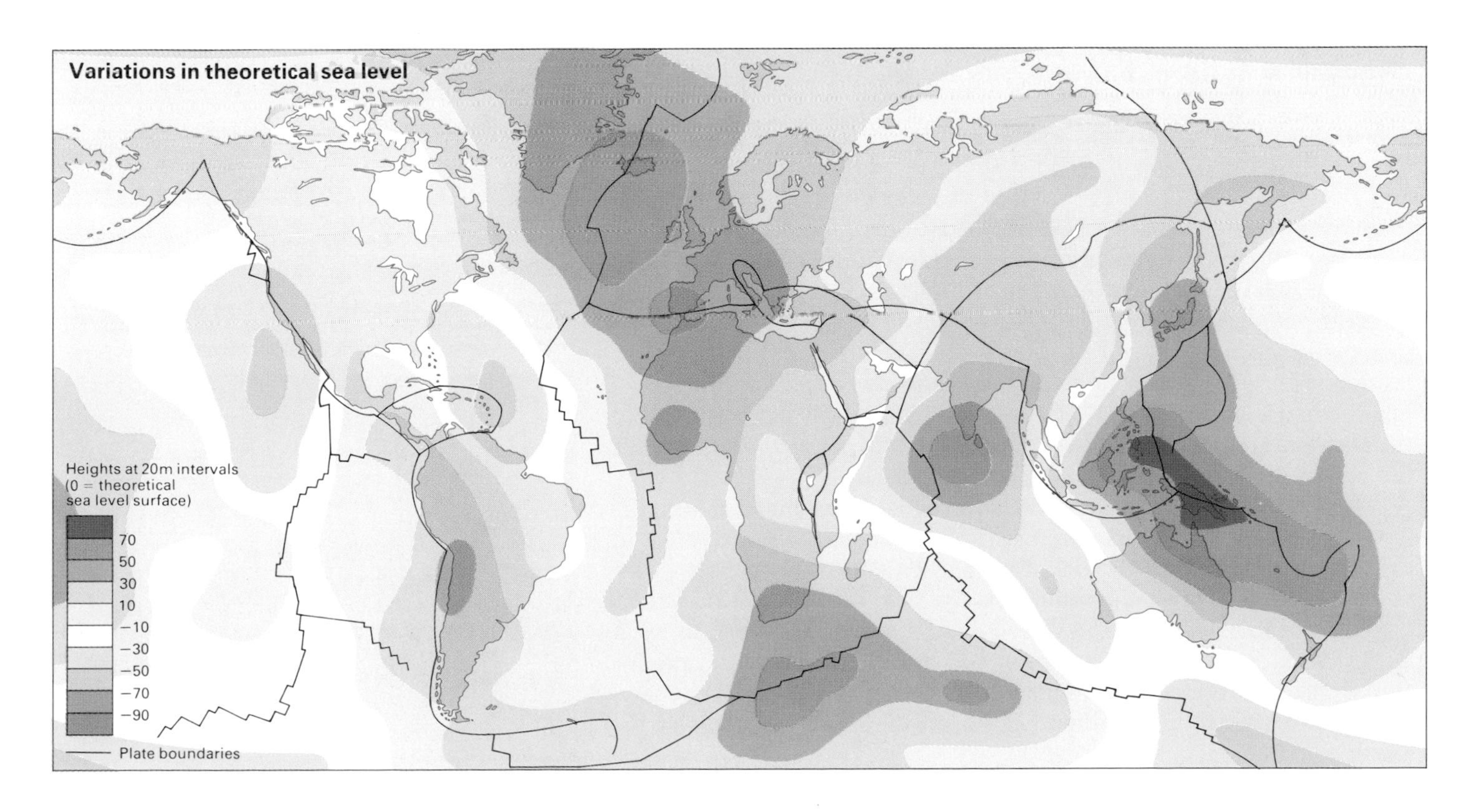

In one year heatflow to the Earth's surface is 1,000 times greater than the annual energy released by earthquakes

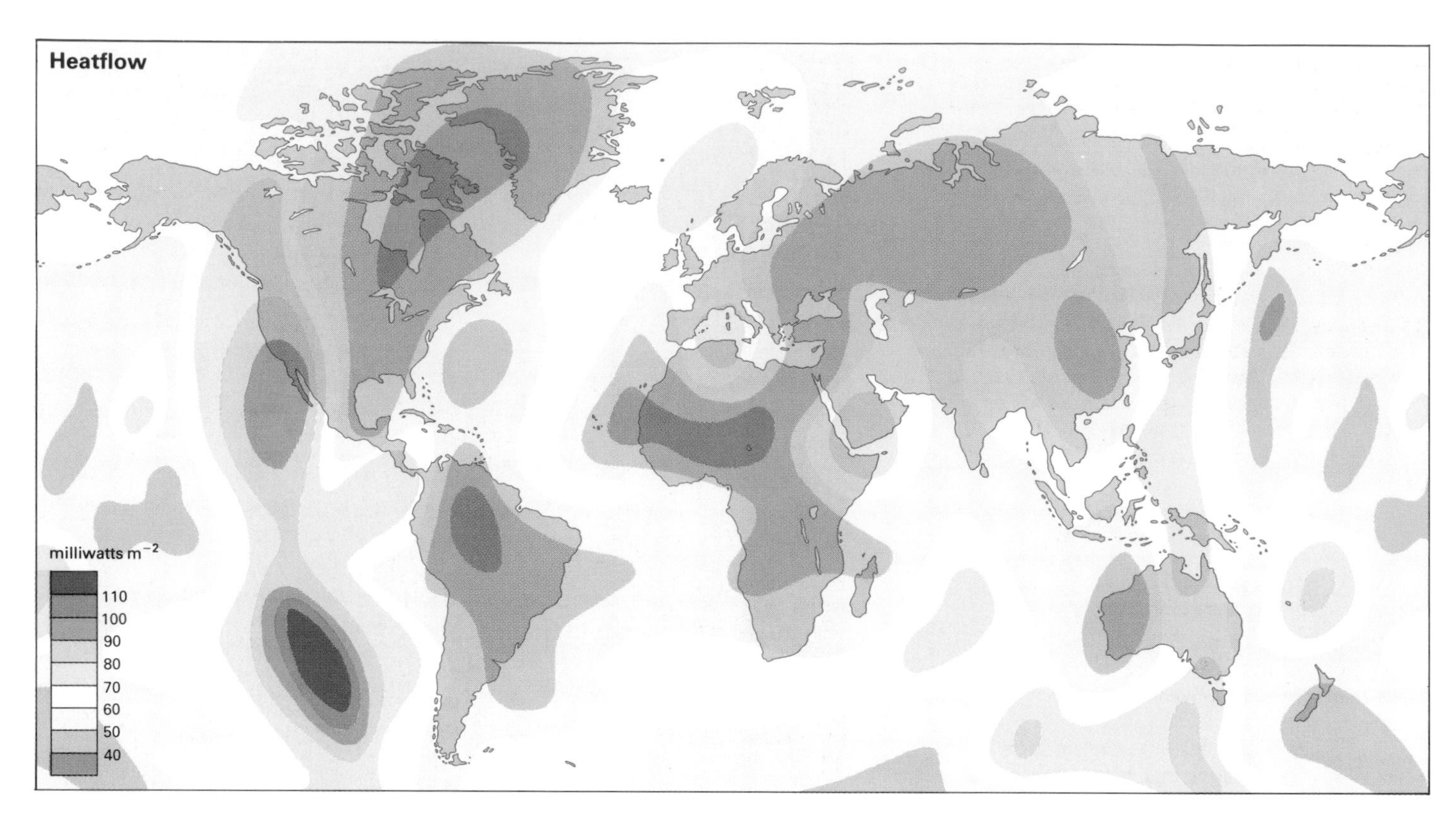

Temperature variations in the Earth

As miners have known for centuries, the Earth gets hotter with depth. Near the surface the average rate of increase in temperature is about 3°C per 100 meters, but in very cold continental crust it can be as low as a third of that and in volcanically active areas it can be very much higher. If this average gradient were to persist throughout the whole Earth, the temperature at the center would be in excess of 190,000°C. The fact that the inner core is solid shows that the temperatures are nothing like that, which must mean that the temperature gradients deep in the Earth are much lower than at the surface.

The major source of heat today appears to be the decay of long-lived radioactive isotopes, chiefly those of uranium (two), thorium and potassium. As the concentration of these elements and their distributions with depth are uncertain, it is impossible to say precisely what proportion of the total heat is generated in this way; it could be anything from 50 percent to almost 100 percent. Nevertheless, some of the heat now emerging at the Earth's surface is almost certainly primordial, generated during, or very soon after, the Earth's formation. Heat travels through the Earth so slowly that an injection of heat near the center would take tens of billions of years to be felt at the surface, and so it is likely that in the 4·6 billion years of the Earth's life not all the primordial heat has been dissipated.

There is no shortage of possible sources of the primordial heat. As the Earth grew by the accretion of small planetary objects, some of the energy (kinetic) of the colliding particles would have been converted to heat. Then there would be the heat of compression generated as further accretion produced higher and higher pressures at depth. Next, gravitational energy would have been released as heat as the Earth differentiated into crust, mantle and core. And finally, there were probably short-lived radioactive elements present at the Earth's formation.

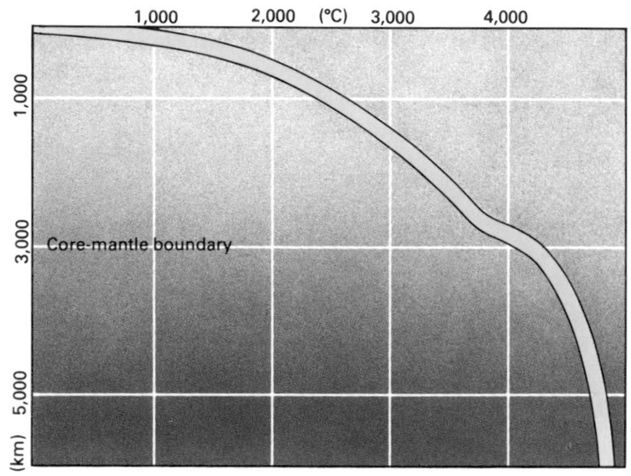

▲ ***The temperature of the Earth increases most rapidly with depth through the outermost layers.***

▼ ***The Earth's heat production through time has been due to various radioactive elements.***

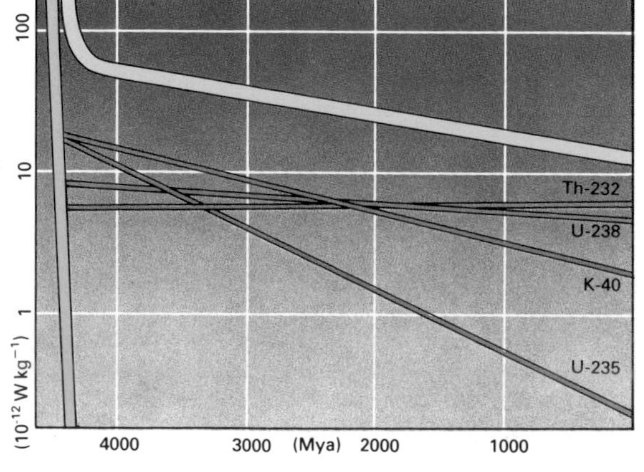

◀ A heatflow map of the world shows that more heat reaches the surface of the Earth from its interior at constructive plate margins than elsewhere.

▶ The heat that comes to the surface of the Earth is most obvious in volcanic areas, and in those regions where hot springs and geysers erupt.

Temperature measurements and inferences

The variation of temperature throughout the Earth is rather speculative, for direct measurements can only be made close to the planet's surface. On the other hand, geophysicists and geochemists can make inferences about the Earth's internal state at various levels. These inferences enable them to place constraints on probable temperatures and in this way they can keep at least some of the plot of temperatures versus depth reasonably close to reality.

For example, as the asthenosphere is partially molten it must be close to the melting point of mantle material. At the relevant pressures, the temperature is therefore probably in the region of 1,100-1,200°C in the depth range of about 100-200km. Corroboration of this comes from the lava emerging from oceanic volcanoes, which has about the same temperature and is thought to originate in the asthenosphere.

Below the asthenosphere the mantle is solid and so its temperature must be below the melting point at the pressures concerned. A more precise "fixed point" comes at a depth of 400km, however, where there is a minor but conspicuous jump in the seismic velocity (♦ pages 57-68). The phase change thought to be responsible for this jump has been reproduced at high pressure in the laboratory where it was found to occur at about 1,500°C. There is also a seismic jump at about 650km. The phase change that geophysicists believe to account for this has not been observed under laboratory conditions, but theoretical calculations suggest that it should occur at about 1,900°C.

Where the core is concerned there is perhaps less agreement on temperatures. The outer core is liquid and so its temperature must be higher than the melting point of core material at the relevant pressures. But what is the core material? If it were pure iron the temperature of the outer core would have to be in excess of 3,900°C at the top and 4,400°C at the bottom. However, the core must also contain a small proportion of a light element such as sulfur, and some geophysicists believe that this could reduce melting points in the core by as much as 2,000°C.

The inner core is solid. This means that its temperature is likely to be around 4,400°C, although the more cautious geophysicist would perhaps say only that the temperature at the center of the Earth is probably somewhere between 4,000°C and 5,000°C.

It is evident from the plot of temperature against depth that the temperature gradient (the rate of change of temperature with depth) decreases with depth. So the temperature at the Earth's center is nowhere near 190,000°C. Just above the core there is a small region where the temperature gradient is slightly higher than expected. This is probably due to excessive heat flowing into the solid mantle from the convecting fluid outer core.

The Earth's thermal history is therefore one based on a huge injection of primordial heat followed by continued heating by a few long-lived radioactive isotopes. When these isotopes decay to nothing in over 10 billion years' time, the Earth will cool until it is completely solid and all plate movement and volcanic activity will stop.

The quantity of the Earth's internal heat now flowing outward by conduction through the surface is large in absolute terms. Over the whole Earth it amounts to about 10^{21} joules a year, which is about 10 times greater than the rate of energy dissipation by the tides in slowing the Earth's rotation and about 1,000 times greater than the rate of energy release by earthquakes. On the other hand, the rate at which heat flows through one square meter of the Earth's surface is very small – but measurable. On average this rate is about 0·06 joules per square meter per second (= 60 milliwatts per square meter).

Many thousands of heat flow measurements have been made around the world, from which geophysicists have been able to construct a contour map. The pattern of heat flow thus displayed correlates remarkably well with plate tectonic processes. Heat flow is high over oceanic ridges, where hot material from the Earth's interior is rising to form new oceanic lithosphere. Moreover, it is particularly high over the East Pacific Rise where the spreading is rapid, but less so over the mid-Atlantic and Indian Ocean ridges where spreading is slower. In reality, the heat flow over the oceanic ridges is even higher than the map indicates. Heat flow measurements record only the heat being conducted through the upper crust, but at oceanic ridges heat is removed additionally by the circulation of seawater within the lithosphere, producing the chemical-rich "smokers" at the ridges.

More detailed studies indicate that as the oceanic lithosphere spreads away from the ridges the heat flow reduces. The lithosphere starts off very hot and then cools and contracts with time and, since it is spreading, with distance from the ridges.

See also
The Dynamic Earth 9-24
Internal Structure 33-44
Internal Composition 57-68

► ***Heat is probably transferred from the Earth's core to the surface by means of convection currents in the mantle. It is not known just how complex these convection currents are. Large currents may circulate through the whole mantle (1), or there may be individual currents circulating in the different mantle layers (2).***

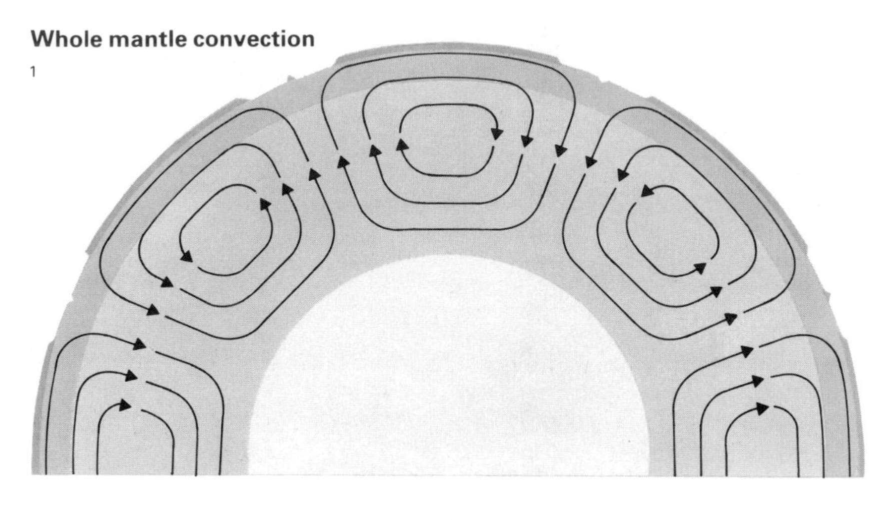

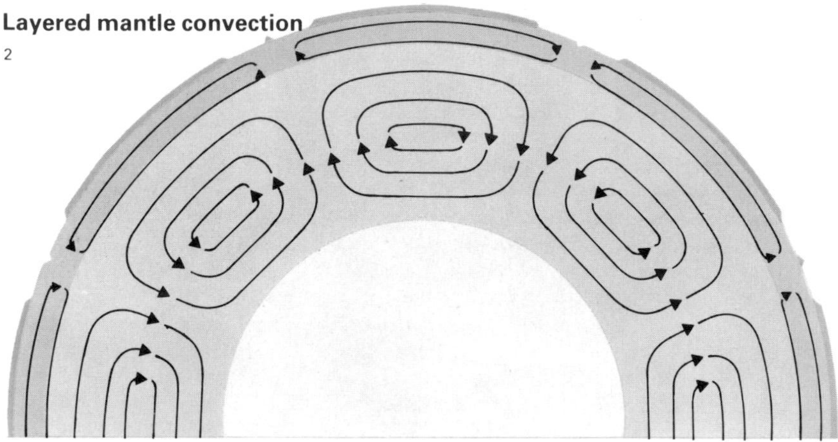

Convection in the mantle? The first ideas

One of the great, long-standing controversies in geophysics concerns whether or not the large amounts of heat in the Earth induce convection in the mantle and, if so, in which part(s) of the mantle. When continental drift began to be taken seriously in the early 1960s, mantle convection was seen as a mechanism for it. The idea was that moving mantle would exert a horizontal force on the base of the continents and thus drag them along. At that stage it was thought that convection currents occurred throughout the whole mantle. A little later, when it became clear that there is a partially molten layer (the asthenosphere) in the upper mantle, some geophysicists suggested that convection took place there only.

By the time continental drift and ocean floor spreading had been combined into plate tectonics, the consensus was that the major force controlling plate motion resulted from the pull of the dense slab of oceanic lithosphere descending at subduction zones. Motion in the asthenosphere, a sort of "return flow" was therefore induced by the motion of the lithosphere above. Though this flow is still rather loosely called "thermal convection" it is not thermal convection in the strict sense, for it is not directly thermally induced, although it would presumably transport heat like any other form of convection.

The lowest heat flow values of all come from the oceanic trenches where the now-cold lithosphere begins its descent into the Earth's interior. To the landward of a trench, however, the heat flow rises to higher values again. This is because friction at the upper edge of the descending lithospheric slab generates heat and magma that rise to produce volcanic activity along the continental edge or in island arcs.

In continental regions the heat flow patterns are more varied because the continental crust is both older and more complex than ocean crust. Nevertheless, as with the ocean floors, it is generally the case that the older the terrain the lower is the heat flow from it. Thus heat flow is higher in young volcanically active continental zones than it is in the Precambrian shield areas. On the other hand, the rate of decrease of heat flow with age is much lower for the continents than for the ocean floors. This is partly because the continents are thicker and thus take longer to cool and partly because of the high concentration of radioactive isotopes.

The average heat flow through continents is not very different from that through the ocean floors but this is an unfortunate coincidence that misled geophysicists for many years. The oceanic crust contains few heat-producing radioactive isotopes; about 90 percent of the oceanic heat flow comes from the mantle. By contrast, the continental crust is rich in radioactive isotopes, and only 50 percent of continental heat flow comes from the mantle. This suggests that the temperature in the mantle immediately below the continental crust should be somewhat lower than that below the oceanic crust, and that the two regions of mantle probably do not have quite the same composition.

The current debate

The debate over convection continues and will doubtless go on for many years. There are now three conflicting hypotheses.

According to the first hypothesis, convection is limited to the asthenosphere. However, it is not always made clear by its proponents whether this refers to thermally-induced convection or motion of the "return flow" kind induced by the motion of the lithosphere above.

The second hypothesis sees convection taking place by means of a single set of convection cells throughout more-or-less the whole mantle. In other words, convection occurs in the asthenosphere and mesosphere, but it does not take place in that layer of the mantle that forms the lower part of the lithosphere.

Finally, in the third hypothesis, convection again takes place throughout more-or-less the whole mantle but there are two independent convecting systems – one located in the upper mantle (the asthenosphere plus the transition zone) and one in the lower mantle (the mesosphere minus the transition zone).

There are things to be said for and against each of these three views, but there appears to be no easy way of settling the issue. No one disputes that the solid parts of the mantle can flow in principle. Disagreement centers on which parts of the mantle have a viscosity low enough to allow convection actually to take place.

Internal Composition

5

Earth and the planets – similarities and differences...Sun's composition – an indicator of the Earth's...Meteorites – their testimony...The crust and mantle – their composition...Minerals – their properties and identification...The core... PERSPECTIVE...Density of the planets...Distribution of elements...Meteorite types...The crust's minerals...Elements of the core

However difficult it may be to determine the physical structure of the Earth's interior, it is even more difficult to determine the planet's chemical and mineralogical composition and how the composition of the various layers differ. Information about structure may be obtained by probing the Earth with seismic waves that are generated and recorded at or near the surface, but there is no comparable way of remotely sensing the types of elements, compounds and mixtures present in the deep Earth. Composition must therefore be inferred from many different sources of evidence.

Very often it is a question of only being able to provide reasonable constraints. For example, the Earth's fluid outer core must be the source of the geomagnetic field and it has a known density; but although these properties are sufficient to exclude a large number of possible core compositions, they are not sufficient to pin the composition down precisely. Similarly, the igneous rocks now at the Earth's surface almost certainly originated in the upper mantle, and so by analyzing them chemically it should be possible in principle to find out something about the compositions of their source regions. But if the source regions are not completely molten, only a part of them will be sampled by the material rising into the upper crust and, besides, different regions of the upper mantle may have different compositions in the first place. So again, the composition of the Earth's surface rocks may put constraints on the overall composition of the upper mantle but not enable geochemists to specify it completely.

Given the impossibility of sampling the Earth's interior below 10-15km, the composition of the deep Earth must ever remain speculative – and that means that there will always be disagreement among geochemists over the true interpretation of the few, and often conflicting, data that exist. Nevertheless, there is broad agreement on the general chemical state of the Earth, and that consensus begins with a consideration of the Earth in the context of our star system, the Galaxy in general and the Solar System in particular.

The Earth in the Solar System

The Earth is just one of nine planets in the solar system, and these planets are of two major types. The outer planets (Jupiter, Saturn, Uranus and Neptune) have comparatively low densities in the range $0{\cdot}7\text{-}1{\cdot}7\times10^3$kg m^{-3} and have masses of 14-318 times that of the Earth. They are thought to consist largely of hydrogen and helium. The inner, or terrestrial, planets (Mercury, Venus, Earth and Mars) have comparatively high densities in the range $3{\cdot}9\text{-}5{\cdot}5\times10^3$kg m^{-3} and consist largely of hard rocks. Mercury, Venus and Mars are all smaller than the Earth in both mass and volume. The odd planet out is Pluto, which is also smaller than the Earth in both mass and volume but has a density comparable to the outer planets and lies beyond them.

Mercury, Venus, Mars and the Moon

*The Earth may have the same bulk composition as the vast majority of the meteorites that have been collected (**▶ page 60**). It might well be expected that Mercury, Venus, Mars and the Moon would also have this composition – and on the same grounds, namely, that all were formed in close proximity and from the same primitive solar nebula at the same time. There is, of course, no more possibility of sampling these bodies throughout their depths than there is of sampling the Earth in like fashion. However, it is possible to carry out a partial test of whether or not all five bodies have the same composition, although this will not reveal whether they have the same minerals or rocks.*

If several planets of different sizes have exactly the same composition, they will not have the same average density. The larger the planet the greater will be the compression towards the center and hence the greater will be the average density. The average density will, in fact, be proportional to the radius, and so a plot of density against radius will be a straight line.

When the relevant figures for the Moon, the Earth, Mars, Venus and Mercury are plotted in this way, the first four do lie more or less on a straight line. However, this certainly does not prove that they have the same composition; still less does it prove that all, or any, of them have the same composition as the meteorites – the only astronomical bodies that can be sampled directly. Nevertheless, it is good circumstantial evidence for supposing that they have the same composition and it strengthens the argument that the composition is something fundamental to at least this part of the Solar System.

But nothing is perfect. Mercury is way off the line, being too heavy for its size. It would appear to have an anomalously large iron core. The decreased ratio of light to heavy elements in Mercury is presumably due to the fact that the planet formed close to the high-temperature region of the Sun.

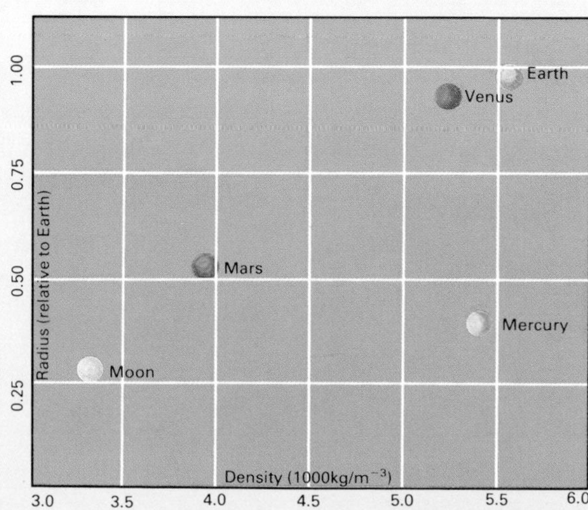

▲ *The fact that the densities of the Moon, Mars, Venus and Earth are proportional to their radii suggests that they have largely the same composition. Mercury, being the exception, must be different.*

Although the Sun is an ordinary star it has elements in common with the Earth

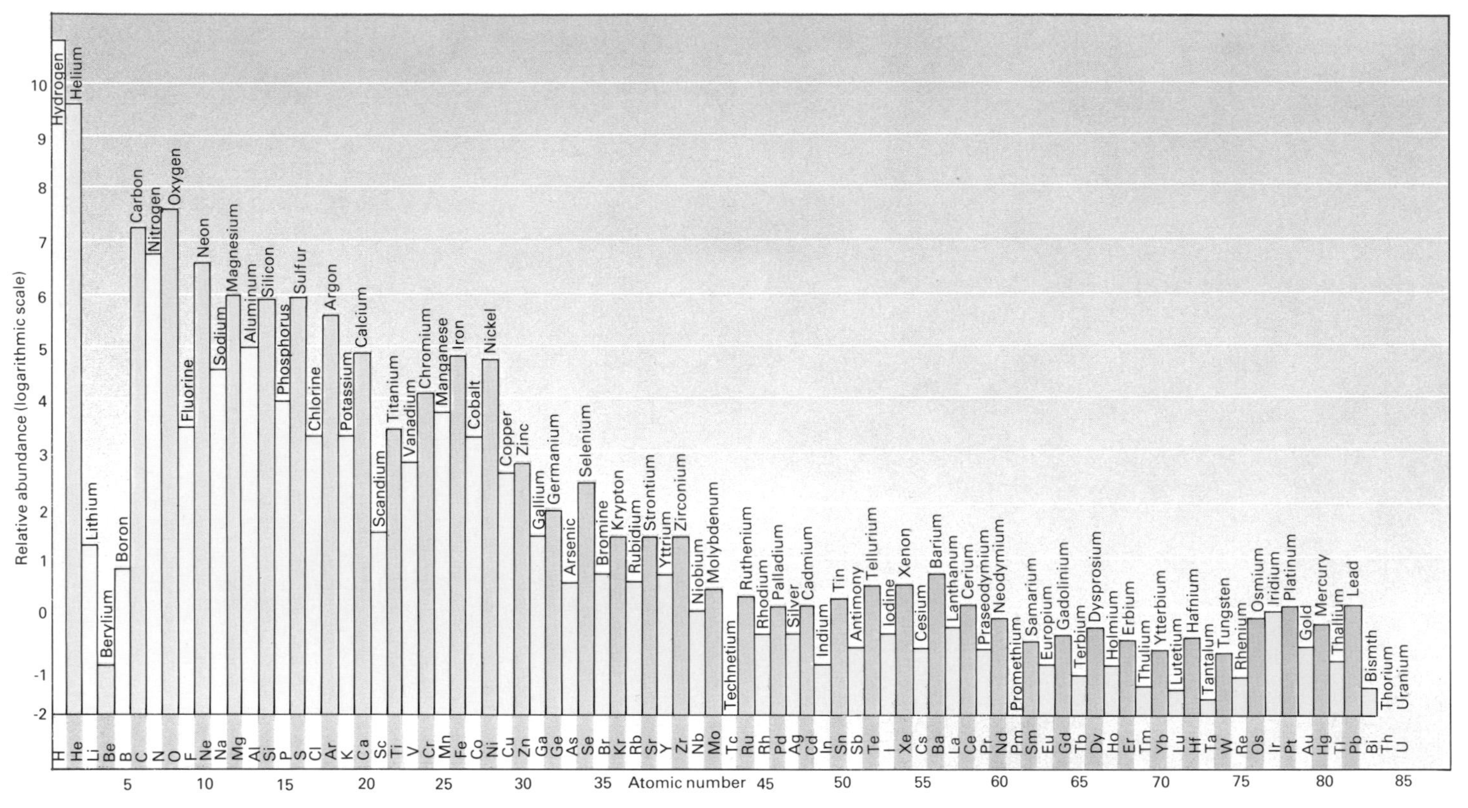

The Sun as a key to the composition of the Earth

The Solar System is dominated by the Sun, which has a mass of about 2×10^{30}kg. This is about 343,000 times the mass of the Earth, but because the Sun's volume is huge (radius about 7×10^5km) its density is low ($1{\cdot}4 \times 10^3$kg m^{-3}). Like the outer planets, the Sun also consists largely of hydrogen and helium. Despite its impressive statistics the Sun is just an ordinary star with roughly the average size, mass and composition of all the stars in the Galaxy, which number about 10^{11}.

This important fact throws light on the probable bulk composition of the Earth. It is possible to determine the elements present in the Sun and stars and their relative abundances by spectrochemical analysis – the study of the wavelengths of light emitted by the chemical reaction inside their interiors. This shows that, although the Sun and stars are composed predominantly of hydrogen and helium (about 98 percent by mass), they also contain many other elements in much smaller proportions. Generally, the proportion of the heavier elements decreases with the elements' increasing atomic mass number (the number of protons and neutrons in the nucleus), but the decrease is not smooth. There are relatively high abundances of elements whose mass numbers are multiples of four – for example, carbon, oxygen, neon, magnesium, silicon, sulfur and iron. This pattern results from nucleosynthesis, probably aided by stellar outbursts. Nucleosynthesis is the process whereby, in the interiors of stars, light elements are combined to produce heavier ones with a consequent release of energy. It is capable of generating elements up to that with an atomic mass of 56 (iron) and is more likely to produce elements with mass numbers that can be divided by four. The bigger a star is, the heavier are the elements it can produce, and so not all stars generate elements as heavy as iron. The Sun, for example, is too small to produce any element beyond helium. As for elements heavier than iron, these require an ***input*** of energy for their formation and thus involve a

Distribution of the elements

The way the elements are distributed in the Earth depends on two main factors. One is the planet's bulk composition. If the overall composition of the Earth were different from what it actually is, the layering of the interior could be different. For example, if the Earth contained only 1 percent of iron by mass instead of the accepted 35 percent, as found in meteorites, it could not possibly have a largely iron core that accounts for about a third of the Earth's mass.

The other factor relates to the chemical affinities of the elements. If all the elements existed as free elements (not in combination with others), then, when the Earth heated up during its early history the heavier elements would presumably have sunk and the lighter ones would have risen. So there would now be an Earth grading from the heaviest elements at the center to the lightest ones at the surface. The reason why this is not so is that most elements form compounds, and it is the properties of the compounds that matter.

For example, sodium, calcium and potassium readily combine with aluminum, silicon and oxygen to form silicates such as the feldspars ($NaAlSi_3O_8$, $CaAl_2Si_2O_8$ and $KAlSi_3O_8$). These have low melting points and are light, and so when the Earth heated up they melted and rose early to form part of the crust. On the other hand, magnesium-iron silicates, which have higher melting points and densities, melted later and rose less far, coming to rest in what is now the mantle. Much of the iron fell towards the center as free iron; but much of the uranium and thorium, which are much heavier than iron, failed to do so because they easily form light oxides. They therefore ended up in the crust. Elements that readily combine with oxygen to form

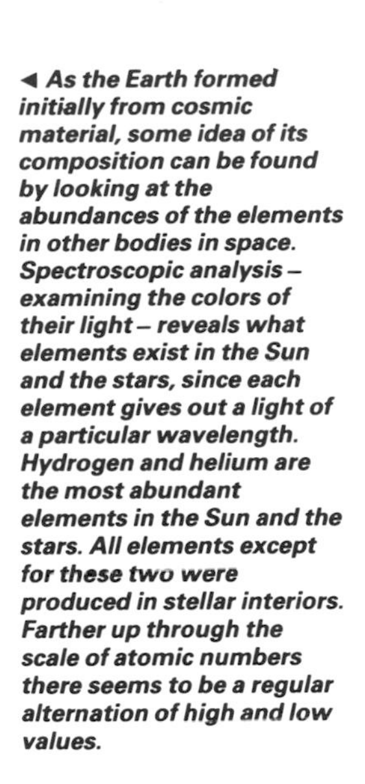

◀ ***As the Earth formed initially from cosmic material, some idea of its composition can be found by looking at the abundances of the elements in other bodies in space. Spectroscopic analysis – examining the colors of their light – reveals what elements exist in the Sun and the stars, since each element gives out a light of a particular wavelength. Hydrogen and helium are the most abundant elements in the Sun and the stars. All elements except for these two were produced in stellar interiors. Farther up through the scale of atomic numbers there seems to be a regular alternation of high and low values.***

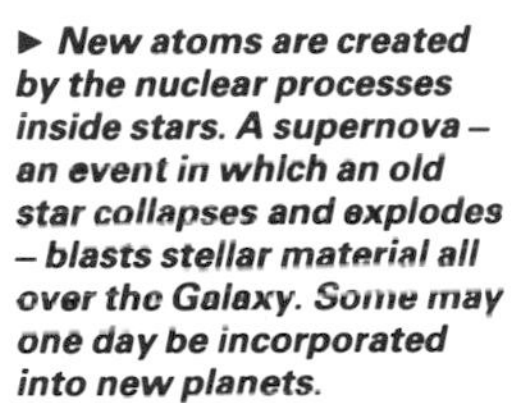

▶ ***New atoms are created by the nuclear processes inside stars. A supernova – an event in which an old star collapses and explodes – blasts stellar material all over the Galaxy. Some may one day be incorporated into new planets.***

oxides or with silicon and oxygen to form silicates are known as lithophile elements. Examples are aluminum, calcium, rubidium and strontium. Elements that have a greater tendency to form sulfides are said to be chalcophile. Examples are zinc, lead, silver and cadmium. Finally, elements that tend to remain as free metals are called siderophile. Examples are nickel, platinum and gold.

Iron is an interesting and important case. It is basically chalcophile by preference, but it is so common that there is insufficient sulfur in the Earth to combine with it. Much iron therefore exhibits siderophile tendencies (as in the core), while a lesser amount acts as if it were lithophile (in the crust and mantle). Much of the iron in the core is probably free metal, but there is a strong case for supposing that some of it is in the form of iron sulfide (FeS).

The fundamental difference between a lithophile, a chalcophile and a siderophile element lies in their ability to attract electrons and become negatively charged ions (anions). The metallic elements cannot do this easily and so tend to form positively charged ions (cations). Those metals with a particularly low "electronegativity" will tend to form ionic bonds (where electrons are transferred) with highly electronegative oxygen, and these are the lithophilic elements. Metallic elements with a slightly higher electronegativity will form a covalent bond (where two electrons are shared) with sulfur of similar electronegativity. These are the chalcophilic elements. The metals that have a high electronegativity tend not to form chemical compounds, and these are the siderophilic group. These are generalizations, and the true picture is rather more complex.

rather more unusual process. For this, astrophysicists invoke supernovae, the explosions of giant stars. When a supernova occurs, the argument goes, it releases some of the iron and elements lighter than iron that were already present in the giant star together with those elements heavier than iron generated either in the supernova itself or late in the life of the giant star. All these elements are then scattered widely throughout interstellar space by the force of the explosion and thus become available for incorporation into new stars.

If this is true, it follows that as time goes by and more and more supernovae occur in the vicinity of the Galaxy, the concentrations in the Galaxy of elements heavier than helium will increase. Younger stars will therefore have higher concentrations of such elements than older ones, which is observed to be the case. However, the *relative* proportions, or abundances, of elements heavier than helium will not necessarily increase, and it appears that they do not. Thus although the Sun is gradually losing hydrogen which is being converted to helium, the elements heavier than helium have, since the Sun was formed, maintained the same relative abundances.

This remarkable consistency has encouraged geochemists to argue that the Earth may also have the same relative abundances as the Sun and stars, for the Earth and Sun were formed from the same primitive solar nebula. Of course, despite their common origin in the Solar System, the Earth and Sun clearly do not have the same relative abundances throughout the complete range of elements. If they did, the Earth would also consist largely of hydrogen and helium. Geochemists suppose that most of the hydrogen and helium that might have been expected in the Earth were either lost as gases at an early stage in the Earth's history or perhaps for some unknown reason not incorporated into the Earth in the first place. The relative abundances of the elements in the Earth are broadly similar to those of the 2 percent of the Sun that is not hydrogen or helium.

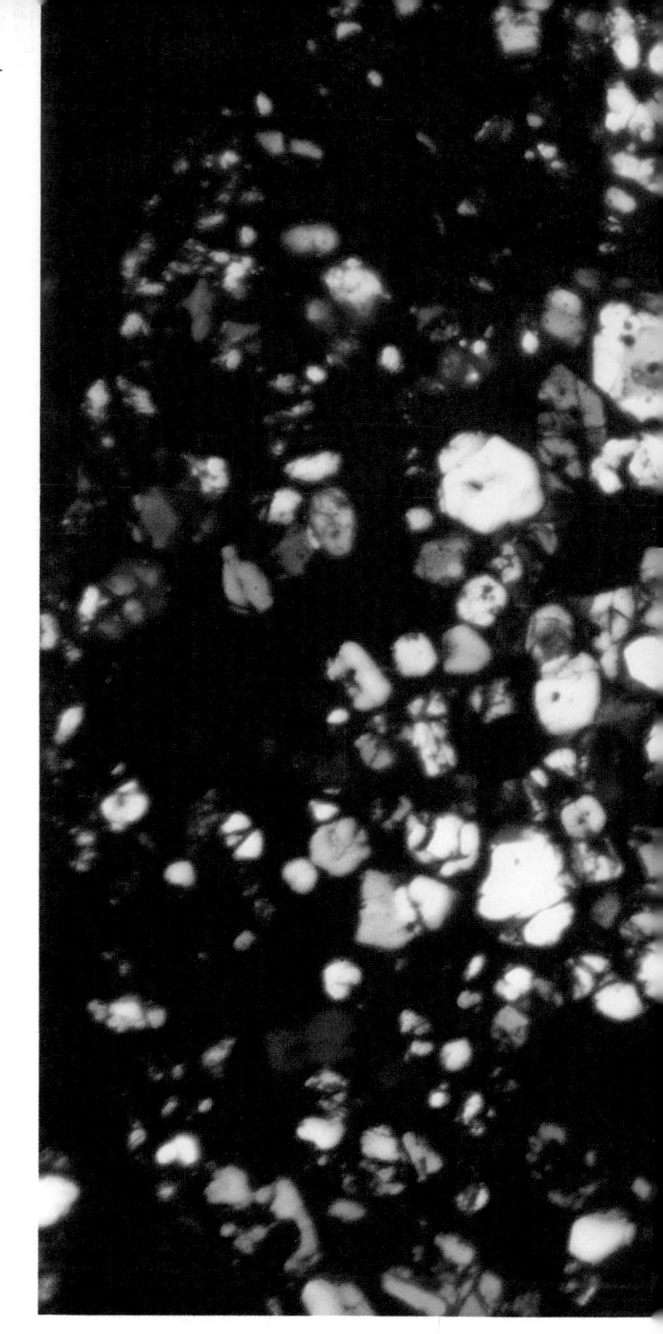

Meteorites and similarities with the Earth

The idea that the Sun may be a key to the composition of the Earth is supported by chemical analyses of meteorites. Meteorites (➧ page 61) are widely believed to be the most primitive materials in the Solar System, for most of them have ages about the same as, or slightly greater than, that of the Earth and many appear to have remained largely unaltered since they formed. Moreover, not only did they originate as part of the Solar System, they concentrated the heavier elements into hard rock (thus at some stage they lost volatiles (➧ page 61) such as hydrogen and helium). In this respect they are more akin to the Earth than they are to the Sun.

The iron meteorites may have come from the core of an asteroid body that had already become layered much as the Earth. Likewise, the achondrites could well have come from the silicate mantle of such a body. The stony-iron meteorites that account for only a few percent of all meteorites, being hybrids, may have come from the core-mantle boundary region. However, the chondrites have no obvious source within any imaginable layered body, but instead appear to resemble most closely the material present in the primitive solar nebula from which the Sun and planets formed. And in the sense that the carbonaceous chondrites are the least changed of all the chondrites, they are probably the closest to primitive matter.

Taking the bulk composition of a carbonaceous chondrite and regarding this as the bulk composition of the Earth would not make a great deal of sense. The Earth would be almost completely lithophilic with a small core of chalcophilic material representing about 20 percent of its mass (➧ page 61). The reason for this is the high percentage of oxygen (32 percent) found in carbonaceous chondrites. However, if some of the oxygen was lost along with other volatiles early in the Earth's formation, this would leave the carbonaceous chondrites' heavier elements, such as iron, silicon, magnesium and sulfur.

If carbonaceous chondrites represent primitive matter, then the relative abundances of their heavier elements are likely to be similar to those of the heavier elements in the Sun – both being made of the same original material. Also these abundances are likely to be similar to those of the Earth. If this were so the proportions of one particular element to another in the chondrites should be the same as the proportions of one to another in the Sun. Spectroscopic studies of the Sun demonstrate that this is more or less so. There are a number of exceptions. The Sun seems to be richer in copper, zinc and cadmium – chalcophilic elements – while the carbonaceous chondrites are richer in such elements as barium, zirconium, lithium, potassium, manganese and aluminum – lithophilic elements. Nevertheless the correlation is good and it can be used as a basis for investigation into the chemistry of the Earth.

The fact that two types of object – the Sun and meteorites – look and behave quite differently and yet have such very similar heavy-element abundance patterns, suggests that the pattern is fundamental to whatever the two have in common. What both the Sun and carbonaceous chondrites have in common is that they were formed from the same source material at about the same time. The Earth was also formed from the same source material at the same time, which is good reason for supposing that it might have the same heavy-element relative abundance pattern. The hypothesis that the Earth has the same relative abundances as carbonaceous chondrites (and hence as the heavier 2 percent of the Sun) is known as the chondritic Earth model.

◄ ▲ **There are two main types of meteorite – iron and stony. Their relative abundances may reflect the composition of the Earth's interior – the iron core and the stony mantle. One particular type of stony meteorite is the chondrite – so-called because it contains chondrules, or once-molten globules of silicate (seen in microscopic view above). Some, called carbonaceous chondrites (left), still contain volatile materials and so cannot have been altered in any way since they formed. They represent the most primitive objects in the solar system.**

Meteorites – the types and classes

Meteorites are fragments of rock that have entered the Earth's atmosphere from outside and have managed to reach the planet's surface without completely burning up. Most, if not all, of them probably originate in the asteroid belt, which lies between Mars and Jupiter and contains thousands of bodies (asteroids) ranging in size from a radius of 500km downwards. This part of the Solar System may originally have contained several small planets that failed to aggregate into a single large one but which subsequently broke up into smaller fragments by collision.

More than 90 percent of all known meteorites are classified as stony meteorites and consist largely of silicates (compounds of silicon and oxygen with one or more metals). They are divided into chondrites and achondrites. Chondrites are so called because they contain characteristic, small, once-molten globules of silicate (chondrules). They consist largely of iron-magnesium silicates in which there are dispersed small grains of iron-nickel alloy and the iron sulfide mineral troilite (FeS). More than 80 percent of all known meteorites are chondrites. Achondrites do not contain chondrules and have far fewer metallic grains.

The so-called iron meteorites, by contrast, are almost entirely metallic. They are predominantly iron-nickel alloy with 4-20 percent nickel. They only account for a few percent of all known meteorites. Stony-iron meteorites, as their name implies, are hybrids, comprising a matrix of iron nickel alloy surrounding fragments of silicate material.

Chondrites are further divided into numerous classes depending on the quantity of free metal in them and the degree of change that has been induced by heating (before they entered the Earth's atmosphere). No chondrites have been melted subsequent to their formation. If they had been, the iron and silicate would have separated and the meteorites would no longer be chondrites. On the other hand, some have been heated to various degrees, producing some changes such as partial recrystallization and the loss of such volatiles as carbon dioxide and sulfur dioxide. The least altered in this way are those known as carbonaceous chondrites. These meteorites are regarded as the most primitive in the sense that they are most representative of the primitive solar nebula in respect of elements heavier than neon.

The composition of the crust

The mantle and core together account for well over 99 percent of both the Earth's mass and volume, so in the context of the composition of the Earth as a whole the crust has little significance. On the other hand, it (or rather the uppermost part of it) is the only part of the Earth that can be sampled directly and thus chemically analyzed. It is therefore of considerable interest to academic geochemists even though they appreciate that it is of limited use. (To applied geochemists, by contrast, it is the whole world, where all the Earth's economically recoverable mineral resources lie.)

There is no reason to expect the crust to be chondritic even if it could be proved beyond doubt that the Earth is chondritic as a whole. On the contrary, there is every reason to suppose the crust not to be chondritic. The Earth is known to be divided into four major layers (crust, mantle, outer core and inner core) with seismic velocities so different that they almost certainly represent different chemical compositions. It is therefore most unlikely that any individual layer will have the Earth's average composition. Nevertheless, it is instructive to compare crustal and chondritic abundances.

The eight most abundant elements in a chondritic Earth comprise more than 99 percent of the Earth's mass, and the eight most abundant elements in the crust comprise more than 99 percent of the crustal mass. But the two lists of eight elements are not identical and those elements that appear in both lists do not exhibit the same proportions. Iron, for example, is much less abundant in the crust than in the Earth as a whole, because much of this element is probably concentrated in the core. Magnesium is similarly depleted because it has moved down to the mantle. On the other hand, silicon, aluminum, calcium, sodium and potassium are enriched in the crust.

In the last analysis it is impossible to prove whether the Earth as a whole is chondritic or not. There is therefore no way of proving beyond doubt that the crustal abundance of, say, magnesium is not actually representative of the Earth's bulk composition. However, in trying to decide where the various elements are situated in the Earth, geochemists take into account factors other than the supposed bulk composition of the Earth and chemical analyses of the upper crust.

Mineralogy of the crust

The chemical behavior of the Earth is frequently better understood in terms of minerals and rocks than it is of elements, but composition is always being expressed indirectly.

In many areas of the continental crust there is a rather diffuse seismic discontinuity at depths of 10-30km (◀ page 38). This is known as the Conrad discontinuity and is regarded as the boundary between the upper crust and lower crust. The upper crust has two parts. The top few kilometers are very variable, and this is sometimes called the sedimentary layer. It accounts for only about 7 percent of the continental crustal mass. The rest of the upper crust (about 45 percent) consists largely of igneous rocks and metamorphic rocks (▶ page 141) derived from them, ranging in type from granite (a coarse-grained rock consisting chiefly of the minerals quartz and feldspar) to diorite (a coarse-grained rock consisting mainly of the mineral plagioclase feldspar and one or more of the minerals augite, biotite and hornblende). Rocks in this range are sometimes called granodiorites.

The lower crust (about 48 percent of the total continental crust) is less well known, but geologists believe it to consist largely of gabbro (a coarse grained igneous rock comprising the minerals plagioclase feldspar, pyroxene and frequently olivine). Although the composition of the lower crust is uncertain, it must be different from that of the upper crust because the increase in seismic P wave velocity from the upper to the lower crust is too great to be due entirely to the increased compression of the lower crustal rocks.

The oceanic crust is quite different from the continental crust. Layer 1 is the thin layer of sediment on the ocean floor. Layer 2 is mainly basalt. Layer 3 is probably gabbro. Although the lowest layers of both the continental and oceanic crust are both thought to be gabbro, they should not be thought of as continuous, for the two crusts have quite different origins. The oceanic gabbro is very much younger than the continental.

▲ Olivine is a silicate mineral. With a chemical formula of $(Mg,Fe)_2SiO_4$ it contains less silica in its molecule than other silicate minerals. Hence it tends to form where silica concentration is low. Heavy rocks of the lower crust and mantle contain olivine.

▶ When lava erupts under the sea, the chill of the water breaks it into fluid lumps about a meter in diameter covered by a cool skin. These roll about on the sea bed, and come to rest draped over one another. Their shape gives them the name pillow lavas.

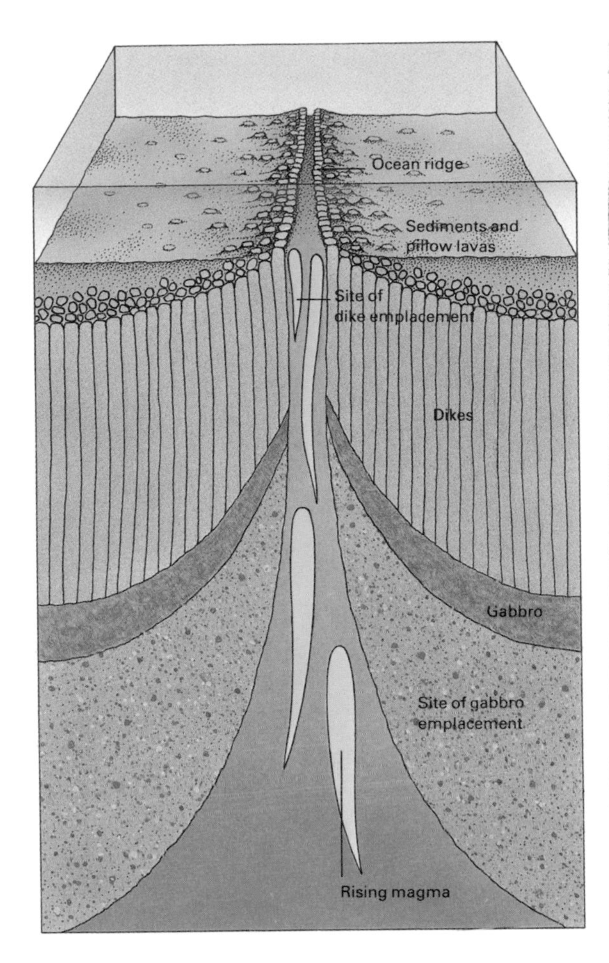

▲ **New oceanic crust is formed at constructive plate margins. As two sections of the crust pull apart, liquid material from the mantle wells up to take its place. This forms a series of dikes along the zone of weakness and volcanoes on the sea bed above. The oceanic crust thus consists of several layers. At the top there is a mixture of ocean sediments and pillow lavas. Below this is a layer consisting entirely of dikes stacked against one another. At the base is a layer of olivine-rich rock solidified directly from the mantle material.**

▲ **Kimberlites are formed in carrot-shaped structures called diatremes. They consists of material that has risen from the great depths of the upper mantle. As the hot fluids rose they expanded with the lower pressure and formed a structure that opened upwards. Now, long after they have solidified, these diatremes are the source of diamonds as well as of fragments of mantle material. The Big Hole at Kimberley in South Africa is a classic example. It is now hollow, having been dug out by diamond miners.**

The composition of the mantle

Geologists think that the upper mantle consists largely of iron-magnesium silicates in the form of the rock known as peridotite. Peridotite is a general name for a group of coarse-grained rocks which consist mainly of the mineral olivine together with a lesser amount of the mineral pyroxene. Olivine in turn is a compound of the minerals forsterite (magnesium silicate) and fayalite (iron silicate) with the general formula $(Mg,Fe)Si_2O_4$. The silicate in stony meteorites is chiefly iron-magnesium silicate in the form of olivine, and so by analogy the Earth's chief silicate layer, the mantle, should also be dominated by olivine. The seismic P wave velocity in the upper mantle is also the same as that in peridotite.

However, there are also several other lines of evidence based on samples from the Earth itself. The present mantle is far too deep to sample, but in various parts of the world there are geological sections that geologists believe to be parts of old oceanic lithosphere that have been uplifted and exposed at the surface during continent-continent or continent-island arc collisions. One of the best known examples is the Troodos mountains of Cyprus which were raised 15-20 million years ago as part of Africa's northward push into Europe.

The Troodos outcrops are known as ophiolite sequences (◀ page 42), and the arrangement of rocks within them parallels the arrangement in oceanic lithosphere. Thus they begin with a thin layer of sediment (the equivalent of oceanic layer 1) which overlies basalt in the form of pillow lavas (layer 2) and sheets of dikes (uppermost part of layer 3). Below the gabbro, however, lies peridotite which appears to correspond to that part of the lower lithosphere that is part of the mantle.

Information of a different kind comes from basalt lavas that form oceanic islands. Some of these contain olivine nodules, pieces of olivine-rich material that have evidently been torn away from the lower sections of the walls of magma conduits as magma rose to the surface.

Finally, there is the evidence from kimberlite pipes, vertical pipe-like bodies that were first recognized at Kimberley in South Africa.

Mineral Properties 1

Cleavage

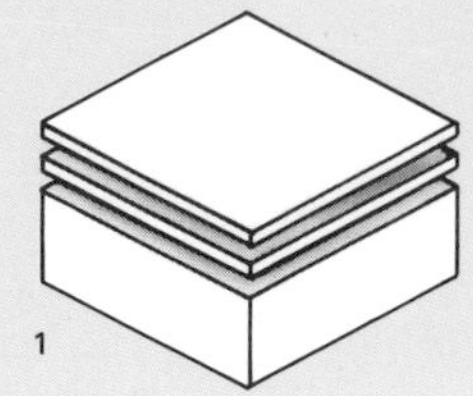

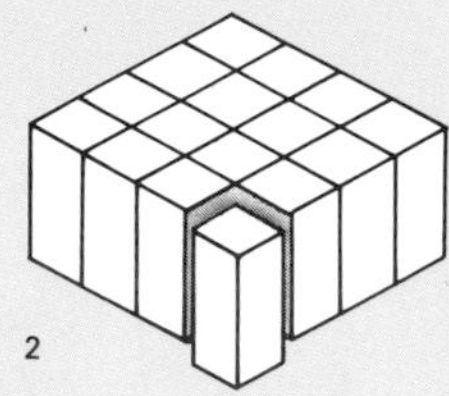

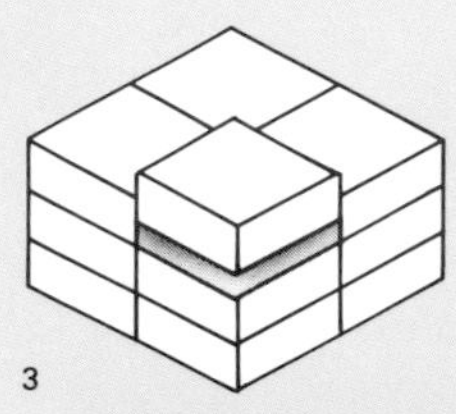

◄ The cleavage of a mineral is, like its crystal structure, related to the arrangement of the atoms in its molecules. Many minerals split along structural planes when damaged, as in the example of mica below. These are known as its cleavage planes. When the mineral has one cleavage plane it splits into sheets (1). Two cleavage planes will cause it to break in columns (2). Three cleavage planes will form regular shapes like crystals (3).

Crystal structure

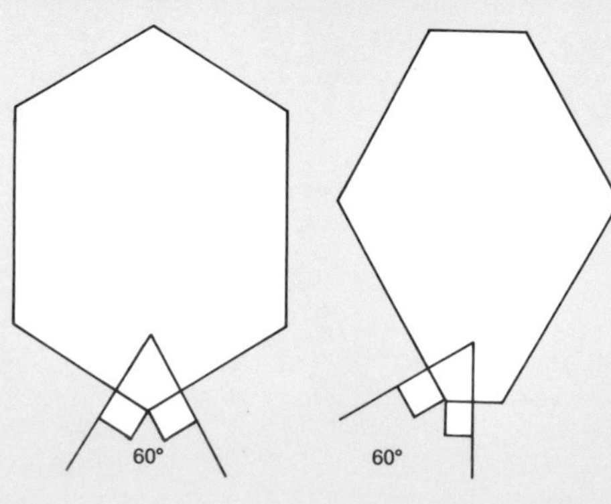

◄ The crystal shape of a mineral is the three-dimensional form it naturally adopts if it is allowed to grow unhindered. The arrangement of the faces and the angles is determined by the positions of the atoms in its molecule. In different crystals of the same mineral, the faces may be of different sizes relative to one another, but the angles between them are always the same.

Specific gravity

◄ ▲ *A useful property of a mineral for use in its identification is its specific gravity, or its weight compared with the same volume of water. This can be measured very accurately in the laboratory, but often an idea of the specific gravity can be gained in the field by hefting the mineral specimen in the hand. A heavy mineral such as galena (left), with a specific gravity of 7·5 will be obviously heavier than a light mineral such as rock salt (above) with a specific gravity of 2·2.*

Fracture

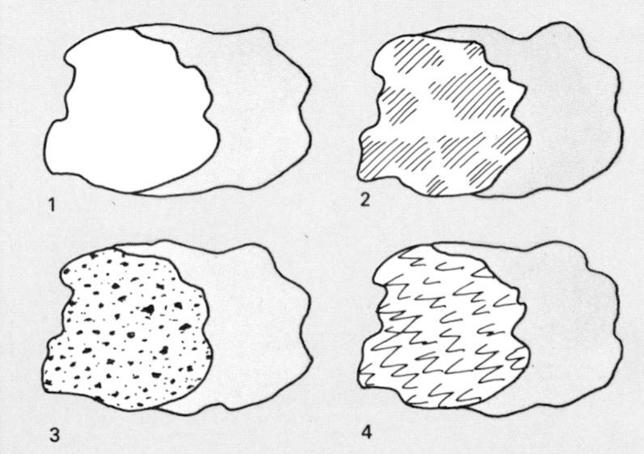

◄ *The fracture is the way in which a mineral breaks, and at times this can give a clue to its identification. The most distinctive fracture is conchoidal, as in flint (above). This produces a curved face with concentric lines, like the growth lines on a shell. Other fractures are defined as even (1), uneven (2), earthy (3) and hackly (4). The latter is often seen in some metals and consists of microscopic jagged spikes.*

Hardness

▲ *A mineral's hardness is distinctive. Any mineral will scratch one that is softer than itself but not one that is harder. A gradation called Moh's scale of hardness is the one that is used by geologists. From soft to hard it runs – talc 1, gypsum 2, calcite 3, fluorite 4, apatite 5, orthoclase 6, quartz 7, tourmaline 8, corundum 9 and diamond 10 (above). A specimen is rubbed against a sample of each of these in turn to see which it can scratch and which can scratch it.*

Mineral Properties 2

Color

1

2

▲ ► The colors of minerals are notoriously deceptive. A mineral is rarely pure, and all kinds of contaminants can enter the crystal structure as it is forming. Quartz, for example, is clear and transparent (and known as rock crystal) when it is pure (2), but more often it will be a milky white (1) with tiny air bubbles, or stained to brown smoky quartz (3) or purple amethyst (4) by mineral inclusions.

3

4

Luster

◄ *The luster of a mineral is the way it shines, the way it reflects light from its surface. There are six types:*
Metallic – like pyrites (below left) and hematite (left)
Vitreous, or glassy – like quartz
Resinous – like opal
Pearly – like feldspar (below)
Silky – like gypsum
Adamantine – like diamond

Streak

► *There is one aspect of a mineral's color that can be useful in identification, especially amongst the ore minerals. If a specimen is rubbed along a white ceramic plate it will leave behind a streak of tiny particles. This streak may be different from the body color of the mineral but it is quite constant from one specimen of the mineral to another. For example, the iron ore hematite is usually black in hand specimen, but will give a reddish-brown streak. Steel-gray specimens of hematite will also give the reddish-brown streak. Gold (right) can be distinguished from "fool's gold" (iron pyrites) by its different streak.*

See also
Internal Structure 33-44
Magnetism, Gravity and Heat 45-56
Origin and Evolution 69-72
Dating the Rocks 75-80
Rocks and Rock Cycle 141-152

The composition of the core

Geochemists are in almost universal agreement that the chief constituent of the Earth's fluid outer core is iron, perhaps with a little nickel and other metals. The evidence for this is strong. The density of the core is approximately right for iron at the relevant pressures, the core material must be a good conductor in order to generate the geomagnetic field, and iron is apparently the only metal available in sufficient quantities. In addition, the existence of iron meteorites suggests that planetary bodies can and do become differentiated into at least one metallic layer. But although the density of the core is about right for iron, it is not exactly right. It is slightly too high, implying that the core must also contain a lighter element up to 5-20 percent by mass, depending on precisely what it is. At this point agreement breaks down, for there are several possibilities.

An obvious one is silicon. Silicon is very abundant in the Earth and is known to be able to form iron-silicon alloys; indeed, small quantities of silicon are used in the manufacture of some steels. Moreover, traces of a natural iron-nickel-silicon alloy called perryite are found in some meteorites, and small quantities of silicon are dissolved in iron in some chondrites. But there are severe difficulties. One is that the production of iron-silicon alloy during, or soon after, the formation of the Earth would have resulted in the release of huge amounts of carbon monoxide, and the mass lost would have represented a large proportion of the early Earth. Other, more technical arguments, all combine to suggest that silicon is not the answer, although it could be present in very small quantities.

Another possibility is sulfur. Sulfur is common in meteorites; indeed, all chondrites contain substantial amounts of it in the form of iron sulfide. Second, the crust and upper mantle appear to be severely depleted in sulfur compared to chondritic abundances. If the Earth is to be considered chondritic overall, therefore, the core would be a convenient repository for the sulfur supposedly "missing" from the upper part of the Earth. Third, the melting point of iron sulfide is considerably lower than that for pure iron. Thus when the Earth was heating up during its early history, molten iron sulfide would have formed first in preference to molten iron. The arguments in favor of sulfur, which would have to account for 9-12 percent of the core to make the density right, are thus strong. Some geochemists argue, however, that the quantity of sulfur in the upper Earth has been underestimated, that the crust and upper mantle are not really depleted in sulfur at all, and that there is therefore no need to look for missing sulfur in the core. Others claim that the Earth is not chondritic in sulfur anyway, and so again there is no need to invoke sulfur in the core. There are also other points to be made against sulfur as the core light element, but for all that it remains one of the most favored candidates.

A third possibility is carbon. Carbon, like sulfur, is abundant in meteorites, can alloy with iron, and is also thought to be depleted in the upper Earth. For some reason, geochemists have given little attention to carbon as a possible outer core constituent. On the other hand, they have drawn attention to the fact that carbon in the solid inner core would, under the huge pressures there, be diamond. In fact, diamond has actually been identified in small quantities in some iron meteorites, but there is some dispute over whether it was there right from the start or whether, more likely, it formed only during impact with the Earth.

▲ **A.E. Ringwood (right) and his team carry out high pressure tests to study the nature of the Earth's core.**

Other possible core elements

One of the latest (1972) entries for the core light element stakes is oxygen. The distinguished Australian geochemist A.E. Ringwood has argued strongly that oxygen could be present in the core in the form of iron oxide (FeO) dissolved in iron. Laboratory experiments have shown that at high temperatures and pressures FeO does dissolve in iron. To account for the outer core's density there would have to be about 40 percent of FeO present, which is equivalent to about 9 percent oxygen. However, this would not make the core any less conducting, for at core pressures the FeO would almost certainly be in a metallic form. One argument against the presence of FeO is that it would raise the melting point and hence require higher temperatures to keep the outer core molten. Another drawback is that there is little evidence of FeO in meteorites. But these arguments are not conclusive, and oxygen remains in the race.

Potassium is not a contender for the role of core light element, but a little may be present anyway. The chief argument in favor of its presence is that it readily combines with sulfur, and so if there is sulfur in the core there is likely to be potassium too. Another point to consider is that if potassium were present in the core its radioactive isotope would generate heat by decay, thereby producing the convection necessary to maintain the geomagnetic dynamo. The possibility of small amounts of potassium in the core also draws attention to the fact that the low density component of the core could be a mixture of light elements. When carrying out their studies, geochemists tend to think of one element at a time, but there is no guarantee that nature works that way, and several elements may be present in the core.

Origin and Evolution

6

The solar nebula...Earth's formation from the nebula – as a ball of dust, or an agglomeration of lumps?...The hypotheses and the evidence – theoretical arguments over something that can never be proved...The origin and formation of the Moon – a lost piece of the Earth, a captured wanderer or Earth's physical brother?

▲ A nebula is an interstellar cloud of dust and gas. It contains a great quantity of material but is spread through a vast volume of space.

The Earth formed as part of the Solar System. Modern theory envisages that the Sun and Solar System evolved from a primitive nebula, an enormous rotating cloud of interstellar gas and dust. About 4·6 billion years ago, for reasons unknown but possibly sparked off by a nearby stellar explosion, this nebula began to contract under the gravitational attraction of its constituent particles. The rotation then speeded up, causing the cloud to flatten progressively into a disk. At the very center, in the densest part of the disk, a proto-Sun formed. Ultimately the proto-Sun collapsed under its own gravitation, its temperature rose to above one million degrees Centigrade, thermonuclear reactions were instigated, and the Sun began to shine.

Meanwhile, the planets, their satellites and the asteroids and comets were also being formed from the nebula. As the material surrounding the infant Sun continued to cool, micrometer-sized grains collided and hung together, building up into a thin disk. More collisions built up larger bodies, which, under gravitational attraction, eventually coalesced into kilometer-sized bodies ("planetesimals"); these finally aggregated into the planets. In the outer regions the temperatures were low enough to attract volatiles such as hydrogen and helium, and these remain part of the outer planets. Closer to the Sun the temperatures were too high for such volatiles. Instead, they were probably removed by an intense solar wind, leaving rocky planets consisting largely of the heavier elements. Once they had begun to form, the planets probably reached more or less their present masses very quickly. One estimate puts the time at no more than about 10,000 years.

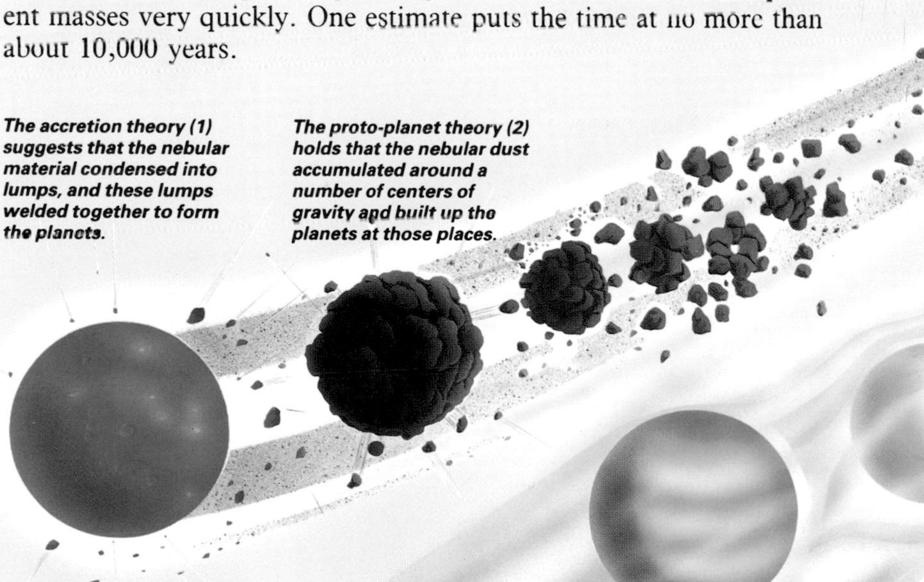

The accretion theory (1) suggests that the nebular material condensed into lumps, and these lumps welded together to form the planets.

The proto-planet theory (2) holds that the nebular dust accumulated around a number of centers of gravity and built up the planets at those places.

▲ The Solar System probably started life as a nebula. At some stage it contracted to form the Sun and the planets. It is not known how this happened. Possibly the cloud of dust aggregated into lumps, and the lumps came together as the planets, or the planets built up from layers of dust.

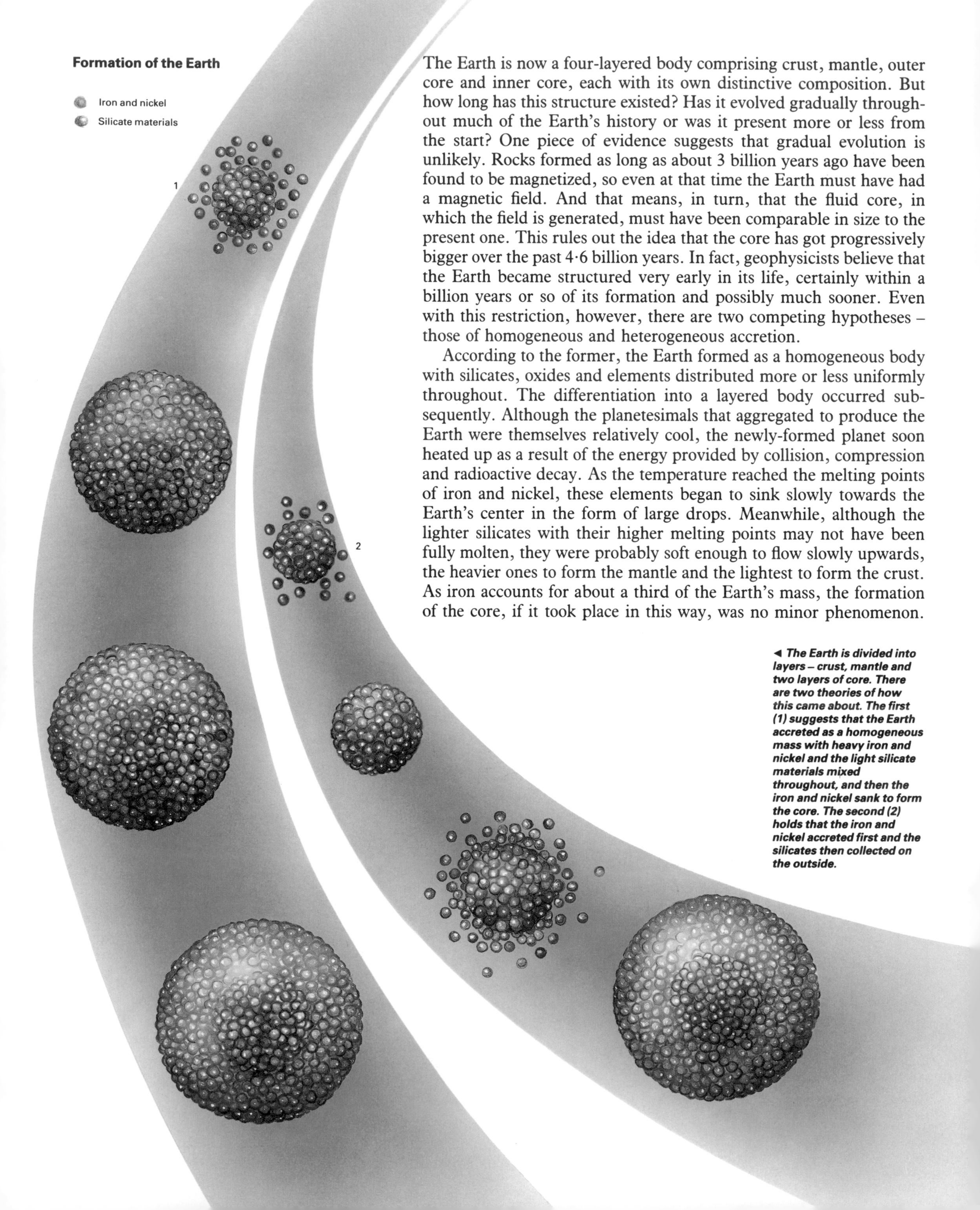

The Earth is now a four-layered body comprising crust, mantle, outer core and inner core, each with its own distinctive composition. But how long has this structure existed? Has it evolved gradually throughout much of the Earth's history or was it present more or less from the start? One piece of evidence suggests that gradual evolution is unlikely. Rocks formed as long as about 3 billion years ago have been found to be magnetized, so even at that time the Earth must have had a magnetic field. And that means, in turn, that the fluid core, in which the field is generated, must have been comparable in size to the present one. This rules out the idea that the core has got progressively bigger over the past 4·6 billion years. In fact, geophysicists believe that the Earth became structured very early in its life, certainly within a billion years or so of its formation and possibly much sooner. Even with this restriction, however, there are two competing hypotheses – those of homogeneous and heterogeneous accretion.

According to the former, the Earth formed as a homogeneous body with silicates, oxides and elements distributed more or less uniformly throughout. The differentiation into a layered body occurred subsequently. Although the planetesimals that aggregated to produce the Earth were themselves relatively cool, the newly-formed planet soon heated up as a result of the energy provided by collision, compression and radioactive decay. As the temperature reached the melting points of iron and nickel, these elements began to sink slowly towards the Earth's center in the form of large drops. Meanwhile, although the lighter silicates with their higher melting points may not have been fully molten, they were probably soft enough to flow slowly upwards, the heavier ones to form the mantle and the lightest to form the crust. As iron accounts for about a third of the Earth's mass, the formation of the core, if it took place in this way, was no minor phenomenon.

◄ ***The Earth is divided into layers – crust, mantle and two layers of core. There are two theories of how this came about. The first (1) suggests that the Earth accreted as a homogeneous mass with heavy iron and nickel and the light silicate materials mixed throughout, and then the iron and nickel sank to form the core. The second (2) holds that the iron and nickel accreted first and the silicates then collected on the outside.***

The gravitational energy lost by the falling iron was immense and, converted to heat, raised the temperature much higher than it would otherwise have been, thus contributing further to the melting and differentiation process.

There are a number of technical arguments against the homogeneous accretion model. A series of complex condensation processes must have occurred before accretion began, to get the contracting nebula into a homogeneous state to start with. For this and other reasons, some geophysicists have been attracted towards a heterogeneous accretion model according to which the Earth's core was present right from the beginning. Indeed, it was the first part of the Earth to form. The argument here is that the iron-nickel would have condensed out of the cooling nebula before the silicates, making metal particles the first planetesimals. As the nebula cooled further, the silicates would then have condensed out and built up on the outside of the metallic bodies already in existence. In short, core and mantle were present right from the start.

But this hypothesis, too, has its difficulties. The iron-nickel may well have condensed out of the nebula before the silicates, but the oxides of calcium and aluminum would have condensed out before either. In principle, then, the Earth should now have a calcium-aluminum oxide core. Even allowing that iron-nickel might be able to sink through the oxides immediately above the core, such a layer would be detectable by seismic methods and it is not so detected.

Unless there is a third hypothesis that geophysicists have not yet thought of, the truth probably lies somewhere between the homogeneous and heterogeneous models. The Earth could well have been partially structured upon formation and then subjected to further differentiation to produce its currently well-defined four layers.

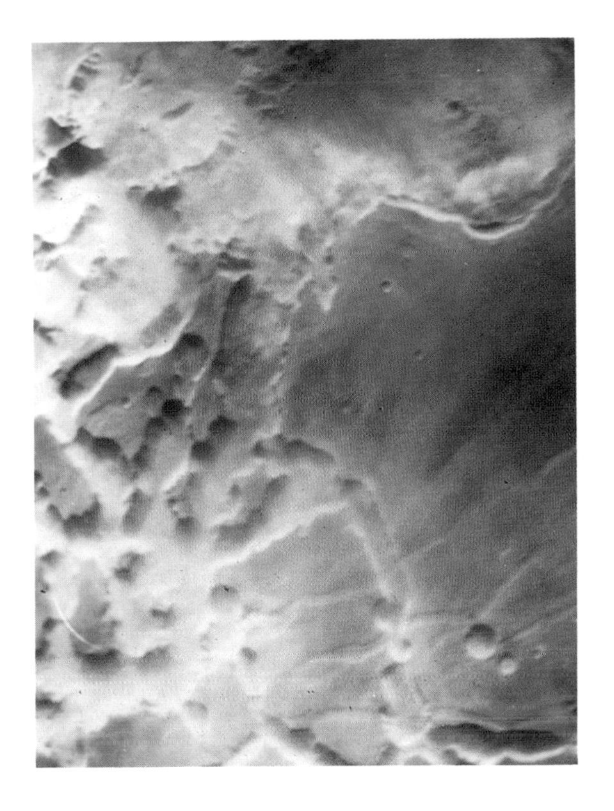

▲ Mars is tectonically dead, but its thin atmosphere can produce erosive features giving a landscape similar to the deserts of the Earth. The crust and mantle are very thick and beneath these lies a low-density core which may be made of iron sulfide. Tectonic activity ended on Mars about one billion years ago and any mobile part of the mantle must be about 300km deep.

▲ Mercury, the closest planet to the Sun, is quite airless and is covered with impact craters like the Moon. The craters were formed by intense meteorite bombardment.

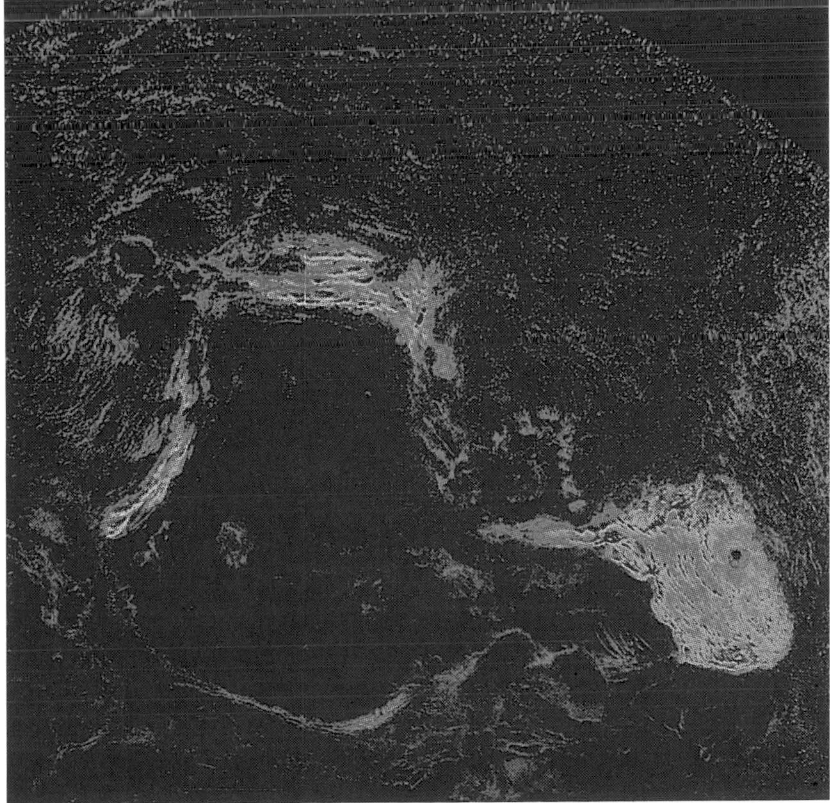

► Beneath the thick poisonous clouds of Venus is a planet that appears to have continents with hot spots and volcanic activity, shown here in false color photography.

See also
Earthquakes and Volcanoes 25-32
Internal Structure 33-44
Magnetism, Gravity and Heat 45-56
Internal Composition 57-68

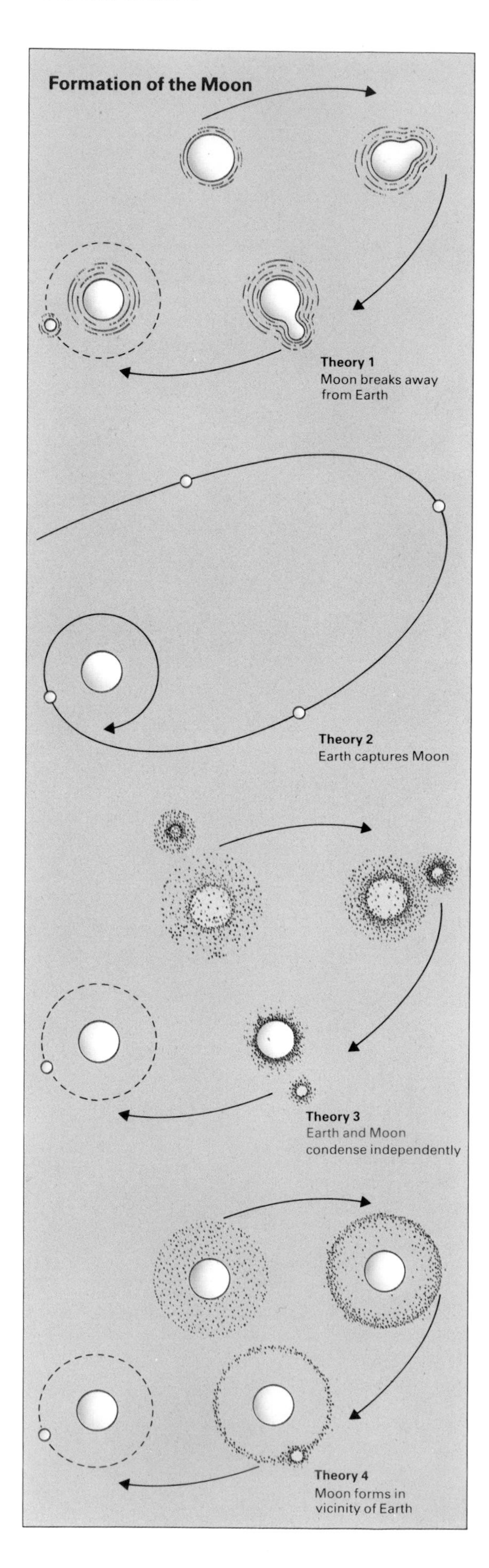

The origin of the Moon has long been controversial. Because of the Moon's close association with the Earth, there seems little reason at first sight to suppose that it formed in a way different from that of its parent planet. However, the Moon is now quite different from the Earth in both structure and behavior. It has no fluid core, it is not tectonically active, and its density is about 60 percent of the Earth's.

On the surface of the side of the Moon that always faces the Earth, there are light and dark areas known, respectively, as highlands and maria (singular, mare). The highlands are rugged and heavily marked with meteorite impact craters. The maria are smoother and flatter regions consisting of lava flows that have filled large depressions in the original, highland crust. They contain far fewer craters than do the highlands, indicating that most meteorite bombardment occurred before the maria formed. Incidentally, there are no major maria on the far side of the Moon, but no one understands why.

Mare rocks returned to the Earth by the Apollo missions all have ages in the range 4·0-3·1 billion years, whereas highland rocks are rather older. Geologists believe that the Moon formed at about the same time as the Earth and then heated up rapidly, causing extensive melting and the formation of a magma ocean. A crust of anorthosite (an igneous rock consisting largely of calcium-rich plagioclase feldspar) then formed on the surface, while the heavier material sank to become the source of the mare lavas that erupted later. During the first 600 million years or so of the Moon's history there was intense meteorite bombardment which produced the rugged topography of the original crust, including the very large craters that were to become filled with mare lava. The early Moon may also have had a fluid field-generating iron core, for lunar rocks are magnetized. But by about 3 billion years ago the core and the outer layers had solidified, leaving a Moon with a lithosphere 10 times thicker than the Earth's and undivided into active plates.

But how did the Moon originate and become the Earth's satellite? There are at least four possibilities. Because the Moon consists mainly of silicate materials (its now-solid iron core is very small), some scientists have suggested that it was originally a part of the Earth's mantle that became unstable and broke away. One problem with this idea, however, is that the Moon's composition is rather different from that of the Earth's mantle. The Moon is relatively depleted in volatiles (materials that vaporize at comparatively low temperatures) such as water, carbon, nitrogen, sulfur, mercury and lead, but is enriched in refractory elements (those not easily melted) such as titanium, aluminum and chromium. Another objection is that the Earth and Moon together appear to have insufficient angular momentum to enable the original separation to have taken place. Other scientists have argued that the Moon formed elsewhere in the solar nebula and was subsequently "captured" by the Earth. This would explain the difference in composition between the Moon and Earth, however, some scientists believe capture to be a statistically unlikely event.

A third possibility is that the Moon condensed from the solar nebula close to the Earth but completely independently of it. But that makes the chemical differences between the two bodies even more difficult to understand. One way out of this problem is to imagine that the Moon formed in a region of the nebula that had already produced the Earth, perhaps in a cloud of silicate material vaporized from the Earth's surface by the high temperatures generated by impacting particles.

Dating the Rocks

The age of the Earth – how is it determined?... Early principles that still hold good...Fossils – ancient life tells a rock's age...Isotope dating – radioactive decay dates ancient rocks... PERSPECTIVE...Relative ages of rocks...The geological time scale...The Earth's great age

Earth scientists seek not only to understand the nature and behavior of the Earth as it is today but also to discover how the planet has changed with time since it was formed about 4·6 billion years ago – and that is no easy task. The record of many events in the Earth's history has been completely obliterated; other events have left only meager traces. Even where the evidence of some past event or process is plentiful, its interpretation may be ambiguous. But whatever the nature and quality of the data available, it is important that they be related to some moment or period of time. Geological phenomena without a clear calendar are of little more use to geologists than are battles without dates to historians.

The discovery of radioactivity by Henri Becquerel 1896 was ultimately to provide geologists with techniques for dating rocks in an absolute sense using the known decay characteristics of naturally-occurring radioactive isotopes (➧ page 80). In recent decades, such radiometric methods have been brought to a fine pitch, enabling geologists to date many, but by no means all, rocks with great precision. Before that, however, geologists had to rely on relative dating, which simply means placing rocks and geological events in the correct chronological order. Using the physical and paleontological methods, relative dating was developed from the late 18th century onwards and by the end of the following century had become a highly developed art. Based on the principle that younger sedimentary rocks overlay older (the Principle of Superposition) the physical method is not so straightforward in practice. Indeed, the unraveling of the sequence of the principal rocks in the industrial countries (chiefly those of Europe) remains one of the great intellectual achievements not only of geology but of science as a whole.

The central problem is that there is nowhere in the world where the deposition of sedimentary rocks has been continuous. Thus a sequence of sedimentation at any one place will represent only a small proportion of the actual geological time span. There will have been intervals when no deposition occurred at all and there will have been intervals when, in addition, sediments already deposited were actually being eroded away. Moreover, only some of those periods of erosion will now be evident. Others will have taken place when the rocks below were still horizontal and hence the break may no longer be discernable at all. So not only is the sedimentary sequence at any given site a rather imperfect record of geological time, the positions of some of the gaps may not even be detectable.

A more mundane reason why the concept of increasing age with increasing depth may be difficult to apply in practice is that there may be a lack of suitable rock exposures. Although all continents contain vast thicknesses of sediments laid down at various times, they now often lie beneath a thin layer of soil and vegetation.

▲ *Nicholas Steno, a pioneer in geological studies.*

Relative dating

Relative dating by physical methods depends on a number of principles, the most basic of which is the Principle of Superposition, also known as Steno's Law after Nicolaus Steno, the Danish court physician who first formulated it in 1669. The Principle states that in a sequence of undeformed sedimentary rocks any bed, or stratum, must be older than those lying above it and younger than those lying below. Today, this may appear to be just common sense and harldly worth the name "principle" at all, but in Steno's time the issue was not so clear because the process of sedimentation was little understood. Sediments are laid down in horizontal layers under the influence of gravity; and only when that was appreciated did it become obvious that sediments can normally be deposited only on top of older rocks and get buried only by younger ones.

The Principle of Superposition may be applied to sedimentary rocks even if they have since been quite severely tilted or folded, for the top and bottom of a sequence may still be apparent. Problems arise only when the strata have been very intensely deformed and perhaps even overturned completely. This is particularly likely to be the case in mountain belts where the collision of plates forces the rocks into very complex arrangements.

It is common to find, for example, that strata that were originally horizontal have since been folded and then partly eroded. New sediments have then been laid down horizontally on the eroded surface. In this case the boundary, or discontinuity, between the old and new sediments is known as an angular unconformity, for the strata above and below it lie at various angles to each other.

"In . . . Earth history we find no vestige of a beginning – no prospect of an end." Pioneer geologist James Hutton, 1785

Unraveling rock sequences – the four principles

Geologists have to rely on exposures of the rock in such places as river valleys, quarries, road cuttings and mines. The problem there, however, is that such sites are usually scattered, and no single one will have exposures of the complete sedimentary sequence available.

To build up as complete a sequence as possible, it will therefore be necessary to study sub-sequences exposed at different sites, and that will mean correlating between the sites. One exposure, for example, may reveal sedimentary layers A to G (bottom to top) and another some distance away may contain layers G to N (again bottom to top), but the two sites can be regarded as providing a complete sequence of layers A to N if G can be proved to be the same layer at each site. This is known as the Principle of Lateral Continuity. As long as the two sites are not too far apart, it may be possible to do this by showing that, at both sites, G has, for example, the same color, grain size and mineral composition. On the other hand, it may not, for a sedimentary layer does not necessarily maintain constant properties along its length. Where physical correlation of this type proves impracticable, it is sometimes possible to make successful use of such phenomena as the geomagnetic polarity-time scale (◆ pages 45-7), fossils (◆ page 76) volcanic ash layers, or those properties of sediments that depend on climatic events.

The Principle of Superposition (◆ page 73) applies specifically to sedimentary rocks, but it may be relevant to igneous rocks in certain circumstances. In some parts of the world, for example, there are plateau basalts – huge piles of more or less horizontal lava flows that were extruded from fissures. Each lava flowed out across its predecessor before solidifying and acting as the base for the next lava flow; and so the flows get younger in the upward direction.

More usually, however, igneous rocks are intruded from below into overlying rocks that are older. In such cases the basic principle of relative dating, the Principle of Crosscutting Relationships, is that the intruded rock must be younger than the rock it cuts through. For example, in the lower crust (although erosion has brought some of them close to the present surface), there are huge masses of granite known as batholiths. These were intruded into the surrounding rock (known as country rock) with which they make sharp contacts. Higher in the crust, intrusions of igneous rock are more likely to be in the form of dikes – thin sheets of rock that formed as magma forced its way into cracks or fissures and then solidified. Dikes, which are frequently vertical or near-vertical but can lie at an angle, cut right through the country rock, with which they are said to be discordant. On the other hand, thin-sheet intrusions can also be horizontal or near-horizontal, not cutting through earlier strata but lying between them. Such intrusions are said to be concordant and are known as sills. Of course, not all geological processes involve the formation of new rocks. The Principle of Crosscutting Relationships also applies to faults (◆ pages 153-6), which must have taken place more recently than the formation of the rocks that are faulted.

Finally, there is one principle that applies to both sedimentary and igneous rocks. This is the Principle of Inclusions, which states that rock fragments included in some other rock must be older than their host. Thus pebbles incorporated into a sedimentary formation must be older than the formation itself. Likewise, unmelted fragments of country rock that get broken off and included in a magma as it rises to the surface must be older than the magma that dislodged them.

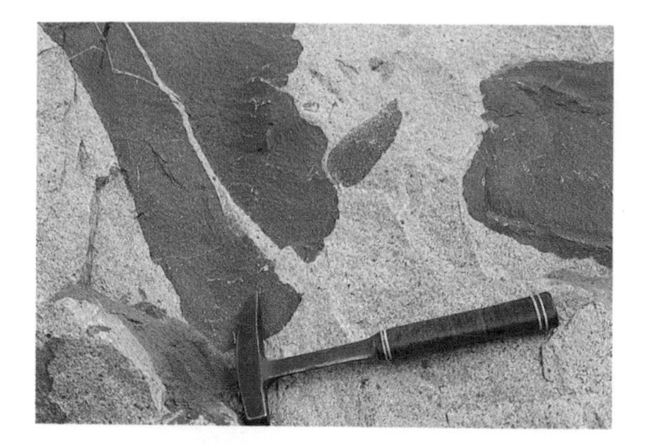

▶ ***The simple idea of the Principle of Crosscutting Relationships is that, if a particular geological structure is seen to cut across or through another, that structure must be younger than the other.***

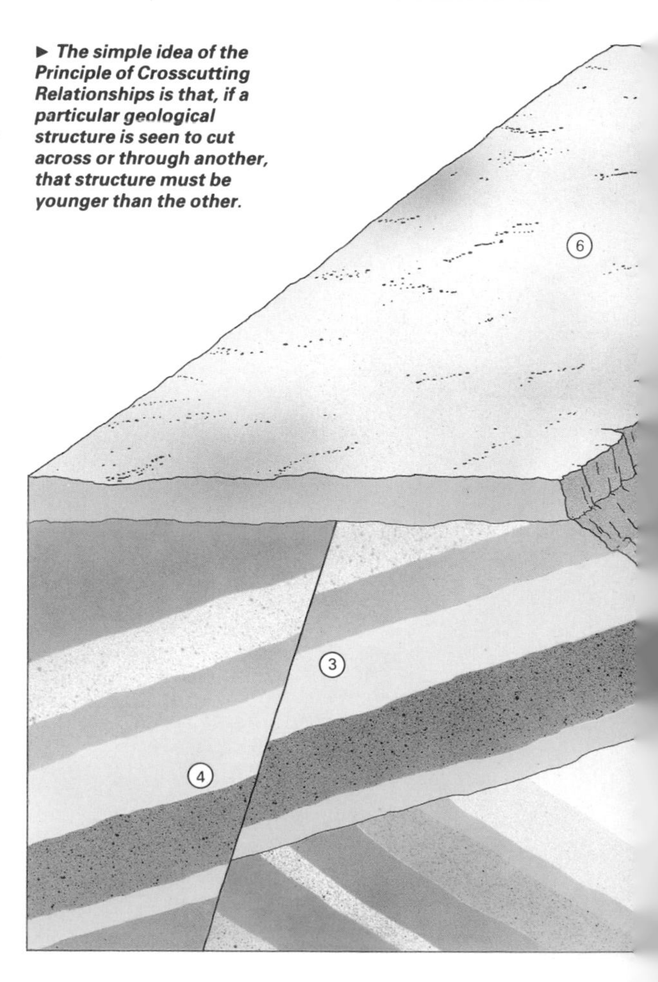

► In an unconformity a set of beds is eroded flat and younger beds deposited on them. Here, in Wyoming, the tilted Triassic beds have Pliocene sediments unconformably laid over them (***♦ page 78******).***

◄ Inclusions, such as these metamorphosed sediments engulfed by granite, must be of an older rock than the rock that contains them. In a case such as this it is obvious which rock is the younger.

◄ Dikes are sheet-like masses of igneous rock that cut through rocks already in existence. The dikes must therefore be younger than the country rock in which they have been emplaced.

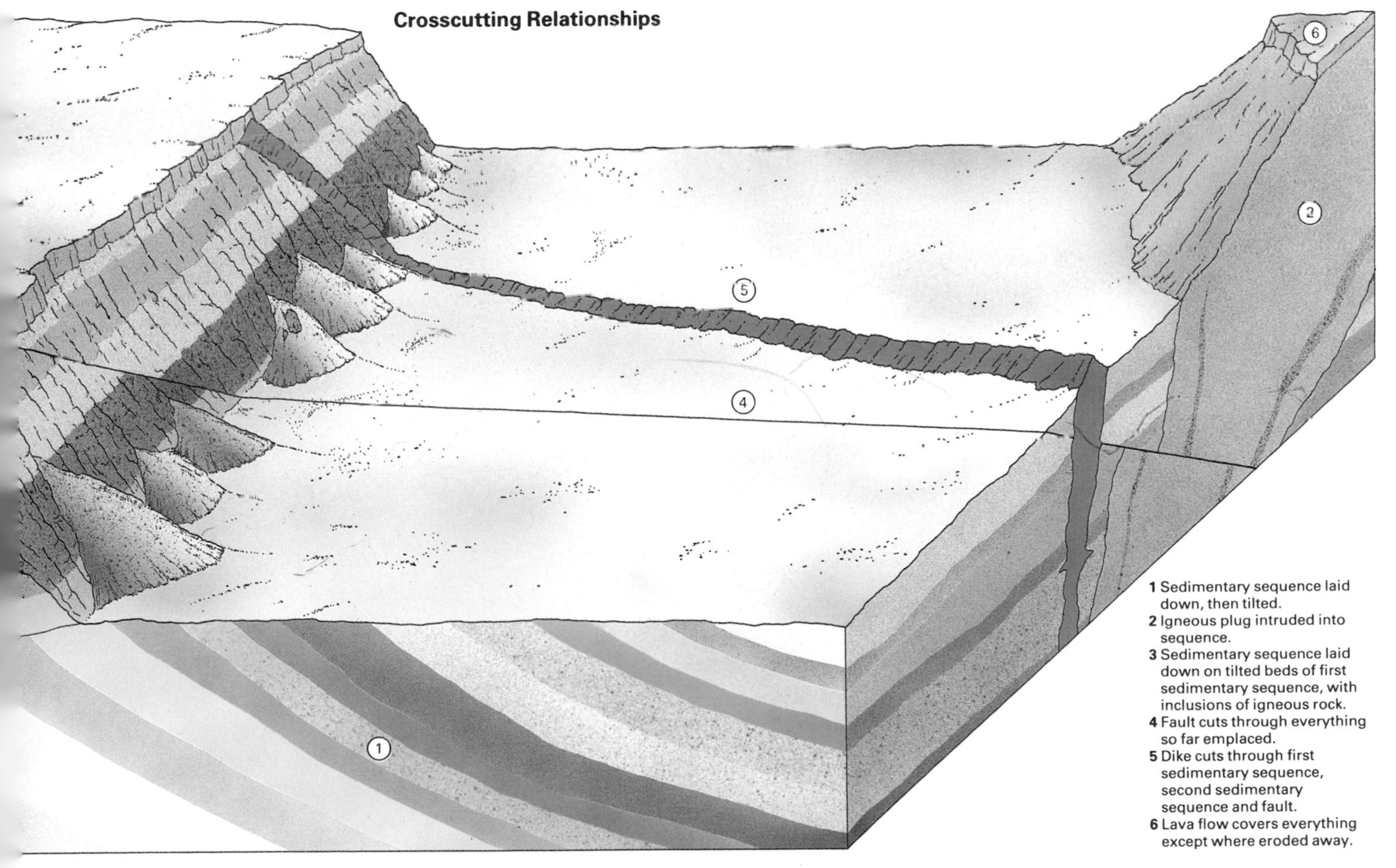

1 Sedimentary sequence laid down, then tilted.
2 Igneous plug intruded into sequence.
3 Sedimentary sequence laid down on tilted beds of first sedimentary sequence, with inclusions of igneous rock.
4 Fault cuts through everything so far emplaced.
5 Dike cuts through first sedimentary sequence, second sedimentary sequence and fault.
6 Lava flow covers everything except where eroded away.

Fossils – the final arbiter

The physical correlation of strata on the basis of rock type is not possible over large distances because sedimentary layers change their characteristics along their length. The method is therefore generally limited in its application to sediments deposited in the same environment, which means over distances of tens or hundreds of kilometers rather than thousands. It certainly cannot be applied across whole continents, still less between continents. Fortunately, however, fossils offer an independent method of correlation which can not only be applied on a worldwide basis, but also enables a global relative chronology to be established.

In the early 19th century the English surveyor William Smith (1769-1839) found that successive sedimentary strata in southwest England contained different assemblages of fossils and that a characteristic fossil assemblage could therefore be used to recognize a particular layer wherever it occurred. Smith had thus discovered what came to be known as the "law of faunal succession", namely, that successive strata can be identified by their particular fossil contents. Geologists soon came to realize that a particular fossil assemblage was a good indicator of the age of the corresponding stratum and that it occurs in rocks of the same age irrespective of their type. In other words, correlations need rely no longer on tracing the same stratum across large distances; instead the same fossil assemblage can be traced without reference to the nature of the particular sediment in which it happened to be. Very soon the correlations of southwest England were being extended to the country as a whole, to Europe and even to North America. A worldwide relative chronology was thus gradually constructed which was independent of sediment type.

Much later it came to be appreciated that fossil assemblages change with time because of the evolution of the relevant fauna and flora. Species arise, expand, get modified, contract and become extinct at different times and at different rates so that at any time there is a characteristic mix of species that represents that time only (although there are also some geographical variations). However, relative dating by fossil correlation does not depend on knowing the cause of the fossil changes with time. It can proceed quite happily without the need to believe in evolution at all and, indeed, did so for half a century before 1859 and *The Origin of Species*. All that is required is information on how (not why) the fossil assemblages have changed with time. William Smith began to acquire this information by studying the fossil assemblages in successive layers and noting the relative ages of the layers using the Principle of Superposition. As a result, he was able to construct the first geological map in 1815.

Because radiometric dating is not generally applicable to sediments, relative dating by fossils is still the chief method of working out the history of the sedimentary crust. The best fossils to use in dating (index fossils) are those that change quickly with time. They should also be fossils of creatures that lived in all the seas of the world. A free-swimming creature will be independent of any sedimentation that is going on beneath, and its fossils should be found in a variety of sedimentary rocks. This type of dating can only be used where there are fossils, which means up to about 590 million years ago. For the vast length of time before this only the physical methods of relative dating can be used. As physical correlations cannot be made over large distances, a worldwide chronology going back 3·8 billion years had to await the development of radiometric methods of absolute dating.

▲ *A particular species of fossil ammonite can be used to date the rock in which it is found.*

▼ *William Smith, an English canal engineer, was first to use fossils to date the rocks.*

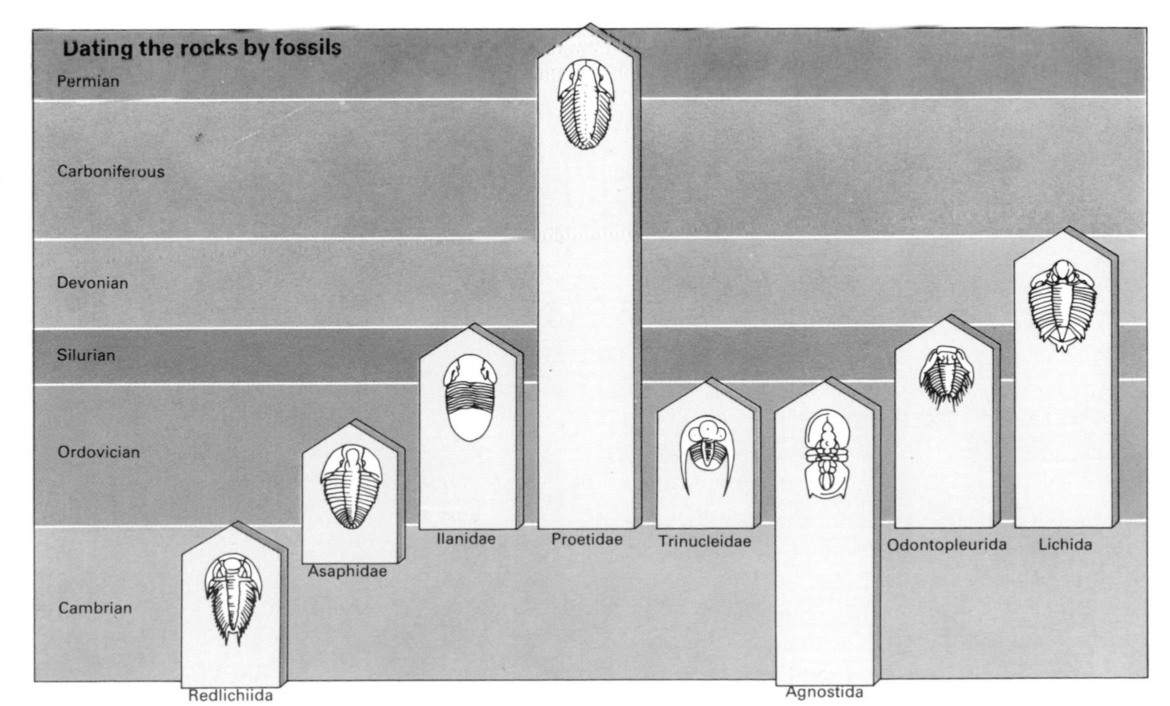

▲ ***William Smith compiled the first geological maps, in 1815, coloring each rock sequence according to age.***

► ***Biostratigraphy – dating by fossils – can only be used with fossils of creatures that evolved rapidy and were widespread. In this generalized example using trilobites, agnostids are known to have lived in Cambrian and Ordovician times, while ilanids survived from the Ordovician to the Silurian Periods (◆ page 78). If a rock sequence contains both agnostid and ilanid trilobites, that sequence must be Ordovician in age, since that is the only period in which the ranges of the two trilobite types overlapped.***

The Geological Time Scale

One of the chief legacies of 19th century geology is the geological time scale; the division of geological time into named intervals separated from each other by major changes in rock type, obvious breaks in the succession, and abrupt changes in fossil groups. The coarsest division is the eon, then come eras, periods and epochs.

Archean Eon
The very ancient eon. Between the formation of the Earth and 2·5 billion years ago. Some geologists place the oldest limit at the time of the oldest rock, 3·8 billion years, and class everything before this as the Hadean Eon.

Proterozoic Eon
The eon of first life. 2·5 billion to 590 million years ago. It is a time when life is known to have existed but left no clear fossils. The 2·5 billion year boundary has been set by the radiometric dating of a number of igneous and metamorphic events. However, it is now evident that life existed for some time before that.

Phanerozoic Eon
The eon of obvious life. 590 million years ago to the present day. This is the time from which good fossils are known, due to the evolution of hard shells and skeletons at the beginning of the eon.

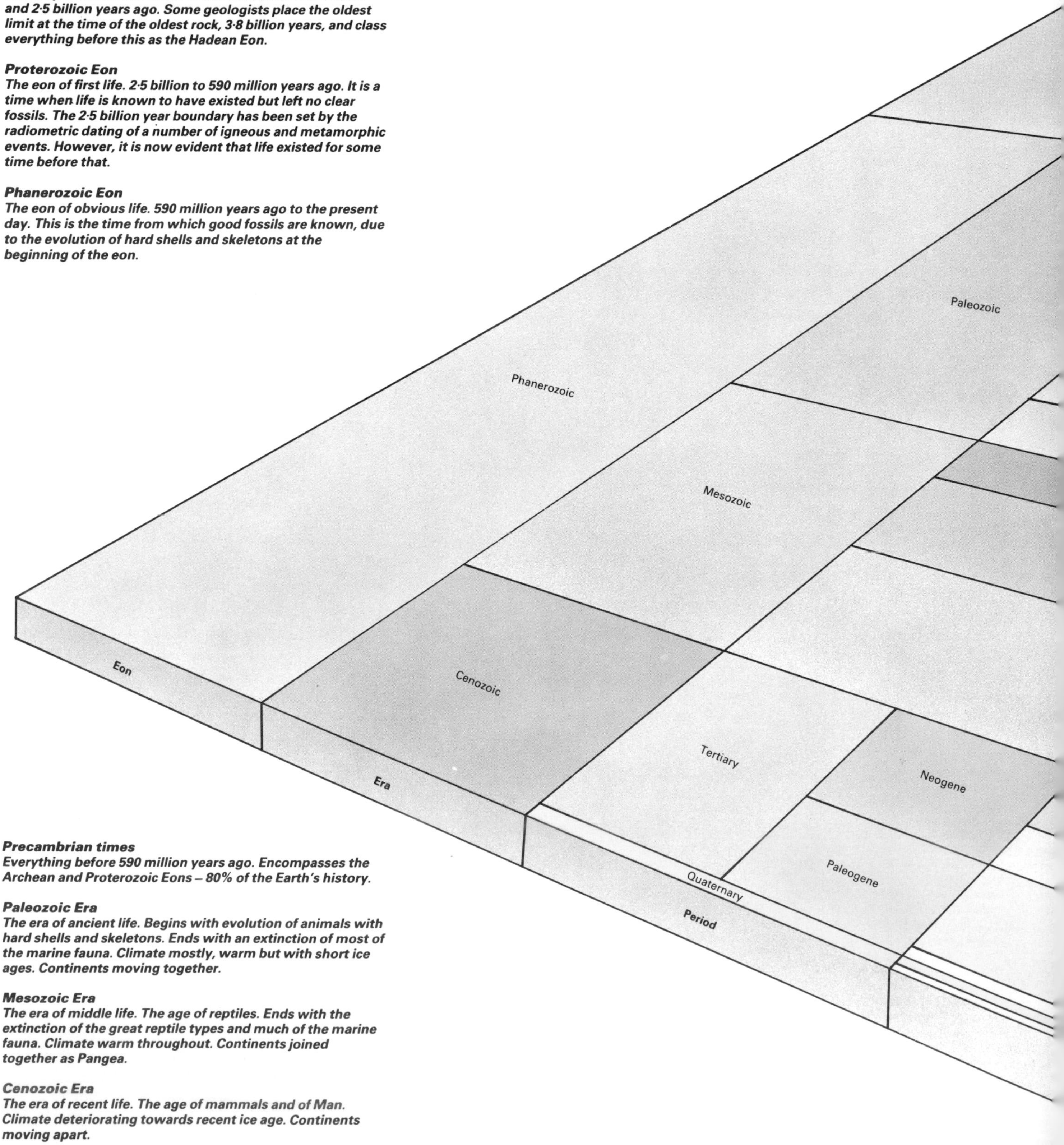

Paleozoic
Phanerozoic
Mesozoic
Eon
Cenozoic
Tertiary
Neogene
Era
Quaternary
Paleogene
Period

Precambrian times
Everything before 590 million years ago. Encompasses the Archean and Proterozoic Eons – 80% of the Earth's history.

Paleozoic Era
The era of ancient life. Begins with evolution of animals with hard shells and skeletons. Ends with an extinction of most of the marine fauna. Climate mostly, warm but with short ice ages. Continents moving together.

Mesozoic Era
The era of middle life. The age of reptiles. Ends with the extinction of the great reptile types and much of the marine fauna. Climate warm throughout. Continents joined together as Pangea.

Cenozoic Era
The era of recent life. The age of mammals and of Man. Climate deteriorating towards recent ice age. Continents moving apart.

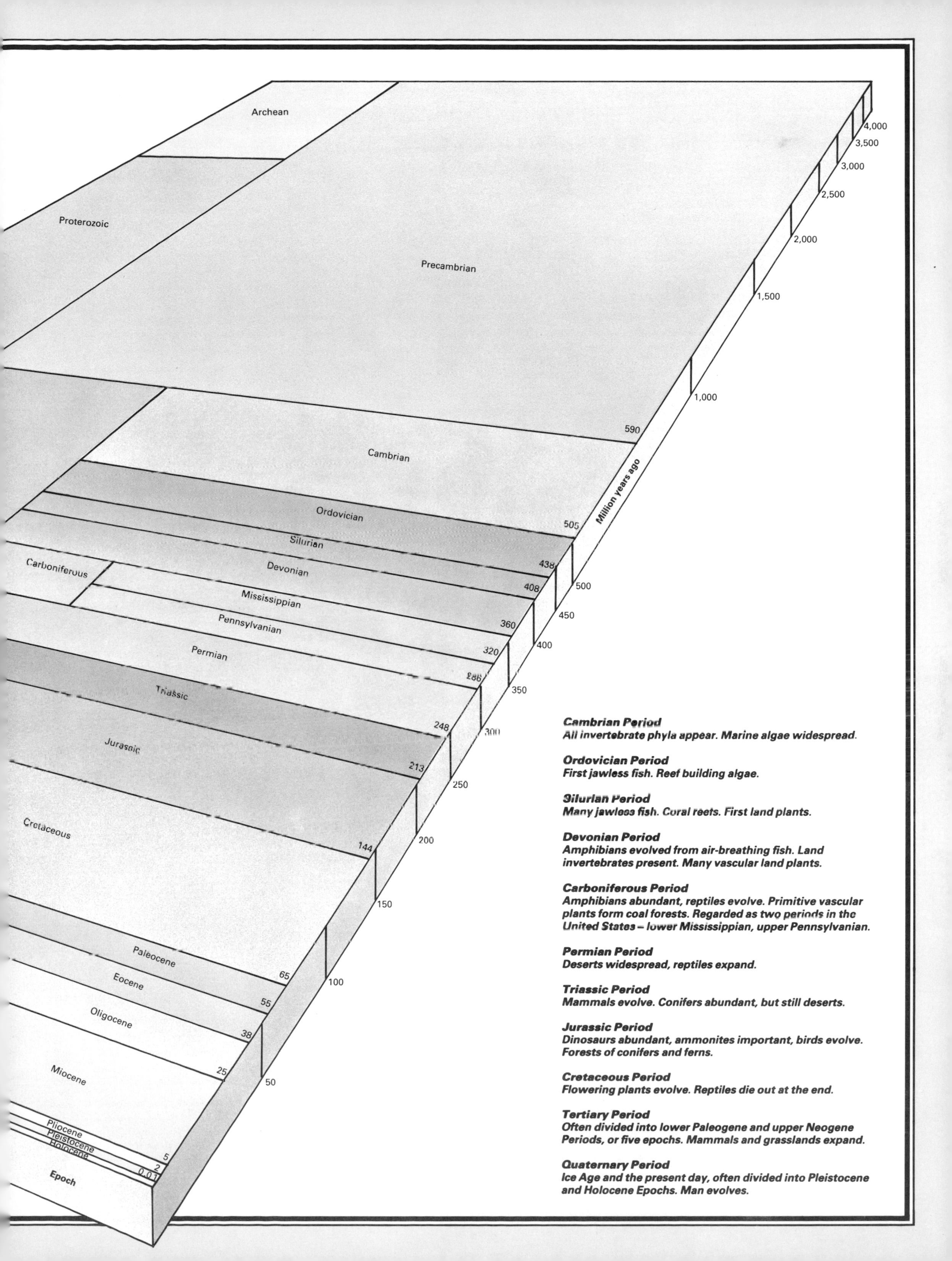

Cambrian Period
All invertebrate phyla appear. Marine algae widespread.

Ordovician Period
First jawless fish. Reef building algae.

Silurian Period
Many jawless fish. Coral reefs. First land plants.

Devonian Period
Amphibians evolved from air-breathing fish. Land invertebrates present. Many vascular land plants.

Carboniferous Period
Amphibians abundant, reptiles evolve. Primitive vascular plants form coal forests. Regarded as two periods in the United States – lower Mississippian, upper Pennsylvanian.

Permian Period
Deserts widespread, reptiles expand.

Triassic Period
Mammals evolve. Conifers abundant, but still deserts.

Jurassic Period
Dinosaurs abundant, ammonites important, birds evolve. Forests of conifers and ferns.

Cretaceous Period
Flowering plants evolve. Reptiles die out at the end.

Tertiary Period
Often divided into lower Paleogene and upper Neogene Periods, or five epochs. Mammals and grasslands expand.

Quaternary Period
Ice Age and the present day, often divided into Pleistocene and Holocene Epochs. Man evolves.

See also
Origin and Evolution 69-72
The Earth Through Time 81-100
Rocks and the Rock Cycle 141-152

Radiometric methods – the key to the oldest rocks

Absolute dating, which means specifying the age of a rock in numerical terms (i.e in years), is based on the decay of the radioactive isotopes of chemical elements that naturally occur in igneous rocks. All radioactive isotopes (parents) decay to non-radioactive isotopes (daughters) at a rate that is constant. The rate of decay is different for each isotope. In other words, in any given year a constant proportion of the radioactive atoms present will decay, although this proportion will not be the same for each isotope. The decay rate is not affected by such environmental factors as temperature and pressure; nor does it depend on the type of chemical compound in which the isotope happens to be. It is governed only by the decay constant for the particular parent isotope, and the decay constants for all radioactive isotopes are now known.

Because a constant proportion of a radioactive isotope decays each year, the actual amount of the parent present decreases exponentially with time and the amount of the daughter increases accordingly. Another way of expressing this phenomenon is in terms of half-life, the time it takes one half of the original atoms to decay. Half the parent isotope decays in one half-life, half of the remainder (or one quarter of the original) decays during the second half-life, and so on. Each radioactive isotope has its own half-life. By measuring the ratio of the amount of the daughter isotope present to the amount of parent isotope (and knowing the decay constant), it is then a simple matter to work back to find out when the decay process began. This is the time at which the rock formed, and the radiometric clock was set at zero and started.

To date igneous rocks, it is best to use isotopes that have half-lives that are roughly the same as the ages of the rocks to be dated. If the half-lives are very much shorter, the isotopes will have already decayed away almost completely. If they are very much longer, the rate of decay will be so slow that the daughter-parent ratio will change little with time. The isotopes chosen must also be abundant and widespread enough. In fact, four decay systems have proved particularly useful. The most important is perhaps the decay of potassium-40 to argon-40, followed by rubidium-87 to strontium-87, uranium-235 to lead-207, and uranium-238 to lead-206.

Radiometric dating may be simple in theory but is not necessarily so in practice, which is why the method has only become reliable in recent decades. For one thing, concentrations of radioactive isotopes in igneous rocks are very low and are therefore difficult to measure with the necessary precision. Moreover, the method will only give reliable results if no parent isotope has been added since the rock formed and none of the daughter isotope has escaped. Sophisticated tests are necessary to ensure that these circumstances apply.

Radiometric dating may also be carried out on metamorphic and sedimentary rocks, but in such cases it does not provide the ages of the rocks concerned. For metamorphic rocks it gives not the age of the original rocks but the time of the event (heat, pressure, etc) that brought about their metamorphism. The metamorphism is, of course, younger than the original formation, but its age is useful all the same. In the case of sediments the ages obtained are not those of the beds (which is what geologists usually need) but those of the fragments that went to form the beds. Clearly the sedimentary fragments must be older, and frequently much older, than the sediments of which they now form part.

The Age of the Earth

The Earth must be older than its oldest rocks, and those have been dated at rather less than 4 billion years. But how much older? There is no way of measuring the age of the Earth directly, but there are three indirect lines of evidence that all point to the same answer. First, almost all meteorites have radiometric ages of about 4·6 billion years. As geologists believe meteorites to be material that formed at the same time as the Solar System, this suggests that the Earth also formed about 4·6 billion years ago. Second, the oldest rocks collected from the Moon, which is also thought to have originated at about the same time as the Earth/Solar System, are again about 4·6 billion years old.

The third piece of evidence comes from lead in the minerals of the Earth's crust. Natural lead consists of four stable isotopes, three of which (lead-206, lead-207 and lead-208) are capable of being produced by the decay of uranium or other radioactive isotopes and one of which (lead-204) is not. Thus all of the lead-204 now on the Earth must have been present since the Earth formed, whereas for the other isotopes, some has been present since the Earth's formation and some has accumulated gradually by decay. If the relative abundances of the four isotopes at the time the Earth formed were known, it would be possible to work out how long it must have taken to get to the present relative abundances. Unfortunately, the original relative abundances are not known directly, although it may be hypothesized that they were the same as those now found in meteorites which contain no lead-producing radioactive isotopes. If the calculation is done on this premise, the answer comes out at about 4·6 billion years.

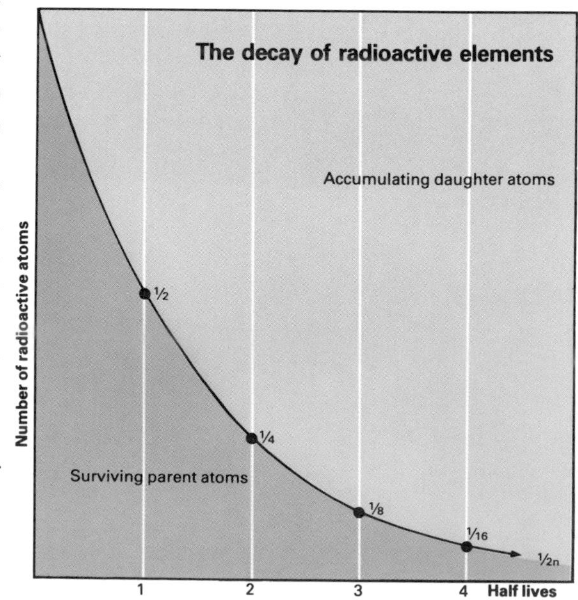

▲ After a period of time, called the half-life, a radioactive element will have decayed to half of its original quantity. After another half-life there will be a quarter left, and so on. The amount of radioactive element in a rock compared with the amount of its decay product gives the number of half-lives that have passed since the rock was formed, and hence the date of the rock.

The Earth Through Time

8

Precambrian – the first 3·2 billion years and the origin of life...Paleozoic – life develops in the sea and emerges on to the land...Mesozoic – the age of reptiles...Tertiary – mammals become dominant...Quaternary – the great Ice Age and the coming of people...PERSPECTIVE... Ancient ice ages...The first life...Oceans of the past... Defining the periods...Extinctions...Ocean sediments... Changing sea levels

The evolution of our planet is contained in the rocks formed at various times, and their preserved fossils. Together they describe the movements of continents and the buildup and destruction of mountains, and the accompanying evolution, development and extinction of the varied plant and animal forms.

The Precambrian – earlier than 590 million years ago

Although comparatively little is known about the geological events and processes of the Precambrian, this is not because of a shortage of rocks of the right age. On the contrary, rocks formed during the Precambrian make up the bulk of today's continents, even though a good many of them are now covered with a layer of sedimentary rocks. The difficulty is that, even in the many areas in which Precambrian rocks are exposed at the surface, description is easy but interpretation is not. The lack of fossils in the sedimentary components prevents the arrangement of the rocks in chronological order. The large areas of exposed Precambrian rock, which exist on all continents are known as shields, and the no less extensive regions in which underlying Precambrian "basement" rocks are covered by sediment are known as stable platforms. Such platforms are stable in the sense that they have not been subjected to large-scale tectonic processes, such as mountain building, since they were formed, which means that they have kept well clear of plate margins.

The older, Archean part of the Precambrian is characterized by two major types of environment – greenstone belts and gneiss terrains. A greenstone belt typically consists of a thick sequence of basalt lavas topped by sediment and intruded by granite. The whole formation is severely folded and deformed. The granite intrusion seems to have occurred at the same time as the folding and deformation, at which time both the lavas and sediments were also metamorphosed (altered by one or more of heat, pressure and invading fluids) in such a way that some of the original minerals in the solid lava were converted to greenish secondary varieties. This explains the name.

Most greenstone belts were produced between 3·5 and 2·5 billion years ago, although their origin is obscure. Many of the lavas are pillow lavas, indicating that they were formed under water. Some geologists have suggested that greenstone belts are the remains of Archean oceanic crust generated during a phase of Precambrian ocean floor spreading either at oceanic ridges or in marginal basins, but this is highly speculative. Greenstone belt lavas are rather different in composition from modern oceanic crustal basalts, but that may simply be because in the Archean the Earth's temperature was higher and hence the mantle more extensively melted. In any event, Archean greenstone belts are of more than purely academic interest. They have an economic value for they have associated with them many economic ores.

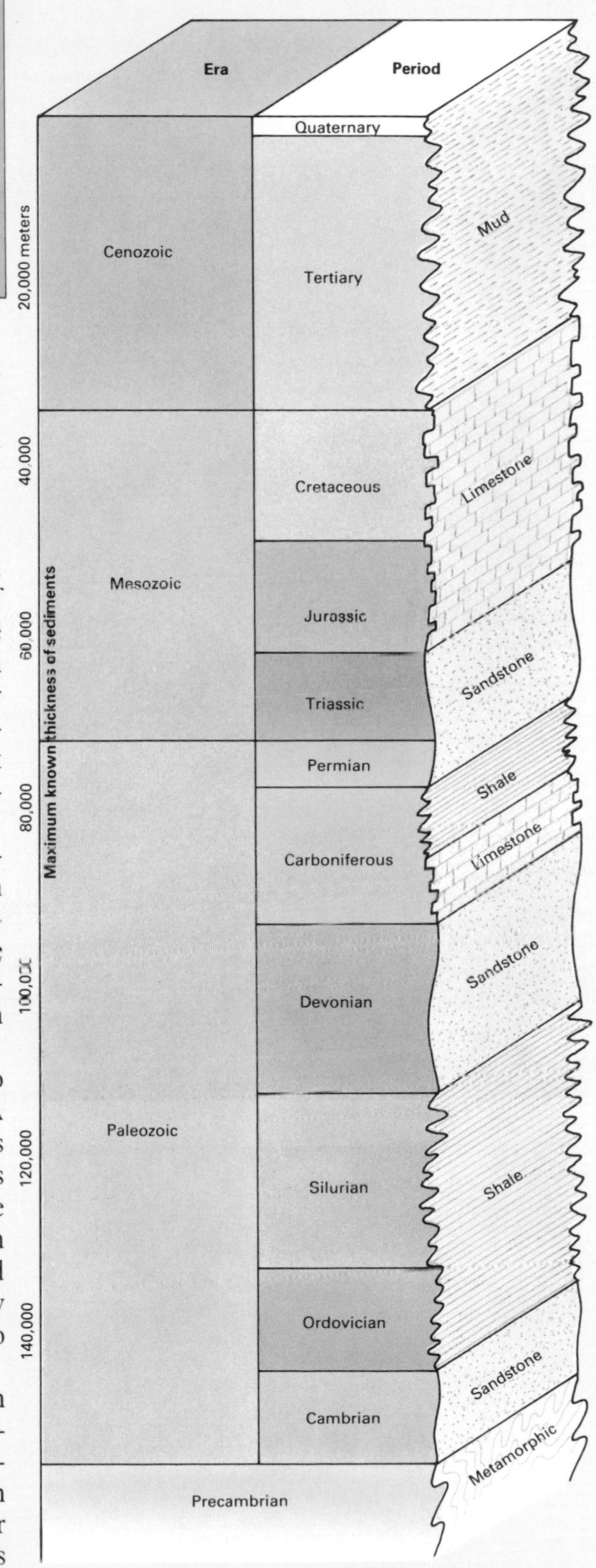

▲ *If a certain spot on the Earth's surface had sediments laid down upon it continuously since Cambrian times there would be a sequence of sedimentary rocks built upon it to a depth of about 160km. Since sedimentation tends to be be sporadic and uneven this kind of thickness is not found anywhere. Earth's history book is read by studying rock sequences from different parts of the world.*

The greenstone belts are distributed within much more extensive gneiss terrains. Gneiss is a banded rock produced by very severe metamorphism, and that of the Archean terrains was evidently formed from original granite. Again, the circumstances in which the gneiss terrains are formed are unknown; but it is interesting that, in addition to highly metamorphosed sediments, gneiss terrains often contain the rock anorthosite. On Earth, anorthosites are not particularly common and are limited to the Precambrian, but they form almost the whole of the Moon's crust (◀ pages 69-72). There may therefore be some significance in the fact that anorthosite appears to be a type of rock formed only very early in the lives of planetary bodies. The world's oldest known rocks are about 3·8 billion years old and come from the gneiss terrains of Greenland.

The Archean was predominantly a time of crustal generation, although the existence of sediments shows that some erosion was already taking place.

The younger, Proterozoic part of the Precambrian, by contrast, was more a time when reworking of the existing crust was occurring in the form of deformation, recrystallization, metamorphism and associated magmatic activity. In other words, some of the Archean crust was subjected to various phases of mountain building, chiefly around its edges.

The Canadian shield is a good case in point, for this well-studied region has a core of Archean greenstone belt/gneiss terrain surrounded by Proterozoic mountain-building zones of different ages. The mountains themselves have long since eroded away, leaving a more or less sea-level relief formed by the now-exposed roots of the ancient mountains. The fact that mountain ranges were being formed during the Proterozoic suggests that some type of plate tectonics was in operation, but what form it took and how similar it was to modern plate tectonics remains unknown.

▲ Most of the exposed Precambrian rocks, such as these gneisses in northern Scotland, are metamorphic. They have been subjected to great heat and pressure in mountain-building activities in the long period since they formed.

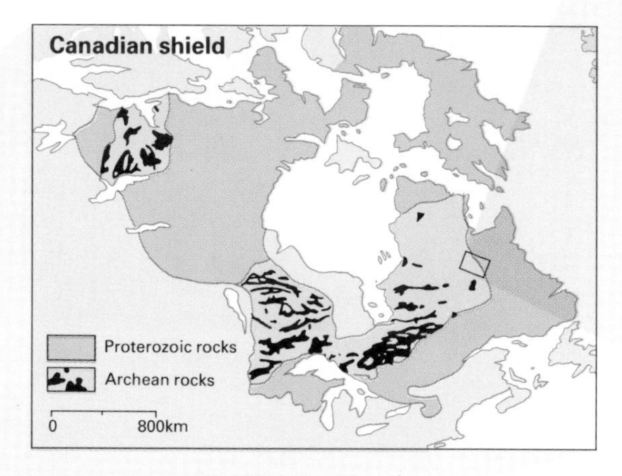

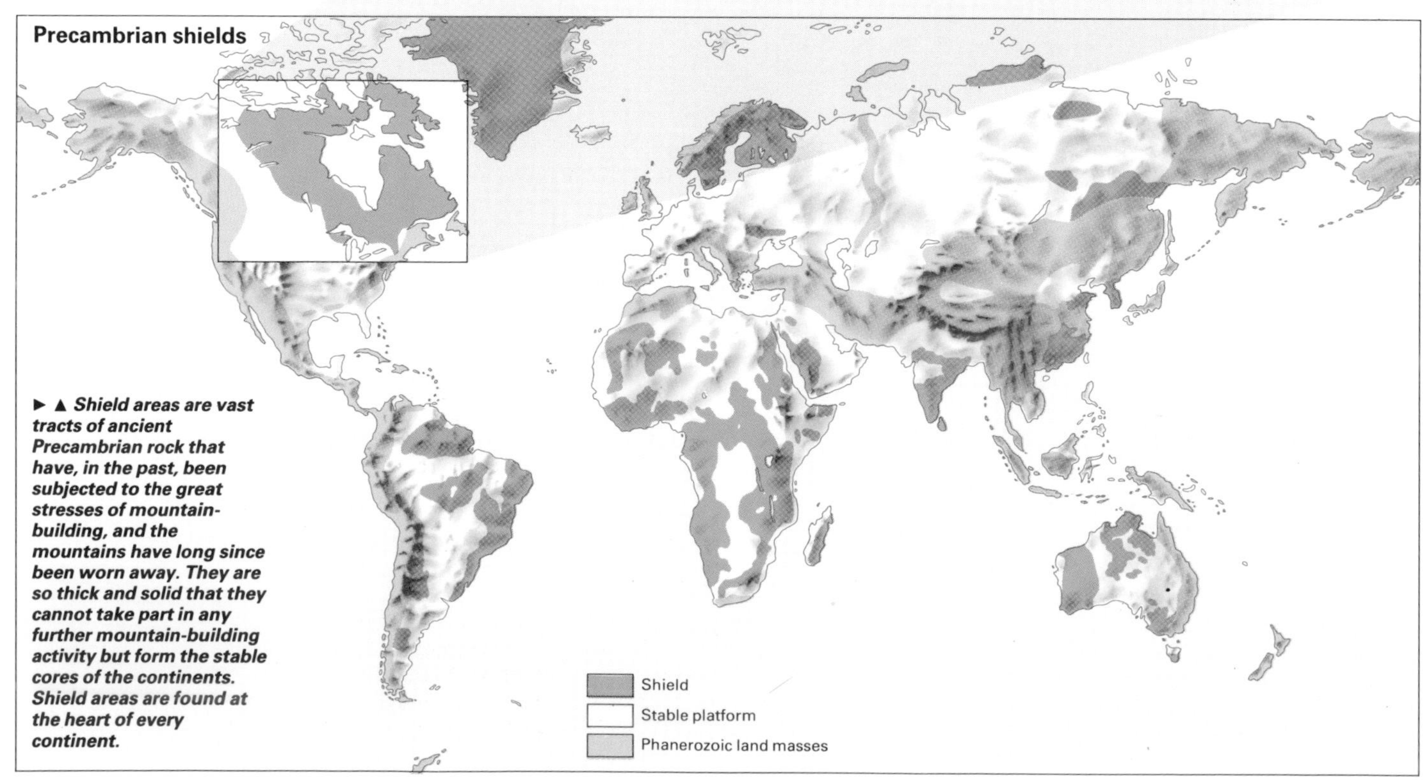

►▲ Shield areas are vast tracts of ancient Precambrian rock that have, in the past, been subjected to the great stresses of mountain-building, and the mountains have long since been worn away. They are so thick and solid that they cannot take part in any further mountain-building activity but form the stable cores of the continents. Shield areas are found at the heart of every continent.

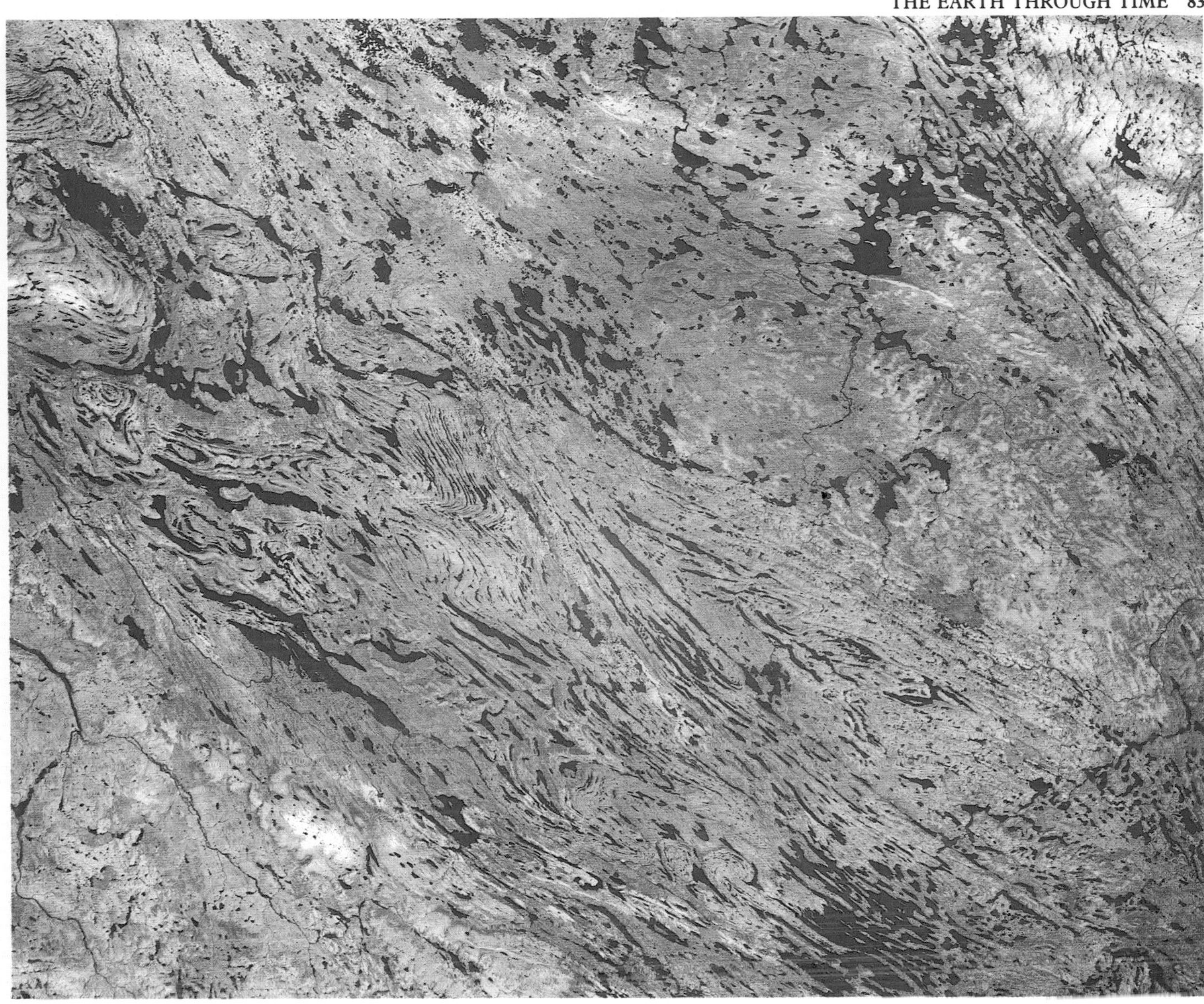

▲ From high altitude the trends of former mountains in a shield area are visible. This is part of the Canadian Shield in Labrador as seen from the Landsat satellite. The harsh climate here restricts the growth of vegetation and the basic rock structure can be seen, with lakes gathering in the hollows. It is a major source of Canadian iron ore.

Precambrian ice ages

When an ice age ends, the retreating glaciers leave behind them masses of unsorted deposits known as till, or drift. The material in till can range from rock flour, consisting of very fine particles produced from rocks that have been fractured or ground together, to boulders of considerable size. The proportions of the various fragment sizes in a till depend on individual circumstances, especially the nature of the bedrock in the regions over which the glaciers picked up their rock load. Till thicknesses can be very great, sometimes as much as 100m.

Geologists have found many tills – or tillites as they are more commonly known in this context – of Precambrian age, suggesting that there must have been widespread Precambrian glaciation. There is one complication here, however, in that unsorted and unstratified sediment mixtures containing boulders are also produced by mudflows – water-saturated masses of mud and rock that in certain circumstances flow rapidly downslope under the action of gravity. Mudflow deposits and till are often hard to distinguish, although in modern times mudflows are comparatively rare events, being associated particularly with volcanic and arid regions with little vegetation. In Precambrian times, on the other hand, when none of the continental areas had any vegetation to speak of, mudflows were probably much more common and extensive. This can give rise to a great deal of confusion. Geologists are therefore not always sure whether individual deposits of unsorted Precambrian material that they are studying are one thing or the other.

Even so, many Precambrian tillite-type deposits on almost all the continents are now accepted by geologists as being genuine tillites, especially those that lie on polished rock surfaces containing glacial scratches. Such tillites indicate that there was an ice age during the Proterozoic around 2·3-2·2 billion years ago and another at the very end of the Precambrian. After the latter, the climate seems to have improved and remained generally warm throughout the Cambrian period and for most of the Ordovician as well.

When Precambrian tillites were discovered during the mid-nineteenth century they came as quite a shock, because at the time geologists believed that the Earth had formed at very high temperatures and had ever since been cooling down. With that premise, the modern ice age appeared a logical consequence of this gradual cooling, and equally cold times as long ago as the Precambrian seemed untenable. In any event, the occurrence of the two oldest known ice ages suggests that surface temperatures during the Proterozoic were much the same as those experienced today.

One consequence of the high relief provided by the Proterozoic mountains was that erosion accelerated, making sedimentation much more important than it was during the Archean. In fact, many of the sedimentary processes and patterns of the Proterozoic closely resemble those of today. On the other hand, there was also extensive sedimentation of a type unique to the Precambrian – the laying down of banded iron formations. These consist largely of iron oxide (or sulfide) and quartz in alternating layers, or bands, of red or black iron-rich material and gray iron-poor material. The circumstances in which these formations appeared is unclear, but some geologists believe that they are limited to the Precambrian because they could only form in an atmosphere with very little oxygen. Be that as it may, such rocks constitute the largest reserves of iron ore in the world.

There is no rock evidence to explain the origin of life on the Earth or provide a clear indication of the atmospheric constituents. However, it is generally accepted that the early atmosphere contained mainly nitrogen with water vapor, carbon dioxide and carbon monoxide. The production of significant amounts of free oxygen had to await the development of life-forms such as green algae. The first living thing would probably have been a complex organic molecule that had the ability to reproduce itself. Evolution would have begun with any change in the molecule that helped this reproductive process. However, the debate about the true origins of life itself is extensive and as yet inconclusive.

Although fossils of creatures with hard parts are rare in the Precambrian and limited to the very late Proterozoic, there is other evidence of life throughout much of the Precambrian. There are two types of "fossil", the most abundant being stromatolites. These are not primary organic remains but distinctive structures of carbonate made by blue-green algae. Stromatolites are produced today along carbonate shorelines in inter-tidal and above-tidal zones which are wetted only occasionally. The sediment surface becomes covered with a layer of filamentous algae cells, and this binds the carbonate particles into a layer known as a "mat". When the tide, or a storm, brings in more carbonate particles they adhere to the mat, get covered by more algae, and so on, building up a characteristic laminated structure. The Precambrian algae themselves have long since decayed, but the laminated mats remain as indicators of their former presence.

The other, less conspicuous and more subtle Precambrian fossils are filaments and spheres which may represent the remains of blue-green algae or other forms of proto-life. They have diameters in the range 10-20μm and are found in cherts, deposits of finely crystalline silica. Whereas the oldest known stromatolites are about 2·9 billion years old, the oldest convincing microfossiliferous cherts are about 3·3 billion years old. However, some geologists claim that there is evidence for even older life. They interpret some marbles, for example, as metamorphosed stromatolites, while other rocks have been interpreted as metamorphosed oil shales. Opinions are divided on the validity of such claims, but if they are accepted it would mean that life was present on Earth as long as about 3·8 billion years ago.

The Cambrian Period – from 590 to 505 million years ago

The most spectacular event of the Cambrian Period was the great outburst of marine animal life with skeletons (hard parts) that were capable of leaving behind fossil remains. Indeed, this remarkable phenomenon is used to define the base of the Cambrian.

▲ **Few Precambrian fossils are known, because animals did not evolve preservable hard parts until about 590 million years ago. The kinds of fossils that do exist consist of trails (upper) and flatworms such as "Dickinsonia" (lower).**

Soft-bodied creatures of the late Precambrian

Although multicellular organisms (metazoa) with hard parts appeared suddenly by the standards of geological time, they nevertheless probably took about 20 million years to develop. Moreover, they also had precursors without hard parts. These soft-bodied forms were unable to leave behind fossils as such, but they did sometimes leave imprints of their bodies that were subsequently preserved. Such fossil imprints have been found in several parts of the world, although the best preserved are those in the sandstones of the Ediacara Formation of southern Australia. Here are to be seen round jellyfish-like forms, elongated worm-like forms, round shell-like forms and a number of others. All the soft-bodied creatures were very thin, presumably to enable them to take in enough oxygen to survive in an environment in which the oxygen concentration was very low.

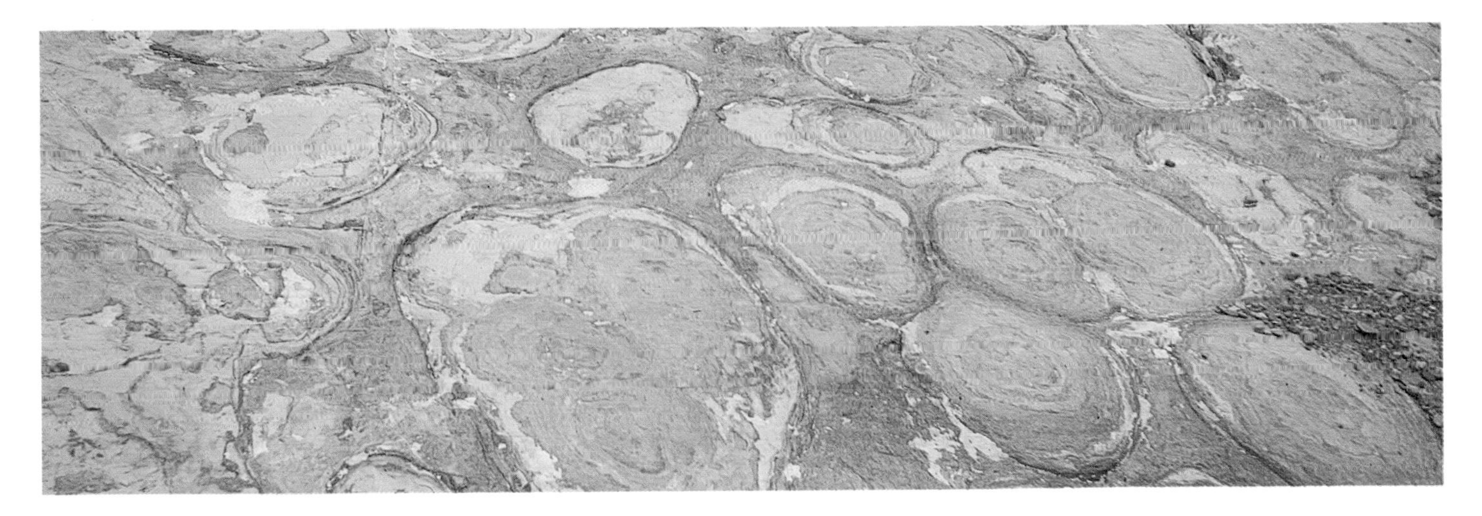

Being thin and planar would have maximized the surface area and hence the exposure to oxygen.

The existence of late Precambrian soft-bodied creatures, though interesting, does nothing to solve the problem of why simple single-celled forms which had been present throughout much of the Precambrian suddenly exploded into a great variety of multicellular life. The oldest fossil impressions are about 700 million years old; so the transition from single cell to a multicellular animal with a skeleton occurred within 100 million years or so – a very short interval on the geological time scale.

The reason for it all remains a mystery. One possibility is that the development of soft-bodied metazoa, like the metazoa's later acquisition of hard parts, had something to do with increasing oxygen levels. In this case, the metazoa themselves presumably needed a certain oxygen level and the construction of hard parts required a somewhat higher one. A second hypothesis relates not to the external physical environment but to changes in the organization of the life-forms themselves. Some paleontologists have suggested that the evolutionary burst of multicellular fauna was the result of the invention of sexual reproduction. Single-celled organisms normally multiply by asexual division, but most multicellular creatures reproduce sexually. Sexual reproduction provides vastly greater opportunity for modifying the genetic characteristics of offspring.

The mechanical process of the evolution of multi-celled organisms is also a matter of differing opinions. One idea is that the single cell became multi-celled by budding – with the new cells failing to break away completely. The other is that a number of separate cells aggregated and then acted as a unit, this unit having some enhanced survival characteristics.

▲ Amongst the earliest pieces of evidence of living things are stromatolites. These form at the present day when mats of algae trap sediment, building up columns of mud on which more algae grow. This happens off the west coast of Australia today (top). Fossil stromatolites lie in many Precambrian rocks, such as the Gunflint chert in Canada, 2 billion years old (bottom).

▲ ▼ *In the Cambrian of North America the Canadian Shield formed the core of the continent. Much of it was underwater and the resulting wide shelf seas collected deposits of limestone, sandstone and shale (below). Fold mountains began to rise along the eastern edge of the continent in the succeeding Ordovician period forming the Taconic mountains of the northwest Appalachians (above).*

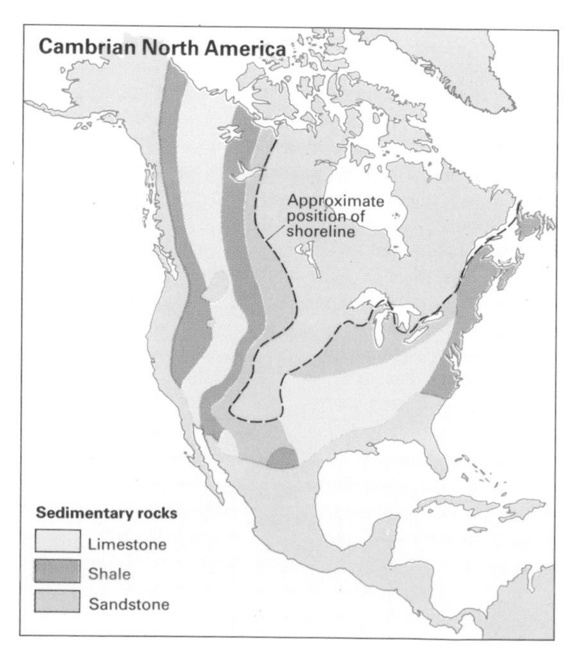

The reason for the sudden development of hard parts of Cambrian organisms is unclear. According to one hypothesis, skeletons were not able to form until the amount of oxygen in the atmosphere had reached a certain level. The Earth's original atmosphere was oxygen-free and the proportion of oxygen gradually built up to its present 21 percent as a result of photosynthesis by plants, especially marine algae. Precisely how the percentage of oxygen has varied with time is unknown; but the critical level could well have been achieved at the beginning of the Cambrian. Oxygen (ozone) in the atmosphere shields the Earth's surface from damaging ultraviolet radiation, thus possibly allowing Cambrian fauna to take to shallower waters near whose surface the primary food source, algae, would have flourished.

A second hypothesis makes use of another event in the Cambrian, namely, the widespread flooding of the continental shorelines by the seas – known as transgression. Minor transgressions occurred during the Proterozoic, but large-scale transgressions do not seem to have occurred before the Paleozoic. The cause of the first major, Cambrian, transgression could well have been a consequence of plate tectonics. Oceanic ridges occupy huge volumes, and so the creation of ridges would displace large quantities of ocean water, perhaps sufficient to flood wide areas of the continents. Transgression occurred slowly throughout the Cambrian at an average rate of 16km each million years. The result was a large increase in the area of warm, shallow marine environments (epicontinental seas) which are likely to have provided many new ecologically favorable sites with adequate food supplies.

▲ *The shellfish of lower Paleozoic seas were mostly brachiopods. These were totally unrelated to modern bivalves.*

► *Graptolites, bryozoans and other communal fauna, both free-floating and encrusting, were common in the lower Paleozoic.*

◄ *Trilobites – many-legged segmented animals – were typical Paleozoic creatures, most important from the Cambrian to the Devonian.*

Among the earliest organisms to appear during the Cambrian were the two-shelled brachiopods, the single-shelled gastropods, the conical-shelled hyolithids, and the sponge-like archeocyathids. The early skeletons were built of phosphates rather than calcium carbonate, although calcareous skeletons ultimately became the rule. The many-legged trilobites came slightly later, quickly multiplying to dominate the Cambrian seas, and the crinoids arrived too. In the Middle Cambrian the graptolites appeared and seaweeds were evidently present. The Cambrian continents were generally deserts.

The Ordovician Period – from 505 to 438 million years ago

The vast epicontinental seas that formed during the Cambrian persisted into the early Ordovician and then retreated, returning large continental areas to dry land. By the middle Ordovician, however, slow invasion by the oceans had recommenced and was to give rise to some of the most widespread epicontinental seas in geological history. There was extensive marine sedimentation on the continents. Today, Ordovician sediments are found either on the stable continental platforms which have remained free of major tectonic activity, or in the linear fold mountain belts where they have been intensely deformed.

Among the many fauna found in Ordovician strata are trilobites, brachiopods, gastropods, crinoids, echinoids (sea urchins), bryozoans and corals. Graptolites are common in the deep sea sediments, and in fact the Ordovician Period was defined by the British geologist Charles Lapworth in 1879 on the basis of the sequence of graptolite evolution. Fish also appeared during the Ordovician Period.

An early Atlantic Ocean

Between the late Precambrian and the Devonian, most of the land areas that were later to become North America and Europe/northern Africa were separated by ocean. Geologists now think that this body of water, now called the Iapetus Ocean, was the result of an early episode of ocean floor spreading. During its first hundred million years or so, the Iapetus grew much as the Atlantic Ocean is growing now. In other words, the ocean floors were spreading from a central ridge and pushing the landmasses away on either side. The continental margins on the opposite sides of the Iapetus were therefore passive (tectonically inactive) just as the present margins of eastern North America and western Europe/Africa are passive.

But in the middle Ordovician this situation changed. At this time the Iapetus Ocean reached its maximum width of about 2,000km and then began to contract. Subduction zones formed at various places around the Iapetus and the ocean floor began to disappear. The precise positions of all the subduction zones are not yet clear, but there was certainly one off the east coast of North America which between the middle and late Ordovician gave rise to what are now the Taconic mountains of eastern New York State (northern Appalachians). This period of mountain building is known as the Taconic Orogeny. Subduction zones are associated with volcanoes. The middle to late Ordovician was therefore a time of extensive volcanism.

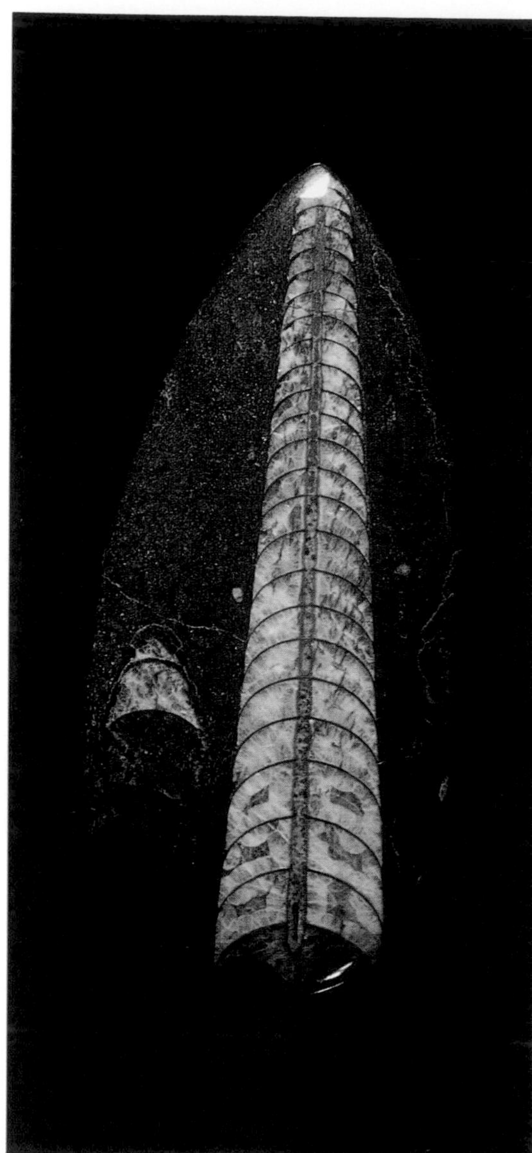

The Silurian Period – from 438 to 408 million years ago

During the late Ordovician the third ice age began, and so the Silurian opened in a state of glaciation (although, of course, not every bit of continent was covered in ice). When the glaciers retreated in the early-to-middle Silurian they left behind tillites which are now found in Argentina and Bolivia. During the middle Silurian the epicontinental seas began slowly to retreat again, going on to complete the second major cycle of transgression and regression. However, the pattern of sea level changes during the late Ordovician and early-to-middle Silurian was not as simple as this would suggest, for worldwide changes in sea level have at least two causes. The appearance and disappearance of oceanic ridges can displace huge quantities of water. Equally, the advance and retreat of extensive glaciers can also affect sea level by removing water from the oceans and then returning it. Sea level changes caused by glaciation probably occur more rapidly than those due to the changes in volume of oceanic ridges, and so the rapid sea level changes resulting from the late Ordovician-early Silurian glaciation were presumably superimposed on the slower ones resulting from the plate tectonic processes.

In any event, for one reason or another sea level rose during the early Silurian. In fact it rose so high that areas of dry land available to supply sand were severely restricted. Silurian sediments produced on the flooded parts of continents are therefore characteristically limestones and muds. Such sediments commonly contain brachiopods, trilobites and corals, and some contain mollusks, bryozoans, crinoids and ostracods. Graptolites are also common in deeper-water sediments. Most Silurian life was marine, although some freshwater fish were present in rivers and lakes. The very late Silurian may also have seen the first colonization of the land by primitive plants, but the main development of plants on land took place during the subsequent Devonian.

In the meantime, the Iapetus Ocean continued to narrow, drawing the ancestral landmasses of North America and Europe ever closer together. As this convergence necessarily involved subduction of the ocean floor, the subduction-related volcanism that began in the Ordovician continued throughout the Silurian.

▲ *Cephalopods, the sea creatures with tentacles, were common in Silurian times. They all had chambered shells, which were usually spiral or curved, but some had shells that were conical.*

► *The echinoderms, the starfish and sea urchins, were present in Silurian seas. Particularly common were the crinoids, or sea lilies, which resembled starfish anchored to the sea bed by a stalk.*

The Devonian Period – from 408 to 360 million years ago

The Devonian was a period of remarkable activity on a number of different fronts. The Iapetus Ocean finally disappeared, having been consumed along its own subduction zones, and the landmasses containing North America and Europe collided. The result was the fusing of the two landmasses and the creation of a mountain range, the remains of which can now be found both in northeastern North America and western Europe (Britain and Scandinavia). Before plate tectonic concepts had come into being, the two manifestations of the orogenic episode on widely separated continents were not seen to be so closely related. In North America the episode was called the Acadian Orogeny and further raised the northern Appalachians that had been formed in the Taconic Orogeny. In Europe it was the Caledonian Orogeny and produced the Caledonides of northern Britain and western Scandinavia. This was the first step in the coming together of all the existing landmasses to produce the supercontinent of Pangea.

Accompanying the Caledonian-Acadian Orogeny were intense metamorphism of the rocks and the intrusion of granite from below. Moreover, the highlands formed in the orogeny subsequently shed huge quantities of eroded sediment both westwards and eastwards. The results were the thick sequences of red sedimentary beds that make up the Old Red Sandstone of Scotland and the Catskill Formation of New York, Pennsylvania and West Virginia. There was a third major transgression and regression of the oceans across much of the land. The Devonian is therefore noted for conspicuous continental sediments and for a wide variety of marine sediments.

As far as life is concerned, the Devonian is often described as the "age of ferns" and/or the "age of fishes", both with equal justice. Vascular plants capable of living out of water began to colonize the land, first as leafless and rootless forms and then as ferns, many of which grew to considerable size, often in forests. Freshwater fish developed rapidly from jawless varieties, which were mostly bottom-dwellers with flat bodies, to jawed types (including sharks) and bony types. In the sea there was an unusual surge of reef growth based on the sponge-like stromatospheroids, ammonoids became much more important than hitherto, and graptolites became extinct.

Adam Sedgwick, who defined much of the lower Paleozoic

The lower Paleozoic controversy

Adam Sedgwick (1785-1873), a British geologist, did much to define the lower Paleozoic rocks through his work in north Wales. At the same time – the early 1830s – another British geologist, Sir Roderick Murchison (1792-1871), worked on similar rocks in south Wales. Sedgwick proposed the term Cambrian to apply to the older sequences, while Murchison called the later beds Silurian. The two men could not agree on the boundary between the two systems. It was not until 1879, when both were dead, that Charles Lapworth resolved the problem by defining the Ordovician and placing it between.

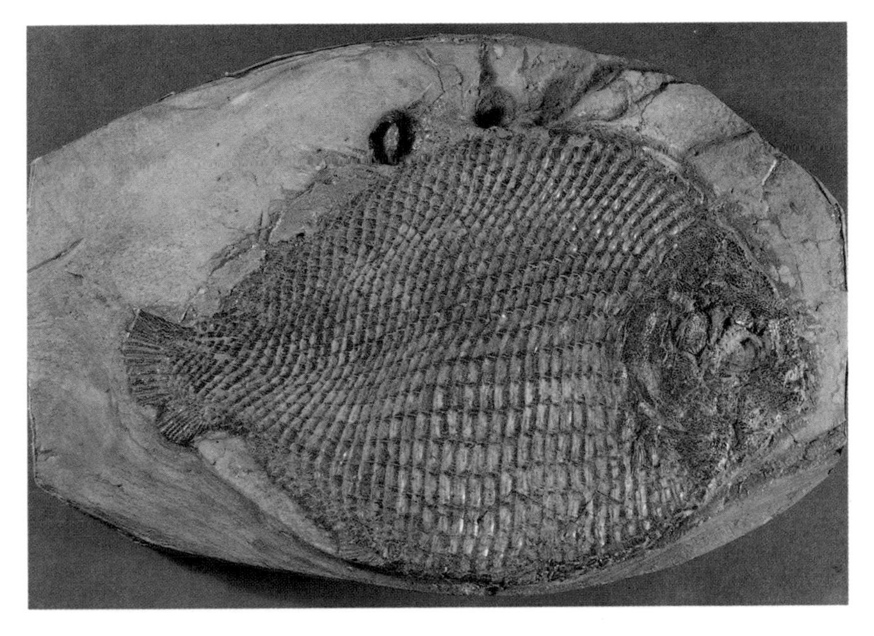

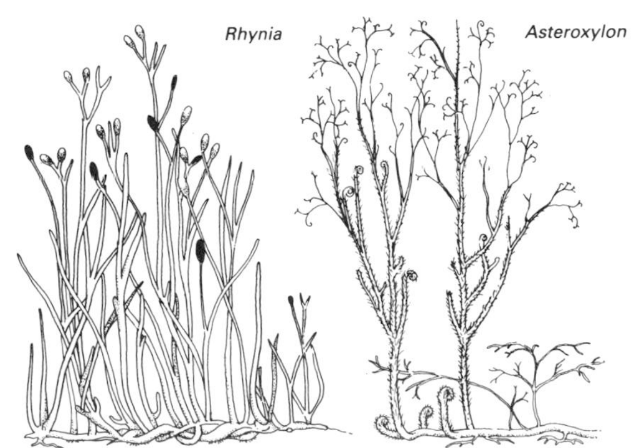

◄ Fish probably first evolved in the Ordovician, but they expanded in the Silurian and dominated the rivers of the Devonian. It was from the fish that the first land vertebrates developed leading eventually to ourselves.

▲ The most important land organisms were the plants. The first land plants were very simple consisting of little more than photosynthetic shoots. The oxygen they put into the atmosphere enabled animal life to leave the water.

1
2
3
5
6
8
10
11

The Carboniferous

From 360 to 286 million years ago

The fourth ice age began around the middle of the Carboniferous, and extensive glaciation developed in the southern hemisphere. On the other hand, those landmasses (including Europe and North America) in equatorial regions enjoyed considerably warmer climates. In these areas, the late Devonian-early Carboniferous saw another rise in sea level, the consequent flooding of low coastal plains, and the formation of wide shallow shelf seas. The results were two-fold. First, there was a change to the deposition of gray and blue-gray marine limestone and shales which contrast strongly with the red continental Devonian sandstones laid down immediately before. Second, along the margins of the seas, erosion products were carried down from the land by rivers to build out deltas and delta swamps. These were then rapidly colonized by tropical forests, the debris from which accumulated to form thick layers of peat. After burial for long periods of time under subsequent sediments, the peat turned to coal. The Carboniferous was not the only period to generate coal, but its coal puts it among the most economically important of all geological periods.

Coal was able to form because in the tropical and sub-tropical regions the flora had undergone substantial development; they included giant tree ferns, seed-bearing ferns and horsetails. By this time also, fish were present in both marine and freshwater environments, amphibians had developed as had (towards the end of the period) reptiles, insects were rapidly evolving on land, and a new evolutionary surge was taking place among marine fauna.

Throughout the Carboniferous, the continents of Gondwana (principally South America, Africa, Australia, Antarctica and India), which had apparently been a single unit thoughout the Paleozoic Era, gradually moved towards the North American landmass. The final collision took place slowly during the late Carboniferous and Permian, resulting in the formation of the southern Appalachians, the Ouachita Mountains and the Marathon. This mountain-building episode has long been known in North America as the Appalachian Orogeny. But at this time North America was already joined to Europe, and Gondwana was also pressing into the latter, causing folding along the southern edge of the continent – an event known as the Hercynian Orogeny. The Hercynian-Appalachian Orogeny represents the coming together of Gondwana and the northern landmass to form Pangea.

In the United States the Carboniferous is regarded as two periods. The Mississippian is the lower part, characterized by limestone deposition, while the later Pennsylvanian is the time of coal formation.

A Carboniferous Landscape
1 *Cordaites* (primitive conifer)
2 *Lepidodendron* (club moss)
3 Tree ferns
4 Horsetail "creepers"
5 Club moss root stock
6 *Sigillaria* (club moss)
7 *Meganeura* (insect)
8 *Eogyrinus* (amphibian)
9 *Arthropleura* (centipede)
10 *Calamites* (horsetail)
11 *Eryops* (amphibian)

The Permian Period – from 286 to 248 million years ago

By the end of the Permian the assembly of Pangea was almost complete. Asia had collided with Europe, forming the Ural Mountains in the process, and only China and its associated islands were perhaps not quite within the fold. The construction of Pangea had a profound effect. One view is that the spreading oceanic crust that had brought the landmasses together continued to exert pressure on the new supercontinent, causing widespread uplift. An alternative view is that the ocean floor spreading stopped for a while after Pangea had been assembled, that the ridges therefore subsided, and that the sea level dropped. Either way, Pangea was widely drained and the wide epicontinental seas receded. This greatly reduced the warm shallow waters available to marine invertebrates and may account, at least in part, for the large number of extinctions that these organisms experienced during the Permian. On the other hand, with the continents now all joined together, the way was open for land fauna to spread with little hindrance.

The ice age which started in the Carboniferous continued into the Permian, but was soon to end. The continents were now well spread out from the South Pole almost to the North Pole across well defined climatic belts. The landmasses near the South Pole were at first extensively glaciated; but as the continents migrated northwards generally, the glaciation slowly disappeared.

As the Permian proceeded, the reptiles continued to gain dominance over the amphibians on land. In the early part of the period the reptile population consisted chiefly of cotylosaurs, which were descended from amphibians, and pelycosaurs, which were the forerunners of mammal-like reptiles. By the end of the Permian, however, mammal-like reptiles had gained supremacy over both amphibians and other reptiles. With the continents spread over such a wide latitude, and hence climatic range, the flora varied markedly. In Euro-America the horsetails declined, while conifers and seed ferns became more common. In Gondwana, by contrast, the flora came to be dominated by the seed plant Glossopteris, a member of a new division of plants that rose out of the almost complete destruction of the previous vegetation by glaciation.

The Permian mass extinctions

A severe crisis overcame life during the Permian. About 30 percent of plant and animal families that were present at the beginning of the period had become extinct by about the end of it. Particularly badly affected were the marine invertebrates. Trilobites became extinct along with certain groups of brachiopods, bryozoans, crinoids and corals that had previously been abundant. On balance, those marine invertebrates that fed on matter suspended in the water were more affected than those that fed on sediment, while those able to cope with ranges of water salinity fared better. But there were no hard and fast rules and, besides, certain plants and reptiles disappeared too.

The cause of the Permian extinctions has long perplexed paleontologists. The coincidence in time makes it likely that the extinctions were related to the final assembly of Pangea, but, if so, the connection is not clear. There are several possible lines of reasoning, however. Uplift of the new supercontinent or lowering of the sea level led to the withdrawal of much of the previously existing epicontinental sea, and this would have severely reduced the space available to inhabitants of warm shallow waters. The coming together of several landmasses, each with its own distinctive mix of flora and fauna, would also have resulted in competition not hitherto experienced. Then the climatic changes induced by the new arrangement of landmasses may have affected some species.

Some paleontologists have suggested that salinity was at the root of it all. Large quantities of salts were deposited as evaporites during the Permian, and this probably depleted the salt concentrations in the oceans quite severely. Known Permian salt deposits are equivalent to more than 10 percent of the salt contents of the present oceans, and the percentage takes no account of either salt deposits that have not yet been discovered or those that were later redissolved. This would explain why those able to cope with wide ranges of salinity preferentially survived.

Life of the Permian Period

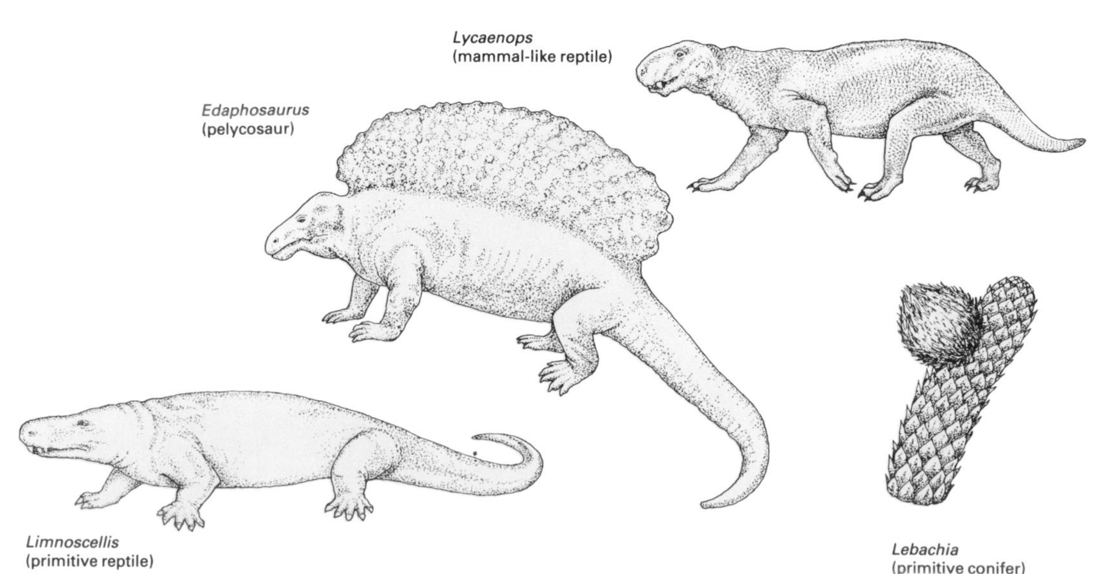

◄ Reptiles evolved in the misty coal forests of the Carboniferous, but it was not until the dry desert conditions spread over the continents during the succeeding Permian that they came into their own. Many different types were successful including crocodile-like forms that were to evolve into the dinosaurs, and the pelycosaurs and mammal-like reptiles that were to evolve into the mammals.

► At the end of the Carboniferous, and the beginning of the Permian, an ice age struck the southern hemisphere. The distribution of glacial deposits is one of the proofs that Gondwana once existed.

The greater ability of sediment feeders to survive is more difficult to understand. Some paleontologists have suggested that there was a worldwide decrease in the amount of plankton available during the Permian, and this would have severely disadvantaged the suspension feeders. But there is no indication of how such a plankton crisis might have arisen.

Another mooted cause of the extinctions is a reduction in the oxygen content of the atmosphere, possibly brought about by the Permian evaporite deposition. According to one estimate, the amount of oxygen in the Permian sulfates alone is the equivalent of 90 percent of that in the present atmosphere. If the hypothesis is valid, those organisms with the greatest dependence on oxygen would presumably have suffered most. A test carried out on the closest modern relatives of Permian animal groups does indeed suggest that the higher the rate of oxygen consumption, the greater the rate of extinction. Unfortunately, however, this test is not definitive. It can only be applied to the descendants of the groups that survived the supposed oxygen crisis.

The true cause of the Permian extinctions will probably never be known. In any case, there may not have been one cause but several. A third possibility is a single basic phenomenon that brought about a slow chain reaction, which would perhaps fit with the knowledge that the extinctions were not sudden but took place over at least several million years. Nor need it be true that the cause of the Permian extinctions was the same as that for other "mass extinctions". The Permian event was the most severe in geological history, but there were smaller crises of the same sort at the ends of the Cambrian, the Ordovician and the Devonian, just as there were to be several more after the Permian. It should always be borne in mind, too, that the extinction of a species or family does not necessarily take place by the deaths of huge numbers of existing individuals. It is much more likely to arise from a failure to reproduce.

Ice sheets in Gondwana

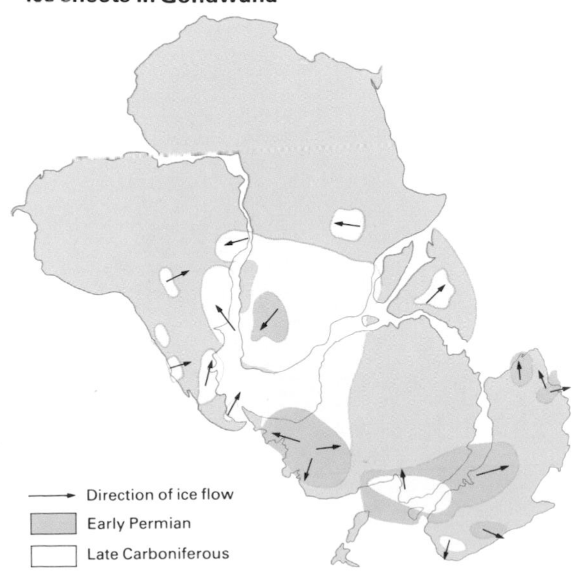

The Triassic Period – from 248 to 213 million years ago

Throughout most of the Triassic, Pangea remained more or less as it was – intact and stationary. The western edge of North America was an active continental margin along which island-arc volcanism was taking place. In fact much of the margin of Pangea bordering Panthalassa, or the "circum-Pacific margin", remained mobile, as did the northeastern margin of the Tethys Sea. Igneous activity took place at very few places within continents, but where it did occur it was remarkably voluminous. Thus in Siberia there are Triassic plateau basalts – layer upon layer of flat-lying lava flows up to 1km thick in total and covering an area of 1,500,000 square kilometers. Similar activity occurred at about the same time in southern Africa where it continued into the Jurassic.

Following the Permain extinctions, the Triassic witnessed a new expansion of life (especially among the marine invertebrates), often involving forms quite different from those of the Paleozoic. The chief vertebrates were still reptiles and amphibians. The dinosaurs became important, and the first primitive mammals appeared. These were small shrew-like creatures that had evolved from mammal-like reptiles, although the transition was so gradual that the precise point of change is often difficult to detect from fossil evidence. Nevertheless, fossils of primitive Triassic mammals have been found in China, South Africa and Wales.

Among the invertebrates, which during the Mesozoic began more to resemble modern forms, mollusks became dominant. The ammonoids became especially important from the stratigrapher's point of view because they are easily recognized and well preserved, and during the Triassic they not only became abundant and widespread but also underwent rapid evolutionary change. Other important invertebrates were lamellibranchs and new forms of coral to replace the many that had been lost during the Permian. Corals are very useful to paleontologists as sensitive indicators of environmental conditions.

Triassic flora were largely ferns, while Glossopteris continued to flourish in Gondwana. In the early part of the period flora were relatively sparse, evidently reflecting poor climate, but towards the end they became much more abundant and varied.

▲ ***Permian volcanoes formed plateau basalts in South Africa.***

The Jurassic

From 213 to 144 million years ago

Pangea may have shown signs of splitting during the late Triassic, but it was during the Jurassic that the dispersal really got underway, with the separation of Laurasia from Gondwana, the start of the opening of the North Atlantic, and the break of Gondwana into Africa-South America, Australia-Antarctica and India. The rise of new oceanic ridges associated with the breakup of Pangea displaced considerable amounts of water. The continents, especially Europe, were therefore gradually inundated until by the end of the period about 25 percent of the total continental area was covered with shallow seas. The result was extensive sedimentation, with the alternating deposition of shales, clays, limestones and sandstones. As clays are formed in deeper water than are either limestones or sandstones, it would appear that water levels, though rising generally, were also fluctuating. It is not clear whether the fluctuations were worldwide or local. In any event, in the very late Jurassic the seas suddenly began to withdraw again from the continents and this process continued into the Cretaceous.

Many of the Jurassic sediments are extremely rich in fossils. Chief among the marine invertebrates are undoubtedly the ammonites (mollusks). From a single ammonoid family that survived the great extinction, the ammonites evolved, developed and spread rapidly during the Jurassic. However, they did not generally live close to shorelines. In very shallow inshore areas foraminifera and ostracods were more abundant. Other important marine organisms were brachiopods, echinoderms and corals. As for vertebrates, again predominant were the reptiles. Dinosaurs developed during the Jurassic into the now-familiar giants, including "Stegosaurus" and "Apatosaurus", which were herbivores, and "Allosaurus" which was carnivorous. There were also the flying pterosaurs and the first bird, "Archaeopteryx". In the sea there were ichthyosaurs and plesiosaurs as well as turtles and crocodiles. But while the giants were stalking the Earth, the mammals, though still very small, were quickly developing. Their time had not yet come.

Conifers, ferns and horsetails continued to dominate the flora, covering much of the land in forests. The trees were gymnosperms (non-flowering types), but angiosperms may also have been about, although they were not to become conspicuous until the following period. Some of the Jurassic forests were to end up as coal – not as important as Carboniferous coal, but economically significant. Another deposit of economic importance is ironstone.

A Jurassic Landscape
1 *Rhamphorhynchus*
2 Conifers and ginkgoes
3 *Brachiosaurus*
4 *Dicraeosaurus*
5 *Megalosaurus*
6 *Kentrosaurus*
7 *Elaphrosaurus*
8 *Dysalotosaurus*
9 Crocodiles
10 Ferns, tree ferns, cycads
11 Horsetails

▲ **Chalk is a particularly pure form of limestone, and it is typical of Cretaceous deposits. The famous white cliffs of the English Channel are made of chalk.**

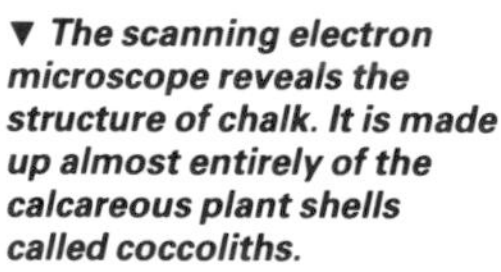

▼ **The scanning electron microscope reveals the structure of chalk. It is made up almost entirely of the calcareous plant shells called coccoliths.**

Calcareous ooze

Because the present oceanic crust is all less than about 200 million years old, geologists have no access at all to older deep-ocean sediment in its original location. Most of the sediment deposited on the deep ocean floor before the break-up of Pangea has long been consumed at subduction zones. As far as deep ocean sediments are concerned, therefore, more than 95 percent of the Earth's history is missing.

The oldest undisturbed ocean floor sediments are of Jurassic age. They are not typical in any way of earlier sediments because the most common deposits by far are derived from organisms of types that did not appear until the Jurassic. These are small carbonate-bearing (calcareous) marine algae called coccolithophores and calcareous fauna known as planktonic foraminifera. Both of them grow near the top of the oceans and, after death, rain down onto the ocean bottom to form sediments known as calcareous ooze. With long burial this ooze becomes compacted and hardened into chalk (limestone). Moreover, as the foraminifera are more easily dissolved than the coccolithophores, the proportion of the latter increases during the squeezing and hardening.

Calcareous oozes have been in production ever since the Jurassic and today cover about 50 percent of the deep ocean floor. They are characteristically deep ocean sediments; and although coccolithophores and planktonic foraminifera are also deposited in abundance on continental shelves, they do not usually form oozes there because they are overwhelmed by land-derived sediments as well as carbonate-bearing organisms that live and die on the sea bed. This makes it most surprising that the vast quantities of chalk laid down during the Cretaceous were deposited from epicontinental seas with maximum depths of about 300m. The Cretaceous shallow seas must have been so extensive and the remaining areas of land so diminished that only coccolithophores and planktonic foraminifera were deposited.

Global sea level

The oceans have been unkind to the continents but the very opposite to geologists. Vast deposits of sedimentary rock are formed in shallow seas – where the edge of a continental mass is flooded. Without the repeated inundations of the land by the ocean waters, the proportion of geological time recorded by sediments now accessible on the dry continents would be far lower than it is. But precisely how has sea level varied through time? J.R. Vail and his colleagues at the Exxon Oil company have attempted to produce a complete sea-level curve for the whole of Phanerozoic time, based partly on the extent of marine sedimentary strata on land and partly on seismic and borehole studies of subsurface discontinuities in the sediments of the continental margins. The marine transgressions of the Ordovician and the Cretaceous are particularly conspicuous, as is the remarkable regression of the Tertiary. The Vail curve is not yet universally accepted by geologists but is a good basis for further discussion.

The Cretaceous Period – from 144 to 65 million years ago

During the Cretaceous, the disintegration of Pangea proceeded apace. The North Atlantic continued to widen, rifting occurred along the line at which the South Atlantic was soon to open, India was involved in its rapid progress northwards, and the Tethys Sea was disappearing as Africa pressed northwards towards Eurasia. The regression of the seas that had begun in the late Jurassic continued into the early Cretaceous, and this led to the replacement of marine sedimentation by non-marine sedimentation resulting from the erosion of exposed continental surface. But before the mid-Cretaceous, the sea level was again rising as new oceanic ridges began to displace more and more water. By the end of the period the continents, including areas that had not been inundated since the Precambrian, were widely flooded in a greater transgression than that of the Ordovician.

The most striking manifestation of this major transgression in Europe and North America is the Cretaceous chalk. Indeed, the very name "Cretaceous" is derived from *creta*, the Latin for chalk. The deposits are of organic origin, the source being marine algae and fauna that reached their peak of abundance at precisely the right time. However, the chalk also contains conspicuous horizons of flint – nodules of silica which originated as sponges and the like but were subsequently altered and thus converted to hard rocky fragments.

On dry land the conspicuous vertebrates were still the reptiles. These were at their peak of size. ***Tyrannosaurus*** had a length of up to 12m. In the sea the plesiosaurs were now up to 10m long, ***Kronosaurus*** had a skull 4m long, while the mosasaurs (a new Cretaceous group resembling "sea serpents") grew to lengths of more than 10m. There were also giant turtles up to 4m long and pterosaurs with wing spans of 10m. The day of the giant reptiles was almost over, however, for at the end of the Cretaceous they were to become extinct along with many other flora and fauna.

Before that happened, the angiosperms (flowering plants) had begun to flourish, often forming forests of deciduous broad-leaved trees, and constituting something like 90 percent of flora by the middle of the Cretaceous. And closely associated with the angiosperms were insects, many of them similar to modern forms.

The Cretaceous-Tertiary mass extinctions

The mass extinction event at the end of the Cretaceous Period is the best known of the several that occurred during the Phanerozoic because it included the dinosaurs. But the dinosaurs were far from being the only form of life involved. Marine, freshwater and terrestrial organisms were all affected to a greater or lesser extent – about 75 percent of all species being wiped out completely.

Geologists have no more idea of why it happened than they have of why the Permian and other mass extinctions occurred. But there is no shortage of hypotheses. Thus the event has at various times been attributed to drastic cooling, to excessive saline oceans, to oceans deficient in salt, to the retreat of the sea from the land, to the breakup of continents, to increased cosmic radiation, to reversals of the Earth's magnetic field, to a stellar outburst, to a change in the concentration of atmospheric oxygen.

In 1980 Luis Alvarez and colleagues at the University of California introduced a new and spectacular hypothesis for which physical evidence exists. Their suggestion was that about 65 million years ago the Earth was struck by a huge asteroid about 10km in diameter. The impact produced a large crater and vast clouds of dust. The dust entered the stratosphere where it spread around the world to form a barrier against sunlight. The lack of light, in turn, suppressed photosynthesis for several years, bringing about the collapse of many food chains, and hence extinctions. The evidence for the supposed impact comes from the sediments at the Cretaceous-Tertiary boundary. At many locations around the world, these sediments have been found to contain very high concentrations of the element iridium – concentrations typical of those occurring in meteorites. The iridium anomalies do not prove that an asteroid impact occurred, but they are consistent with it. Not everyone is convinced by the asteroid impact hypothesis, but ever since 1980 it has proved a lively topic of debate.

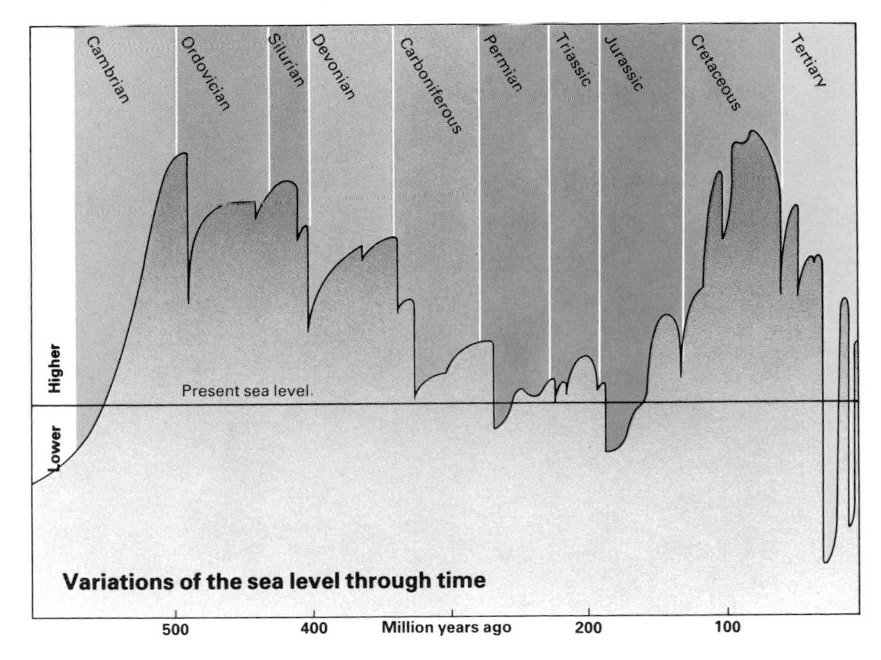

◄ Sea level has risen and fallen throughout time. In the upper Cretaceous it was particularly high, giving vast areas of shallow shelf seas.

▲ This bed of clay between the Cretaceous and Tertiary contains iridium that may be from a meteorite. This may account for the Cretaceous extinctions.

The Tertiary

From 65 to 2 million years ago

The Tertiary is the first of the two periods of the Cenozoic Era, the era of recent life and the "age of mammals". With the giant reptiles gone, the way was open for the mammals to take over the Earth. Early in the Tertiary the mammals were primitive, but they rapidly developed until by the end of the period there were "man-apes" in Africa.

In the Paleocene Epoch, India had not yet reached Asia and Australia had still not separated from Antarctica. The South Atlantic was about three-quarters of its present width. The North Atlantic was only a few percent away from its present width, but Europe and North America were still joined in the far north. This explains why the Paleocene fauna in the two regions are practically identical. On the other hand, the north polar seas and the Tethys were linked by a north-south epicontinental sea (the Uralian Sea) which hindered migration between Europe and Asia. Sea level had dropped to a much lower level than it had been in the Cretaceous, and the sedimentation accordingly changed from largely chalk to land-derived sediments.

Sometime during the Eocene the far northern link between North America and Europe was broken, although later there was still to be communication via the Bering land bridge. Australia also largely separated from Antarctica, except in the region of Tasmania. Various new mammals appeared in the Eocene, including rodents, elephants, whales, artiodactyls and perissodactyls, and primitive mammals became extinct. Sea level glaciation appeared in Antarctica.

By taking up water, this glaciation contributed to a worldwide regression of the oceans during the Oligocene, when sea level fell to the lowest point ever. Oligocene sedimentation was sparse. During the early and late Oligocene the climate cooled leading to the southward migration of high-latitude calcareous plankton. India finally collided with Asia, beginning the main phase of the building of the Himalayas. Australia finally broke from Antarctica at Tasmania.

By the Miocene, the Tethys Sea had become much reduced in size by the movement of Africa towards Europe and was more or less recognizable as the Mediterranean. The continued pressure of Africa against Europe led to the Alpine Orogeny. The Himalayan Orogeny was also continuing as India was forced farther against Asia. The overall result was the vast Alpine-Himalayan chain running across the whole of southern Europe and Asia. Sea level was still low as the Antarctic ice continued to grow; indeed, during the Miocene the ice sheet became larger than it is today. During the Miocene the mammals continued to progress as pigs, deer, elephants, antelope, monkeys, cats, horses and many other species thrived.

During the short Pliocene, glaciation appeared in the north polar regions. The Isthmus of Panama was uplifted, cutting off marine communication, and hence the interchange of marine life, between the Pacific and the Atlantic for the first time for over 100 million years. On the other hand, this allowed the exchange of mammals between North and South America.

A Tertiary Landscape
1 *Amebeledon* (elephant)
2 *Procamelus* (camel)
3 *Neohipparion* (three-toed horse)
4 *Pliohippus* (one-toed horse)

3
4
1

See also
The Dynamic Earth 9–24
Dating the Rocks 73–80
The Growth of Continents 101–8

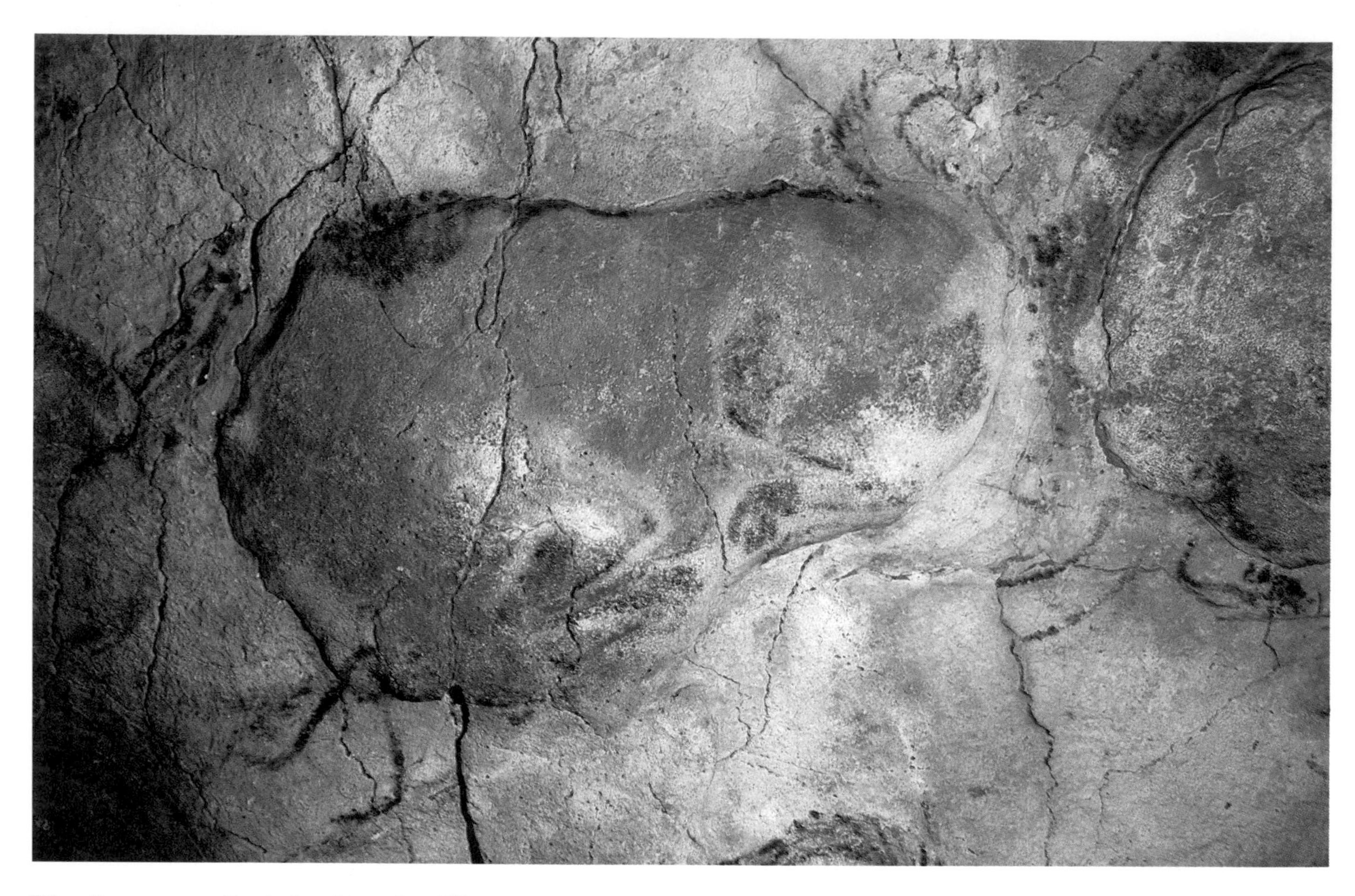

▲ *We have an eye-witness account of the animal life of the Quaternary. Ice Age Man was an accomplished artist and painted the animal life of his home area on cave walls. This may have been done for religious or magical reasons. This painting of a bison is on the roof of the cave of Altimira in the Spanish Pyrenees.*

The Quaternary Period – from 2 million years ago to the present
Geologists take the start of the Quaternary to be the start of the current ice age. The polar ice sheets originated in the Tertiary, but it was at the beginning of the Quaternary that the polar ice began its forays out to lower latitudes. These advances are the "ice ages" of popular parlance, and there were many of them during the Pleistocene Epoch. The ice age last retreated to polar regions about 10,000 years ago, marking the beginning of the Holocene Epoch.

Parts of the Earth's crust/lithosphere have been repeatedly depressed and uplifted as they reacted to the weight of the ice by adjusting isostatically (◀ page 50). These fluctuations have been superimposed on worldwide changes in sea level caused by the continual withdrawal of oceanic water by the glaciers.

During the last "ice age" there was a world drop in sea level of about 90m, which increased the total area of land. On the other hand, most areas of northwest Europe, the northern USSR and North America were covered in ice and thus not available to fauna or flora. Temperature gradients in middle latitudes were increased resulting in much more stormy weather. In the northern hemisphere there were wide permafrost zones to the south of the ice (extending to the southern edge of Europe) and forests to the south of those. In short, zones of vegetation were shifted something like 2,000km southwards. In the permafrost were cold-region fauna such as reindeer, woolly rhinoceros and mammoth. Because of the rapidly changing habitats vertebrates underwent great evolutionary changes during the Quaternary. By contrast, the flora tended to change latitude with the moving climatic zones without great evolutionary development.

▼ *The spectacular fauna of the Ice Age has not been long extinct. Parts of it are sometimes found in an excellent state of preservation. This baby mammoth is 12,000 years old, and has been preserved in frozen mud in the Magadan region of the Soviet Union. Mammoths are often found "fossilized" in this way.*

The Growth of Continents

9

Continental crust – 36 percent of the Earth's surface...Components – shields, platforms and mountain ranges...The first continents – ancient crust or accumulation of volcanics?...Continental growth – the buildup of mountains round the edge... Microplates – little scraps of continent... PERSPECTIVE...The shapes of continents...Ancient plate tectonics...Continental drift before Pangea

The continents today cover about 36 percent of the Earth's surface (29 percent exposed, 7 percent under water) and vary in crustal (not lithospheric) thickness from about 20km to 90km with an average 35-40km. They exist because the average density of the materials of which they are made is lower than both that of the mantle below and that of the surrounding oceanic crust. They therefore lie on top of the mantle, much as scum lies on top of water or molten steel, and stand higher than the ocean floors (◀ pages 34-5). Mountain regions apart, continents are large flat slabs whose surfaces lie on average but a few hundred meters above sea level. Their overall average elevation (including mountains) above sea level is about 0·9km, whereas the surface of the oceanic crust lies on average about 3·7km below sea level.

The continents differ from the ocean floors not only in elevation but also in their range of ages. Whereas the oceanic crust is all less than about 200 million years old, some continental rocks are 3·8 billion years old. It is hardly surprising, therefore, that the continents are the more varied and complex. Floating on the asthenosphere and being pushed around by spreading ocean floors, they are continually being deformed, split, added to and moved.

The shapes of continents

The rocks of the continental areas are subjected to all the tectonic and erosive processes that take place at or near the Earth's surface. They are continually exposed to weathering and are repeatedly being flooded by oceans, they have undergone numerous episodes of erosion and sedimentation. The net result is a complicated mixture of igneous, sedimentary and metamorphic rocks (▶ pages 141-52). However, it is still possible to describe continents in terms of only three basic structural components – shields, stable platforms and young mountain belts.

A simple continent can be thought of as an approximately concentric arrangement of these three elements. At its center would be a shield – a flat plain of ancient metamorphosed rock. Around this would be the stable platforms – areas of the shield that are now overlain by sequences of sedimentary rock. This arrangement would be surrounded by mountain ranges – the oldest being closest to the shield, and the youngest near the edge of the continent.

This, however, is a generalization. Usually ancient simple continents have become fused together – giving more than one shield and stable platform arrangement separated by ancient mountain ranges in the middle of the continent, and sometimes the continent has split apart, giving shield and stable platforms close to the newly formed continental margin.

▼ *Continents all have the same general structure. They have shield areas at their cores, surrounded by regions in which the stable shield is overlain by thick sedimentary rocks. Fold mountains lie on the margins.*

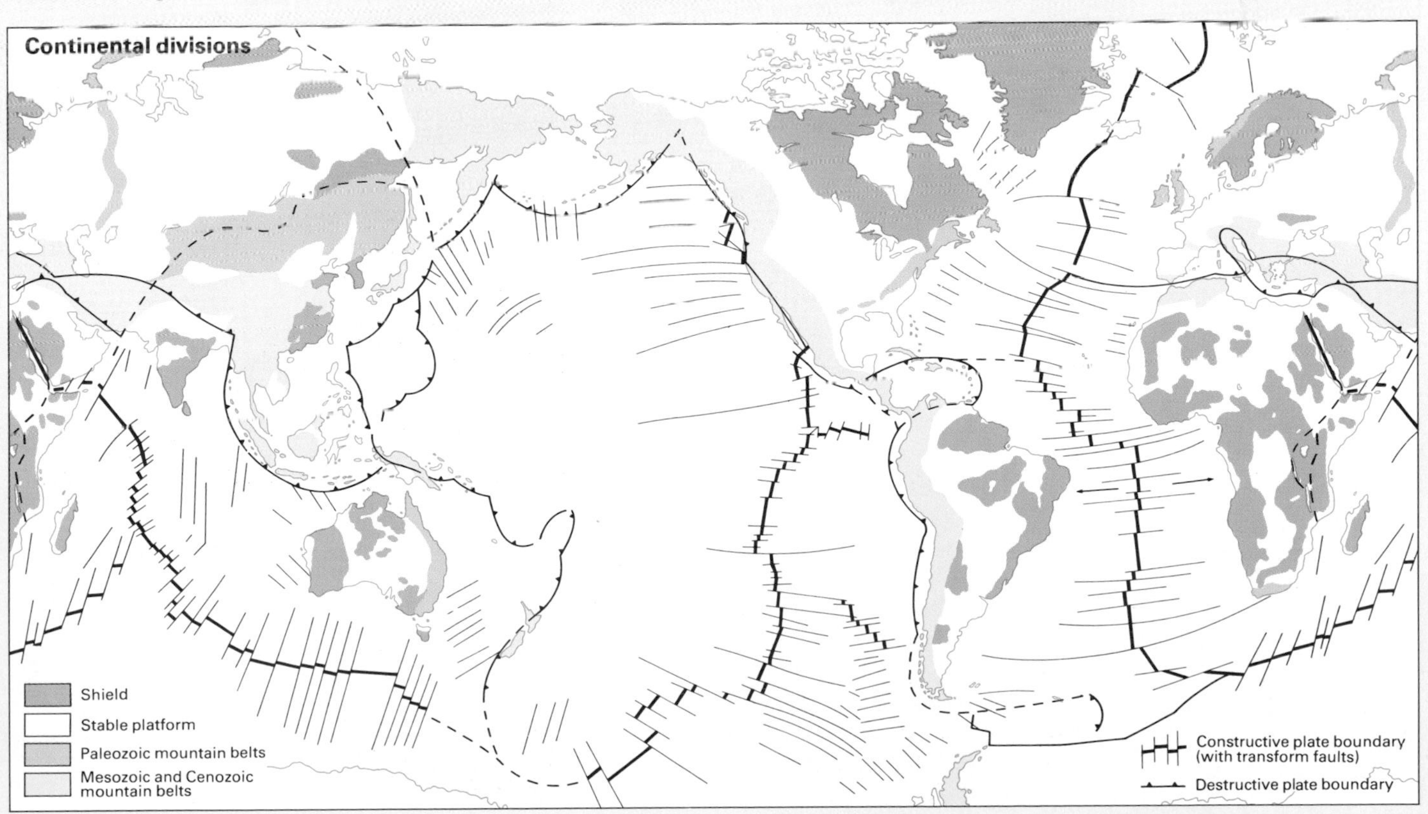

The basic components of a continent

The three basic structural components of the continents – the shields, the stable platforms and the orogenic belts – each have their characteristic features. The shields are large areas of exposed, low-lying rocks of Precambrian age which form the nucleus of each continent. On a regional scale they appear relatively flat, the only features of relief being resistant rock formations that stand a few tens of meters above their surroundings. When examined closely, however, they are seen to be structurally complex mixtures of faulted and jointed igneous and sedimentary rocks that have been intensely deformed and metamorphosed, intruded by granite magmas, and finally eroded almost to sea level. Most, if not all, shields appear to be the roots of mountains that eroded away long ago, and they have remained stable (not subjected to mountain-building episodes) ever since. Not all shield provinces are of the same age, however.

Large areas of the shields are not exposed, but are now covered by largely horizontal sediments. These are known as stable platforms and, like the shields themselves, have not been subjected to orogeny since Precambrian times. The stable platforms are the world's lowlands – the plains and steppes – covered in marine sediments deposited during oceanic transgressions.

Finally, there are the younger orogenic belts – huge linear regions of compressed, folded mountains formed as a result of continent-continent, or ocean-continent collisions. Some of the mountains involved (such as the Andes and Himalayas) are still being formed; others (such as the Appalachians and Urals) have ceased to form but have not yet been eroded away.

The making of a continent

Geologists know what the continents are like today. They also know that, despite changing shape and position, the structural components have followed the same pattern throughout most of geological time. However, the beginnings of the continents are more obscure and contentious. As the oldest known rocks are about 3·8 billion years old, parts of the crust must be at least that age. But they could be considerably older. For one thing, there may be in existence rocks older than 3·8 billion years that have simply not yet been discovered. Rocks could have existed before 3·8 billion years ago and have been reworked at that time (metamorphosed or partially melted, for example), thus resetting the "radiometric clock". Evidence for this comes from geochemical studies, which appear to suggest that the oldest Precambrian rocks were derived by the partial melting of earlier crustal material of a somewhat different composition. Most geologists therefore envisage that a "primordial crust", or "protocrust", formed very early in the Earth's life, perhaps as the direct result of the major differentiation that led to the formation of the mantle and core.

The controversy begins with the interpretation of the role of this primordial crust. Some geologists, but probably a decreasing number, argue that the primordial crust was substantial, accounting for the bulk of the present-day continents. On this argument, because of the highly mobile conditions at this early stage of the Earth's history, the continental crust that first formed would have probably undergone rapid reworking and hence a change in volume. In other words, most of today's continental crust was formed very early. Since then it may have altered its shape and distribution and have been modified by tectonic activity, but the total quantity has not changed appreciably.

Section through a continent

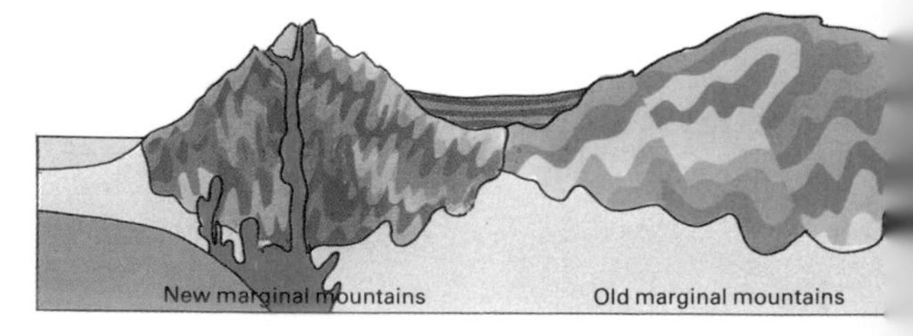

▲ ▼ *North America has the concentric pattern of shield, stable platform and fold mountains. In the east the impact zone shows where it once combined with another continent, and the coastal rifting shows where it later broke away.*

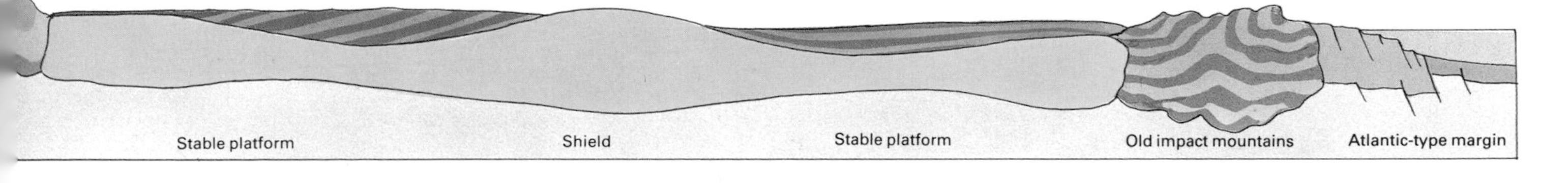

The alternative hypothesis, which has become increasingly popular in recent years, is that the primordial continental crust was very unstable, partly because of the vigorous convection that must have been taking place and partly because it was subjected to perpetual meteorite bombardment. Since it was not very substantial, after reworking it was able to form only quite small pieces of continental crust. The important implication of this hypothesis is that the present continents were still incomplete by about 3·8 billion years ago. They began as small nuclei and have grown larger through time by the gradual accretion of material derived from the upper mantle.

One view, for example, envisions the development of a subduction zone along one edge of a continental nucleus. As the oceanic lithosphere descends beneath the continental fragment, it begins to melt as it encounters higher and higher temperatures, and magma rises to form both volcanoes at the surface and intrusions beneath. New material is thus added to the original continental mass; and because the lower-density minerals tend to melt first, this new material is light and buoyant. In this way, material that originated in the mantle and rose at oceanic ridges to form the spreading ocean floor finally has its lighter components removed and incorporated into a piece of continental crust, which therefore grows bigger.

A variant model does away with the original continental nuclei altogether, arguing that the primordial crust developed solely as a result of ocean floor spreading. If spreading begins in oceanic crust, a subduction zone must develop somewhere to compensate for the material rising at the oceanic ridge. The result of the melting of the descending slab will again be volcanism and intrusion, but in this case the light volcanic and intrusive components will not add to a pre-existing piece of continental crust but will form an island arc and its substructure. The island arc and its substructure thus become, in effect, the first piece of continental crust which then continues to grow.

Another mechanism for creating a continent

Modern hypotheses of continent formation tend to rely on plate tectonics (◆ pages 9-24), but there is at least one that does not. This looks instead to the effect of the many meteorites that must have been bombarding the Earth during the first 600 million years or so of its history. Evidence from the craters on the Moon suggests that the Earth was showered by meteorites of up to 50–60km in diameter right up to 3·9 billion years ago. In the early days this bombardment would perhaps have inhibited the formation of the primordial crust by perpetually breaking it up and remixing it into the mantle; but once the crust had formed, the meteorites could well have built the sort of material now present in the shields.

A meteorite of 60km diameter would penetrate the Earth and almost immediately explode, forming a crater a thousand kilometers wide and a hundred kilometers deep. The instantaneous removal of the pressure of overlying rock and the heat generated by the impact would cause partial melting of the material beneath the crater and result in volcanic activity. Mantle material would rise to fill the crater from below, whilst sediment and volcanic products would enter from above. The consequence would be isostatic subsidence of the material in the crater, its metamorphism at depth, and the intrusion of granite from below. The end result, when all had settled down, would be a stable mass largely comprising material less dense than the mantle material below.

The continental nucleus thus formed could then grow by subduction along one of its edges. Even here, however, plate tectonics are dispensable. The continents could have grown simply by the accretion of meteorite-derived nuclei.

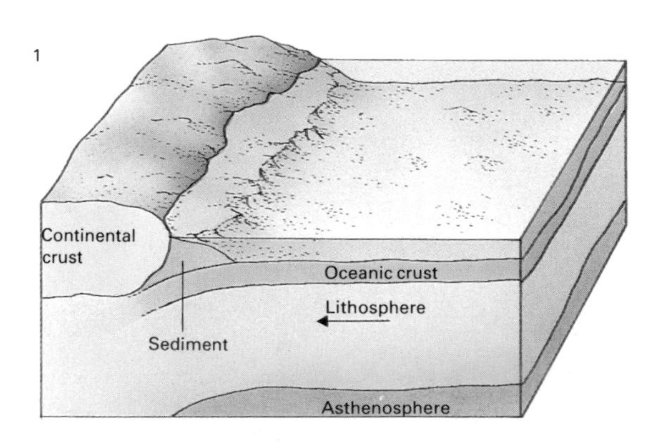

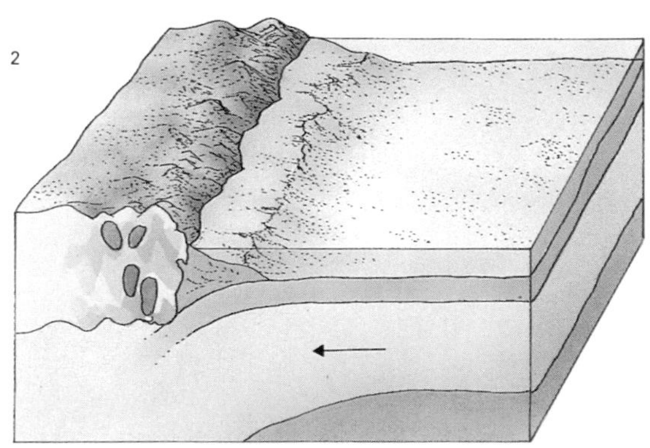

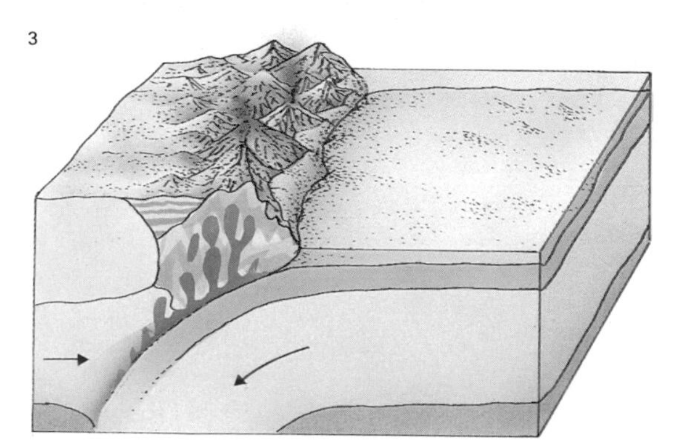

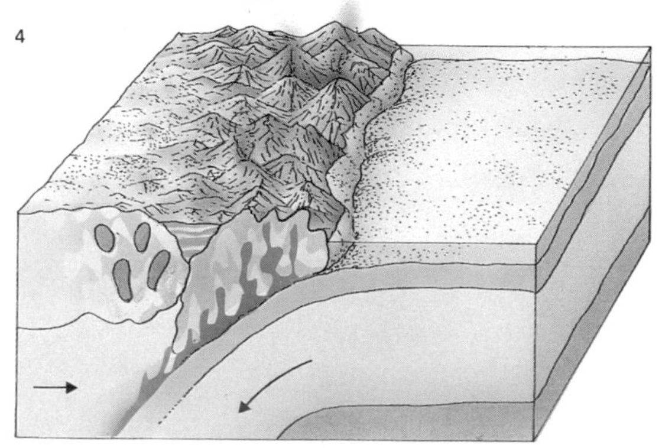

***►* The simplest scheme of continent formation envisages the almost continuous growth of a continent by repeated phases of mountain-building. Once the original patches of continental crust formed, they became caught up in the plate tectonic processes (1). When they settled over the subduction zones the mountain ranges subsequently formed were welded on to their margins (2). Continued plate tectonic activity would have produced increasingly younger fold mountains on to the outsides of these (3) and eventually the continents would have reached their present sizes (4).**

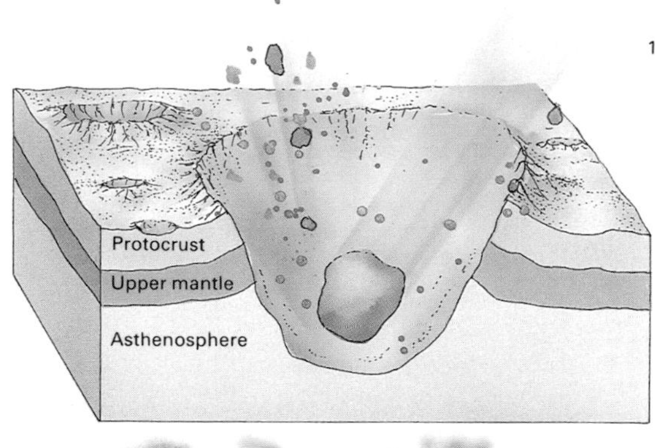

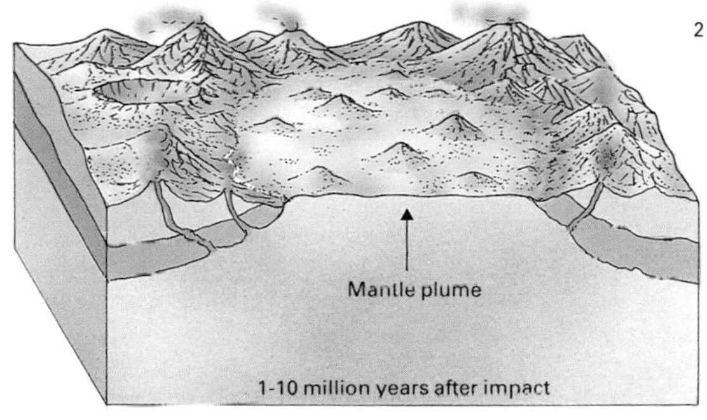

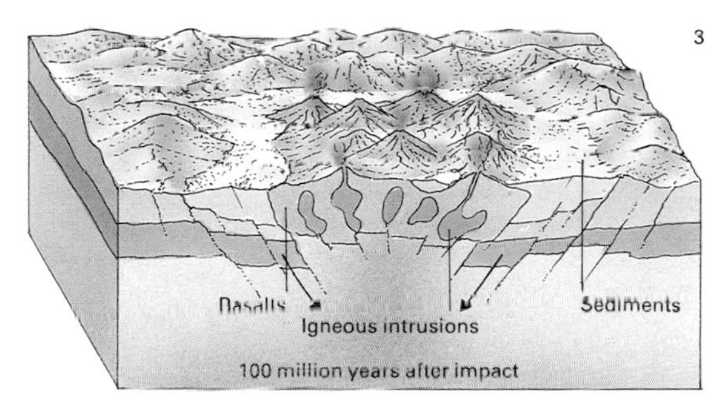

Continental growth

Whichever hypothesis is considered an acceptable explanation for the making of a continent, it is clear that a continent develops and grows. It does so, moreover, in the context of a particular type of mountain building, for intense compression takes place at convergent plate boundaries. (A good modern example of this type of mountain building/continental accretion is the formation of the Andes.) This explains why shield areas appear to be the eroded roots of mountains; they probably accreted in orogenic environments generated adjacent to subduction zones.

This kind of accretion was not necessarily continuous, however; it was more likely to have been episodic. In other words, periods of accretion/mountain building were probably separated by times when, because ocean floor spreading had temporarily ceased or moved elsewhere, the mountains were being eroded. The sediments produced then became part of the next compression/accretion phase. Some geologists believe that this sort of episodic activity during the Archean is what gave rise to the greenstone-gneiss terrains that lie at the heart of the shields (◀ pages 81-2). Subsequent accretion/mountain building then continued around the edges of the greenstone-gneiss terrains during the Proterozoic.

Evidence for the episodic nature of continental accretion comes in part from an analysis of radiometric ages of Precambrian rocks. These suggest that about 10 percent of the continental crust was generated during a phase of accretion/mountain building that occurred between 3·8 and 3·5 billion years ago. A further 60 percent was then added during a second phase between 2·9 and 2·6 billion years ago. The remaining 30 percent was created during two subsequent Proterozoic phases (1·9-1·7 and 1·19 billion – 900 million years ago) and during the Phanerozoic. In short, continental growth has occurred throughout geological time but far from uniformly.

▲ According to one theory the nuclei of the continental masses were formed early in the Earth's history when the new planet was still being bombarded by huge meteorites (1). A meteorite that pierced the crust would have disturbed the topmost layers of the Earth so much that extensive volcanic and plutonic activity would have resulted (2). The result would have been a mass of rock that was different from its surroundings (3).

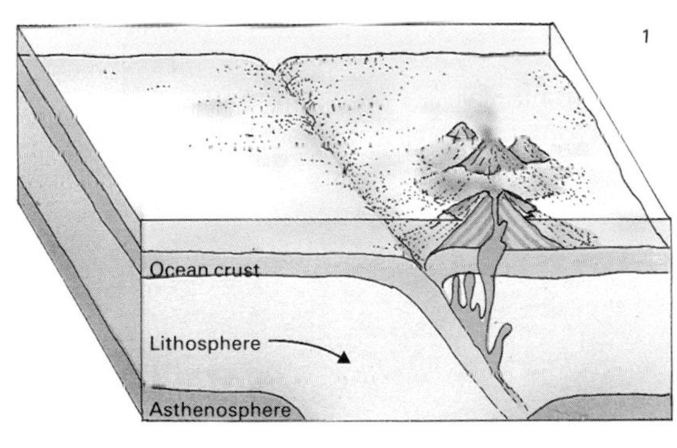

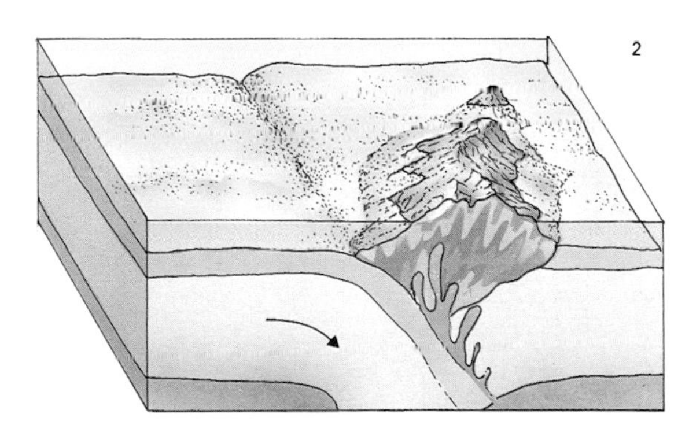

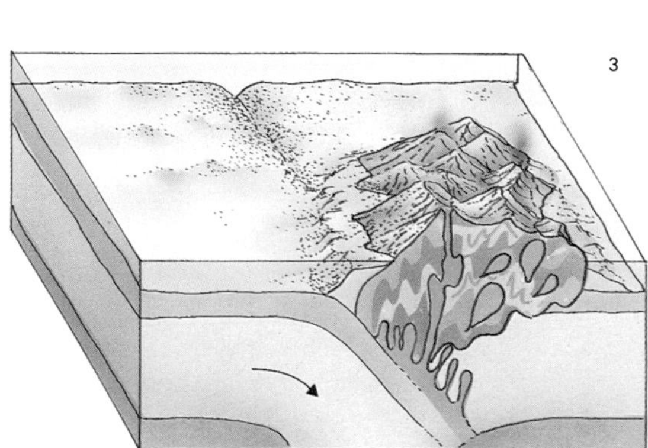

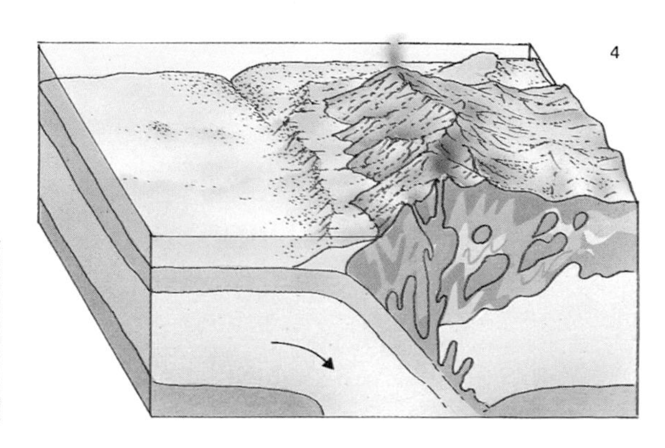

► Another theory maintains that there need never have been continental nuclei to start off the growth of a continent. Today arcs of volcanic islands are seen forming along the subduction zones of ocean-ocean plate boundaries (1). Such island arcs could be added to by subsequent plate tectonic activity (2) and continent could eventually form simply by the accumulation of volcanic and fold mountain material (3,4).

Similar geological processes were at work before Pangea as those occurring today

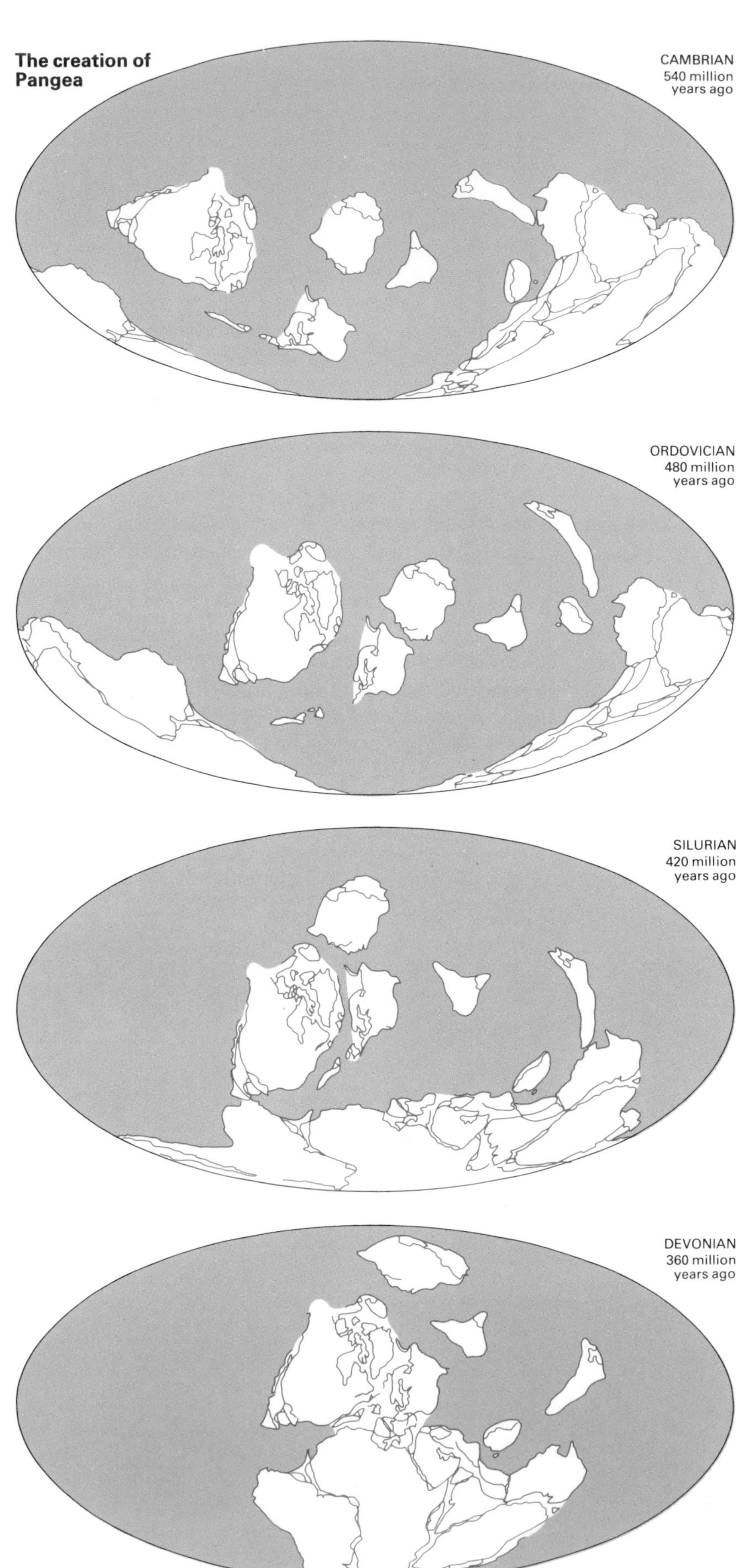

◄ It is often assumed that plate tectonics has only produced continental drift since Pangea broke up. This is quite erroneous, since Pangea itself was the result of the coming together of isolated continents that existed in former times. The shapes of these ancient continents would have been unfamiliar, and their actual positions at various times during the earlier periods are not known with certainty. The farther back in time, the more ambiguous becomes any geological evidence. Between these continents there would have been oceans but no trace of these survives to the present day.

Precambrian plate tectonics

Plate tectonics is, by definition, the theory of global processes that have been taking place over the past 200 million years. It involves, among other things, continental drift, ocean floor spreading, lithospheric plates floating on the asthenosphere, and collisions along plate edges. But were those same processes occurring in earlier times? Geologists do not really know. They hypothesize that Precambrian fold mountains were the result of plate collision, but there can be no certainty that they were. The early Earth must have been different from the present one in a number of respects.

In the absence of definite information to the contrary, most geologists tend to assume that the Precambrian Earth was subject to plate tectonic-style processes, but modified by the conditions of the time. During the early Precambrian, for example, the Earth was probably much hotter than it is now. Convection in the mantle would therefore have been more vigorous, the lithosphere could well have been much thinner, and the lithospheric plates are likely to have been much smaller. If processes of the plate-tectonic type were operating then, some of their expressions would have been somewhat different from those of today.

Thinner plates would have been less rigid than modern ones and thus perhaps susceptible to buckling rather than fracturing and faulting. Higher thermal gradients in the Earth would have inhibited the chemical, and hence density, changes that occur in the slabs of descending oceanic lithosphere. The slabs would have therefore descended less steeply, and the resulting volcanism at the surface (for example, in island arcs) would have taken place over a much broader band. At divergent plate boundaries, by contrast, processes would have been much as they are today, although ocean floor spreading is likely to have been much faster.

Continental drift before Pangea

There is more than enough information from the various sources to allow the production of paleogeographic maps showing fairly precisely how the continents have changed their positions over the past 200 million years. But for older times geologists do not have this luxury. There are paleomagnetic data, of course, although these become less and less reliable the older and older the rocks. Unfortunately, however, there are no magnetic anomaly data (◄ pages 45-56) to be gleaned from the oceanic crust, for all the oceanic crust is less than 200 million years old. The only way of determining the relative positions is therefore to rely on geological information. This comes in principle from a wide range of sources –

from fossil distributions, from the alignments of ancient mountain ranges, from climatic indicators, from sedimentary arrangements, and so on. But geological data for older times would often be no less ambiguous and conflicting than those relating to the past 200 million years, even if they were as complete, which they are not. Moreover, because particular geologists do not necessarily have detailed knowledge of all fields of the subject, they do not always use all the information available when they try to produce paleogeographic maps for periods of geological time before the Jurassic.

The result is that there are now several different published versions of how the Earth's landmasses might have moved prior to the breakup of Pangea. The definitive version has not yet been produced. On the other hand, there is a growing, if limited, consensus on a number of points. First, Pangea was a very temporary state of affairs. It probably lasted as a single supercontinent for less than 100 million years, having come together from a number of independent landmasses during the Permian. Second, the landmasses that coalesced to form Pangea were not generally the familiarly-shaped regions of today. The modern continental shapes are characteristic only of the post-Pangean period. Third, plate tectonic processes must have been going on throughout at least Phanerozoic time, although it is far from clear to what extent, if any, those of pre-Pangean times were identical to those of post-Pangean times.

▲ One of the proofs that Pangea was the result of the merging of more than one continent lies in the highlands of Scotland and Norway. These were formed, like the Himalayas today, by the coming together of two continents. In Silurian and Devonian times the North American continent collided with the European continent, crushing up the mountain range in between. Since then the two continents have separated again.

See also
The Dynamic Earth 9-24
Earthquakes and Volcanoes 25-32
Internal Structure 33-44
The Earth Through Time 81-100
An Expanding Earth? 109-112

If continental growth occurs gradually along the landward edges of subduction zones, it is presumably going on now, especially around the edge of the Pacific where most of the world's subduction is taking place. But such gentle accretion is apparently not the only type of continental growth possible. During the 1970s it became clear that continents can also grow by the addition of discrete fragments that are carried along by the spreading oceanic crust and, on reaching a subduction zone, are forced against the adjoining continent and become "plastered" to it.

Microplates – fragments that build continents

For many years geologists had been puzzled by the fact that most of Alaska (especially) and a strip of several hundred kilometers width along the west coast of Canada and the United States appear to comprise a "mosaic" of hundreds of disparate segments. Adjacent regions consist of quite different rocks and rock segments and the boundaries between them are sharp. Adjoining segments also frequently contain distinct fossils and fossil groups, indicating that they were formed at quite different latitudes. Some of these "suspect terranes" or "exotic terranes", as they have come to be called, are apparently continental microfragments. Others are relics of dead island arcs. Still others are oceanic plateaus, pieces of oceanic ridges, and sedimentary platforms. The sizes of the fragments are not constant. They can range from small outcrops a few kilometers square to large areas of many thousands of square kilometers.

For some time the geologists concerned resisted the obvious conclusion because it seemed too incredible, but in the end they had to accept it. Paleomagnetic studies began to show that some of the exotic terranes had traveled from thousands of kilometers away. Fossil analysis showed that some of the terranes must have originated in environments different not only from those of their neighbors, but also from any that the main bulk of the North American continent could ever have experienced. The only conclusion possible was that the various terranes had once been scattered throughout much of the Pacific and had gradually been propeled towards North America by the spreading ocean floors. Once there, they had often been compressed, uplifted, rotated and tilted as part of the normal processes of deformation and mountain-building at convergent plate boundaries.

Similarly, the mountainous area of Antarctica comprising Byrd Land and the Antarctic Peninsula – the side of the continent that borders the Pacific Ocean – now seems to consist of several small individual plates. However, the perpetual cover of snow and ice in this region makes study difficult.

The behavior of exotic terranes has come to be called microplate tectonics, even though the fragments involved are not tectonic plates in the strict sense. The histories of the hundred or more terranes discovered so far are still being worked out during the 1980s. Nor is it yet clear precisely how a fragment that finally reaches a subduction zone is actually transferred to the adjacent continent. Two things are clear, however. One is that exotic terranes exist not only along the margins of North America but around almost the whole of the rim of the Pacific. The second is that most, if not all, of the terranes reached their present resting places during the current phase of plate tectonics that began about 200 million years ago. Whether a similar phenomenon occurred during earlier phases of plate tectonics is a completely open question since no evidence has been found.

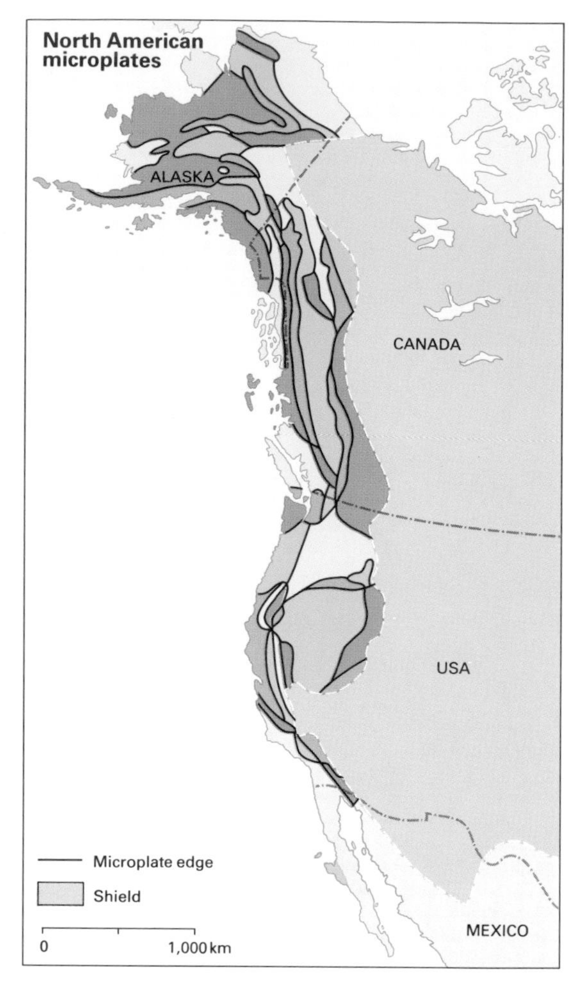

▲ ***Western North America has areas of continental crust that seem to have been brought from afar.***

▼ ***These masses have sharp boundaries, showing that they are different from the rocks surrounding them.***

An Expanding Earth?

10

Did Pangea cover the whole Earth?...Paradoxes needing explanations – Antarctica and Africa are both stationary but moving apart, the Pacific is getting smaller but stretching...Plate tectonics rethought – no subduction zones on an expanding Earth... PERSPECTIVE...Old theories of an expanding or shrinking Earth...The difficulty of finding proof... The speed of any expansion

◄ *One quaint idea attempted to prove that the distribution of continents and oceans on the Earth's surface was due to shrinking. A deflating ball develops broad dimples and ridges and tends to form a four-sided body called a tetrahedron. The theory held that, on a shrinking Earth, the ridges would have formed the continents and the dimples the oceans.*

About 200 million years ago, all of today's continents were joined together in a single supercontinent Pangea. Most geologists believe that Pangea was surrounded by a single world ocean (Panthalassa) equal in area to all the present oceans put together. A few argue, by contrast, that this huge ocean never existed and that, instead, Pangea covered the whole of the surface of an Earth that was once smaller than it is now. For it is a curious fact that if one imagines Pangea to be bent round into the shape of a thin spherical shell, it completely covers a globe with a radius about half the Earth's present radius. And that raises an obvious question: Is the remarkable fit merely a coincidence or has the Earth really doubled its radius?

Although supported by a small minority, the arguments in favor of an expanding Earth are extensive. In the 1970s, for example, the chief proponent of the hypothesis, Professor S. Warren Carey of Tasmania, published a volume of no fewer than 500 pages putting the case in considerable detail. Unfortunately, however, much of the evidence is also complex, very technical and, as most geologists would argue, highly ambiguous and perhaps even wrong. Nevertheless, expansion enthusiasts can point to a number of simple phenomena which apparently defy explanation by plate tectonics alone.

The Earth's changing size

Throughout the history of geological thought a number of theories have been put forward that suggest the changing size of the Earth as the reason for observed phenomena.

Perhaps the most obvious noted that when an apple rots it shrinks. The skin then wrinkles up on the surface. A shrinking Earth could similarly produce wrinkles in its crust and these wrinkles would be today's mountain ranges.

Another curious theory involving shrinkage was the tetrahedron theory. When an air-filled ball deflates it tends to form broad dimples, usually four, separated by ridges. The geometrical shape so produced is also visible on the Earth's surface. The four dimples are the Atlantic, the Indian, the Pacific and the Arctic Oceans. The ridges are the continents separating them.

There are less exotic theories. Growth rings on Devonian corals show that the days were shorter 400 million years ago, in other words the Earth was turning faster. A smaller Earth could account for this, but it is now generally believed that tidal friction is the real reason.

▲ *Professor S. Warren Carey was one of the most assiduous champions of the expanding Earth theory in the 1970s.*

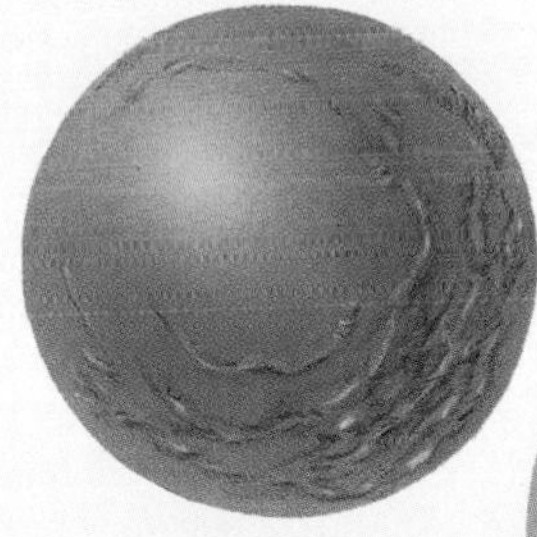

Radius 3,500km (51%)
4.5 billion years ago

Radius 4,400km (69%)
2.8 billion years ago

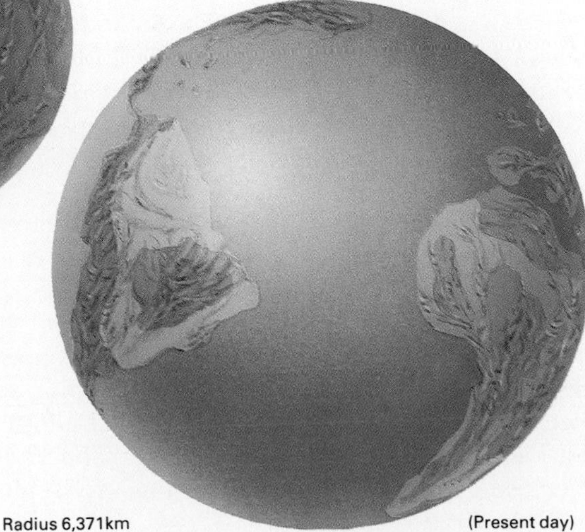

Radius 6,371km (Present day)

► *If the expanding Earth hypothesis were taken to an extreme it would suggest that 4·6 billion years ago the Earth had a radius of 3,300km and was entirely covered by continental crust. Expansion then split the crust into continents.*

Paradoxes in need of an explanation

Antarctica's behavior is a good case in point. This continent is almost completely surrounded by active oceanic ridges, and so newly created lithosphere is continuously spreading towards it. But where does this lithosphere go? It is not being subducted because, with the possible exception of a small region opposite South America, there are no subduction zones around Antarctica. Nor is there any evidence that it is compressing Antarctica and thus forcing the area of the continent to decrease. The only possible explanation, assuming the Earth to have a constant radius, is that the encircling ridges are migrating away from Antarctica at just the right rate to accommodate the new lithosphere being laid down between the ridges and continent. Antarctica is therefore stationary and all other continents are moving away from it. In itself that is quite feasible and not inconsistent with plate tectonic principles. The problem arises in the case of Africa, which is also largely surrounded by ridges and thus by the same token also stationary at the center of a diverging plate system. On an Earth of constant radius, Africa cannot be both stationary and moving away from a stationary Antarctica, but on an Earth of increasing radius it can.

If this is known as the Africa-Antarctica paradox, at the opposite pole of the Earth there is the rather different Arctic paradox. Paleomagnetic data show that all the continents except Antarctica have a northward component of drift, which means, if the Earth has a constant radius, that the Arctic Ocean is getting smaller. But other evidence suggests that the Arctic zone is one not of compression but of extension, implying that it is increasing in area. These mutually exclusive sets of evidence can be reconciled easily only if the Earth's surface area is increasing.

Similar circumstances apply in the Pacific. The continents are generally converging on the Pacific which, again assuming that the Earth has had constant dimensions, must be getting smaller. Yet ocean floor spreading data suggest that each pair of adjacent landmasses around the Pacific are moving farther apart. The Pacific is thus apparently decreasing in area whilst increasing in circumference, which is yet another paradox.

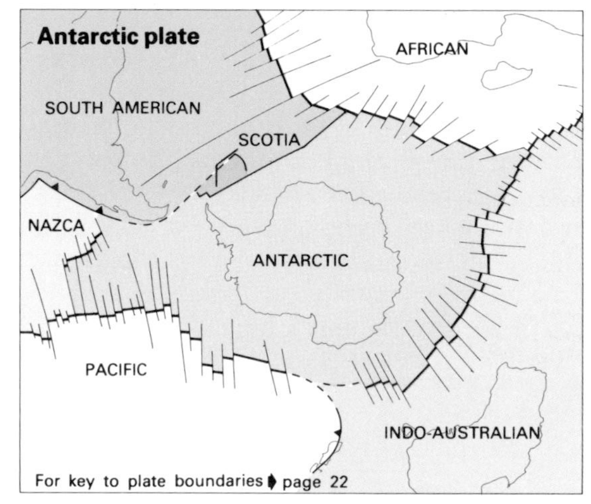

Testing the expansion hypothesis

There are a number of ways in which Earth expansion can be put to the test in principle, but it is notoriously difficult to obtain a convincing result one way or the other in practice. For example, if expansion has occurred, the value of gravity at the Earth's surface will be lower now than it was when the planet was smaller (assuming there has been no increase in mass), because gravity decreases with distance from the center. But how can gravity in the past be measured? There are many phenomena that depend on gravity but such things also depend on many other factors, and the influence of gravity cannot be isolated.

The geomagnetic field which has been recorded in many rocks (➧ pages 45-56), also depends on the distance from the Earth's center. All that is required to determine the radius of the Earth in the past are the directions of magnetization in rocks of the same age from two well-separated sites on the same ancient magnetic meridian. Unfortunately, however, the paleomagnetic data available are simply not accurate enough to detect differences

Expanding Earth

1 Pangea 180 million years ago (assuming an Earth of modern dimensions)

1,000m isobath and modern coastline

Regions of discontinuity

Oceanic crust

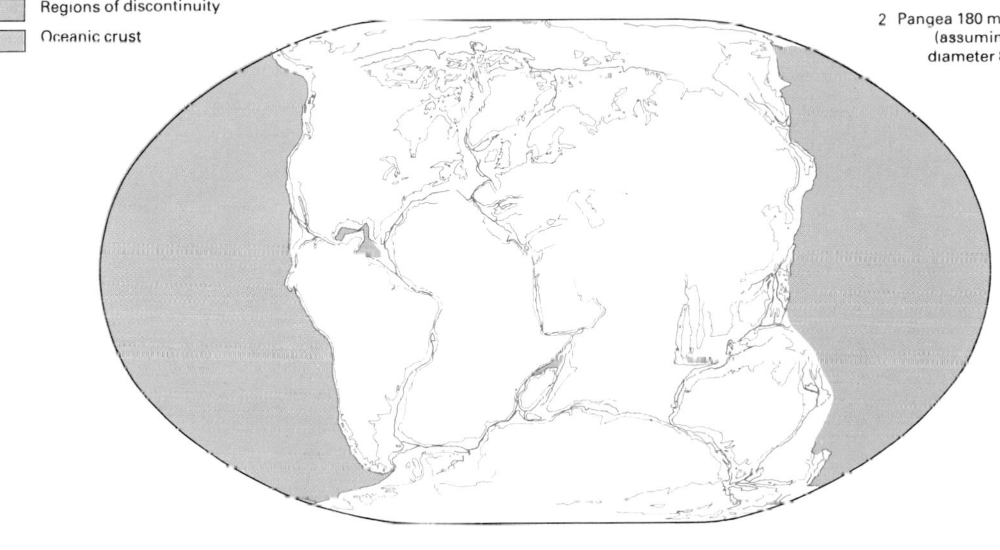

2 Pangea 180 million years ago (assuming an Earth with diameter 80% of modern dimensions)

◀ ▲ *The lonely continent of Antarctica is almost entirely surrounded by constructive plate margins. This means that all other continents are moving away from it. On the next plate, Africa is also surrounded by constructive margins. This is taken as evidence that these two continents are moving away from each other because of the expansion of the Earth.*

▲ *When Pangea is reconstructed on a globe the size of the present day Earth (1), there are a number of gaps where the continents do not quite fit. According to H.G. Owen this is because, at the time of Pangea, the Earth was only about 80 percent of the present diameter. When Pangea is reconstructed on this smaller globe (2) the mismatches are eliminated.*

from the present radius even as large as 30-50 percent.

A quite different approach is to examine possible causes of Earth expansion. The first hypothesis came to light in 1933 when the German geologist O.W. Hilgenberg demonstrated the near perfect fit of the continents on a smaller Earth using a papier-maché model. Soon after, it received more publicity when, quite independently, the British physicist, P.A.M. Dirac, suggested that the universal gravitational forces between particles are weakening, and hence all bodies, including the Earth, will be expanding. But all attempts to detect small variations in gravity in recent times have failed, and there is apparently no evidence of rifting on the tectonically inactive bodies Mercury, Mars and the Moon, which there would surely be if these bodies had also increased their radii by a factor of two.

H.G. Owen of the British Museum (Natural History), a strong supporter of expansion, has suggested that, although the Earth's inner core reacts to seismic waves as if it were solid, it may not really be solid at all, but a plasma – material in which the atoms have been stripped to their basic particles and are thus chemically unrecognizable. The properties of a plasma in that position are uncertain; but it is quite conceivable that, as the Earth cools and the plasma of the inner core gradually converts to atomic material (mainly iron) of the outer core, it expands, thus forcing the Earth as a whole to expand.

For the time being, that remains speculative. Also beyond reach, but perhaps not very far, is the exciting possibility of testing Earth expansion directly using laser beams to measure very precise distances between the Earth and the Moon. In the meantime, the most impressive single set of data supporting Earth expansion comes from H.G. Owen himself. Using ocean floor spreading data he has meticulously constructed a series of maps showing how continental positions have changed over the past 200 million years. The fit of continents and ocean floor segments is better when plotted on an Earth of increasing radius than on an Earth of constant radius.

See also
The Dynamic Earth 9-24
Internal Structure 33-44
The Earth Through Time 81-100

An all-embracing hypothesis?

The few proponents of Earth expansion argue that it is necessary to invoke expansion in order to eliminate such paradoxes. Most go even further, claiming that expansion is the basic cause of continental drift and ocean floor spreading. Thus they envisage that the buildup of strain resulting from the pressure to expand was what caused the breakup of the Earth-covering Pangea in the first place. The continents then moved apart from each other simply because, as the Earth's surface increased, the space between them increased. Molten material rose, and rises, at oceanic ridges only in sufficient quantity to fill those spaces with new oceanic lithosphere. But in that case, what of subduction zones, which on a constant-dimension Earth are needed to compensate for the generation of lithosphere at ridges? They do not exist, say the expanders. The evidence for them has been totally misinterpreted. No lithosphere is being consumed at all.

The majority, who do not support expansion, argue, by contrast, that the refusal to accept the existence of subduction zones is absurd. The evidence for descending slabs is perfectly clear. If the Africa-Antarctica, Arctic and Pacific paradoxes exist at all (which some geologists dispute), they are merely the results of gaps in existing plate-tectonic theory which will be filled sooner or later without the need to invoke anything as radical as Earth expansion. For example, the evidence of stretching between the continents on the margin of a Pacific Ocean that is undoubtedly shrinking, may be due to the continents moving apart as Pangea broke up. And there, for the time being, the matter rests.

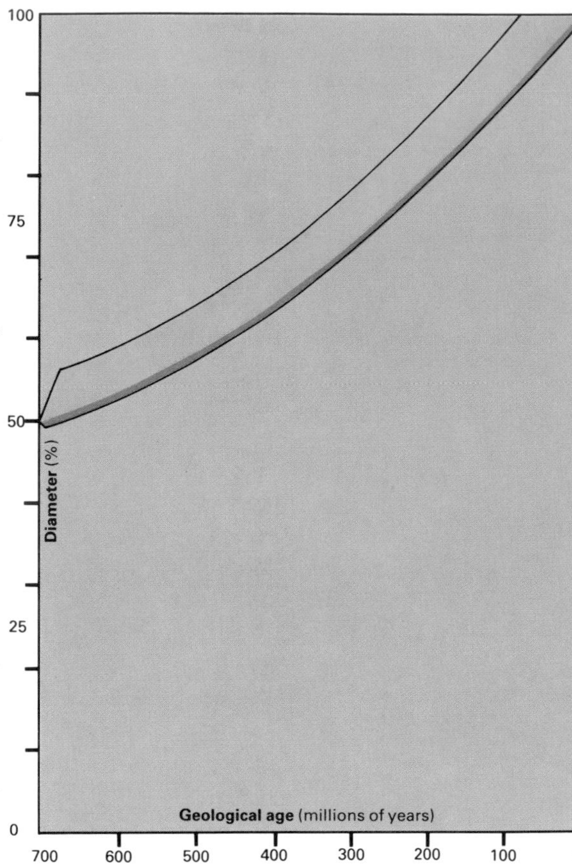

▲ ***If the continental crust formed a complete shell in late Proterozoic times, and the Earth were 80 percent of its modern diameter at the time of Pangea, then our planet would have been expanding at the rate shown on the graph. This graph assumes that the Earth has retained its present shape throughout its expansion.***

What rate of expansion?

If the Earth has been expanding at all, it must presumably have been doing so at a rate that lies between two extremes. If the radius has increased by a factor of two throughout the Earth's entire history, the average rate of increase would have been about 0·7mm a year. On the other hand, if, as Professor Carey himself once suggested, all the expansion has occurred over the past 200 million years (since the breakup of Pangea), the rate of increase in radius would have been as high as 16mm a year.

Many figures between these extremes have been suggested in the past, but the most plausible is perhaps that deduced by H.G. Owen from his maps of continental displacement. He has concluded that during the past 180 million years there has been an exponential (not uniform) increase in radius from 80 percent to 100 percent of its present value. Extrapolating backwards, that implies that the process began about 700 million years ago. This represents an average rate increase in radius of about 4·5mm a year.

◄ ***Subduction zones are an embarrassment to those who support expanding Earth theories. If continental drift were due entirely to the expansion of the Earth, then there would be no need to postulate destructive plate margins. In that case an alternative reason must be found for the existence of destructive plate margin mountains and volcanoes.***

Development of the Atmosphere

11

The ocean of air – the mixture of gases that surrounds us...Atmospheric structure – different layers with different properties..The history – changing atmospheres throughout time...Recent changes – the influence of ice ages and civilization...PERSPECTIVE... Atmospheres of the planets...Escape of gases into space...The greenhouse effect on early Earth... Volcanic gases and their contribution

We live at the bottom of an ocean – an ocean of air. All around us this ocean, called the atmosphere, presses in upon us and affects us in everything we do. We breathe its gases and they keep us alive, it enables fuels to burn, vibrations through it enable us to communicate by speech, particular layers in it shield us from harmful radiations from the Sun and even at heights of tens of kilometers it is thick enough to arrest the flight of meteorites and cause them to burn up with friction before reaching the Earth's surface. It is colorless, tasteless and odorless, but it enables us to exist. To us it is a vast ocean reaching away several hundred kilometers above our heads, but on the world scale it is the thinnest of envelopes – no thicker in proportion than the skin of an apple.

The atmosphere is a mixture of gases. The main one is nitrogen, but the next most abundant, oxygen, is the one that is most important to us, enabling us and all other land animals to breathe. The other gases are present in quite small amounts. Another extremely important component, but one that varies from time to time and from place to place, is water vapor. It is this that determines the climate and the weather in the bottom few kilometers. It is affected by the evaporation of water from the ocean surfaces, by the movement of prevailing and local winds, and by the relief of the underlying ground.

Wind is merely a movement of the atmosphere. Sometimes this can be barely perceptible, but sometimes it rips along at over 300 kilometers per hour. It has great power, and is a major agent of erosion and distribution of sediments and other surface materials.

▲ ***Clouds in a blue sky – the feature of Earth's atmosphere.***

Atmospheres of the inner planets

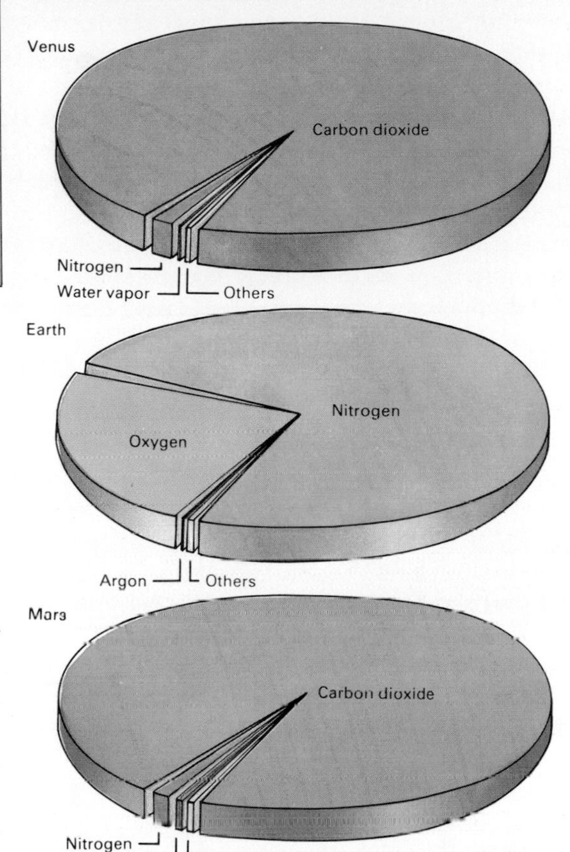

Earth, Venus and Mars compared

The composition of the Earth's atmosphere varies with altitude, but on average it has 75 percent by mass of nitrogen, 23 percent oxygen, 0·05 percent carbon dioxide and 1·28 percent argon. There are other inert gases such as neon and helium in minute amounts. It also contains water vapor in variable quantities ranging from 0·01 percent to 3 percent. Another variable component is sulfur dioxide – a dangerous pollutant that causes acid rain. There may be 10 million tonnes of this substance in the atmosphere at any particular time. At a height of 15 to 50km there is a layer of ozone – a form of oxygen with three atoms per molecule rather than the usual two. There may be about 4 billion tonnes of it.

The Earth's atmosphere is quite unlike that of the other inner planets. For one thing the atmospheres of Venus and Mars contain overwhelming amounts of carbon dioxide, about 95 percent in each case. The other 5 percent of the Venusian atmosphere consists of nitrogen with slight traces of oxygen and sulfur dioxide, and a minute amount of water vapor. Atmospheric pressure on the surface of Venus is about 90 times that of the Earth.

On the other hand the Martian atmosphere is very thin, with a surface pressure about 1 percent of the Earth's, but this can vary depending on the season. As well as carbon dioxide there are small amounts of nitrogen, argon, oxygen and carbon monoxide.

Most of the atmosphere lies within 11km of the Earth's surface

The structure of the atmosphere

For convenience the atmosphere can be divided into various layers, depending on its physical properties, although there are no sharp boundaries. Variations in temperature and pressure, which distinguish the layers, result from the distribution of solar heating.

The lowermost 11km is known as the troposphere. This is the most important layer, where we live and breathe, and where all the weather and climatic activities affect us. Being at the bottom, the pressure here is far greater than anywhere else in the atmosphere, and the compression of gases is such that over 80 percent of the mass of the atmosphere is found here. Near ground level visible and infrared radiation is absorbed and the temperature is high. With increasing height the temperature decreases until it reaches the tropopause.

Above the tropopause, and up to about 50km in height is the stratosphere. In this region lies the ozone layer that absorbs the Sun's harmful ultraviolet rays. As a result the temperature in the stratosphere rises with height until it reaches a maximum – about 10°C – and begins to decrease again. This maximum temperature marks the stratopause – the upper limit of the stratosphere.

The mesosphere – the next layer up – extends from the stratopause up to a height of about 80km. Temperatures in the mesosphere drop gradually to a minimum of about −120°C. The density of air in the mesosphere is very low, but it is enough to burn up most of the meteorites that enter the Earth's atmosphere. Their fiery trails across the night sky indicate where these extraterrestrial objects have come to grief by the friction of their passage through this layer.

Above the mesopause lies the thermosphere which absorbs ultraviolet radiation. The temperature again increases with height, reaching several thousand degrees, although, because of the thinness of the air, very little heat energy is available. This is the source of what is known as the ionosphere. Ultimately there is the rarified exosphere which gradually gets thinner and thinner and trails off into the vacuum of space over 700km above the Earth's surface. Short-wave and long-wave radio transmissions are reflected at various layers within the ionosphere – a circumstance that is important in telecommunications. The whole atmosphere, however, is transparent to very-high-frequency radio waves, and radio signals from distant stars can be received on the Earth's surface.

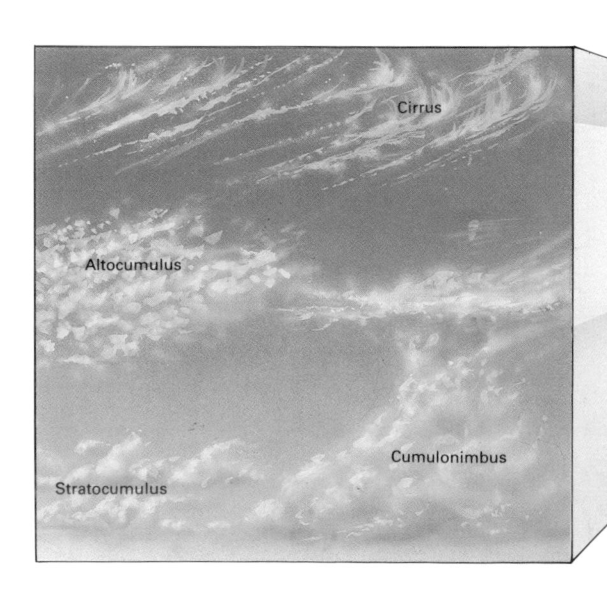

◄ ▲ The structure of the atmosphere is quite well defined: the bottom 11km is the troposphere – the most important layer. This is the layer that we breathe and in which all the important meteorological events occur. Above this the various layers are defined by temperature ranges and other physical properties until, at a height of several hundred kilometers the atmosphere fades off into space.

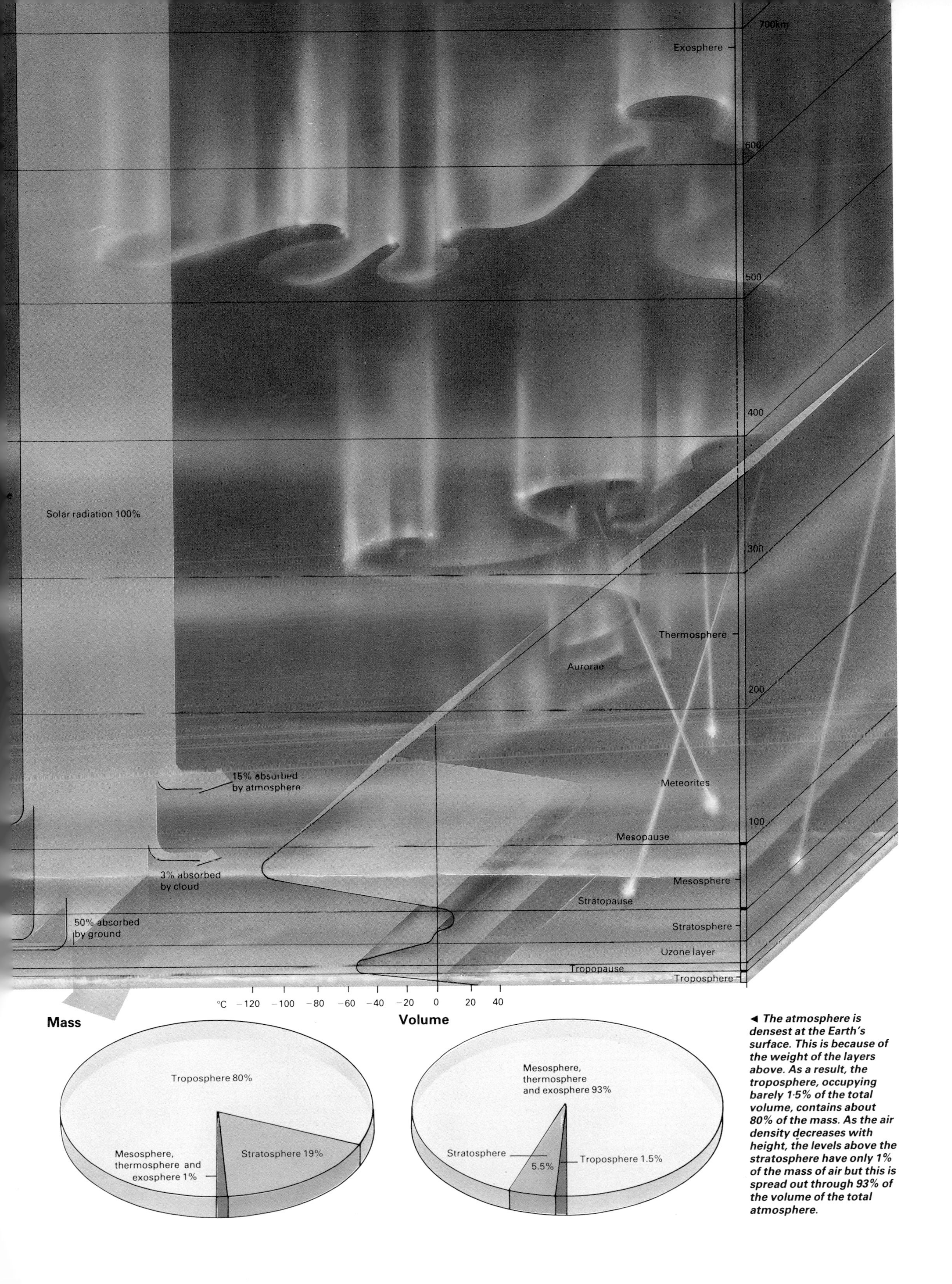

◄ ***The atmosphere is densest at the Earth's surface. This is because of the weight of the layers above. As a result, the troposphere, occupying barely 1·5% of the total volume, contains about 80% of the mass. As the air density decreases with height, the levels above the stratosphere have only 1% of the mass of air but this is spread out through 93% of the volume of the total atmosphere.***

Had we existed 2,000 million years ago, we would not have been able to breathe the air

The development of the atmosphere

Only in recent decades has the history of the Earth's atmosphere been pieced together from theoretical and empirical studies. Much is still uncertain. The development of the atmosphere can be divided into distinct stages. In the early stages of its existence, the nature of the atmosphere was determined by the fundamental processes involved in the formation and early evolution of the Earth itself. Later, more subtle changes are thought to have been induced by the swings between glacial and interglacial conditions, climatic change, and geological processes. Throughout much of its lifetime, there has been a close connection between the biosphere and the atmosphere; the one influencing the development of the other. This continues today as society itself, through pollution and deforestation, affects the chemical composition of the atmosphere. Underlying all these processes is the position of the Earth within the Solar System, the distance of the Earth from the Sun, and, hence, its equilibrium temperature.

Heat from the processes that formed the solid Earth resulted in the outgassing of those elements at or near the Earth's surface that most easily evaporated. It is now believed that this early atmosphere contained nitrogen, carbon monoxide, carbon dioxide, water vapor, hydrogen and inert gases, making up the original cloud of cosmic dust and gas (the primordial atmosphere). The vigorous solar wind may have removed much of this primitive atmosphere during the first billion years of the Earth's lifetime.

Some elements are as abundant on Earth as they are in the Solar System as a whole, but others are notably depleted. The inert gases,

A leaking atmosphere?

The retention of the gases that make up our atmosphere depends on the balance between gravity and temperature. The molecules that make up the gases are characterized by very rapid motions, the mean velocity of which corresponds to a specific temperature.

Up to the thermopause, the boundary between the thermosphere and exosphere, collisions between molecules are sufficiently numerous to provide a complete distribution of particle velocities and thus a specific temperature. Above the thermopause collisions are less and less frequent and the retention of particles depends solely on the gravitational and magnetic fields.

Because of the pull of gravity the denser molecules, with slower velocities, are greatest near the Earth's surface and the lighter, with greater velocities, float to the outer layers of the atmosphere. As a result, in the exosphere a light atom such as hydrogen can escape from the atmosphere because of its velocity – four times that of atmospheric oxygen, which remains permanently subject to the gravity field.

The characteristic escape time of atomic hydrogen is a few hours and of helium around a century, which makes it difficult to trace their history. The evolution of heavier molecules, however, cannot have been affected by their escape time – atomic oxygen has an escape time longer than the age of the Earth.

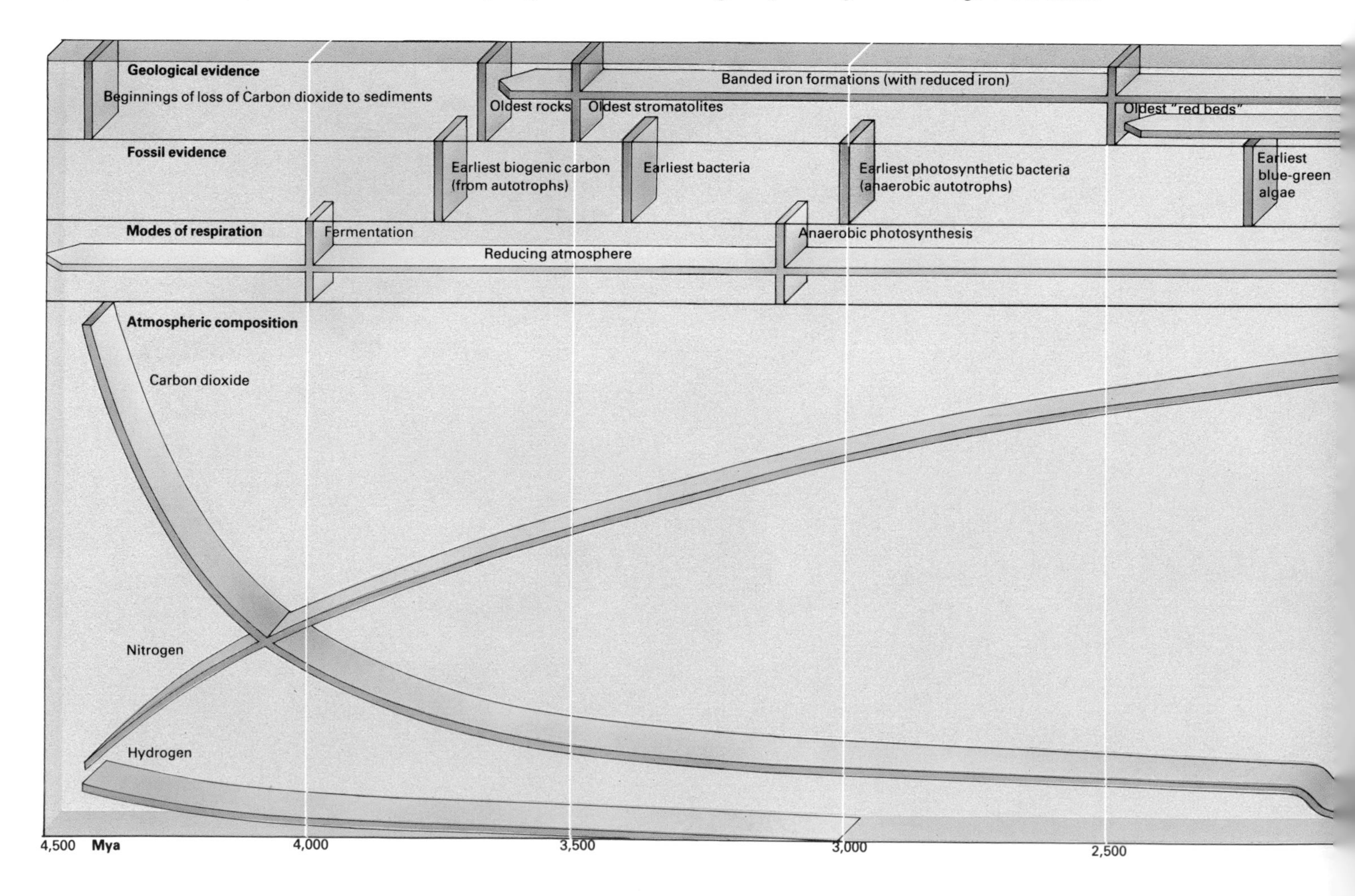

A positive "greenhouse effect"?
Water vapor and carbon dioxide produce what is known as a "greenhouse effect". In the same way that glass traps heat in a greenhouse, these gases trap radiation from the Earth. This is then reradiated, much of it back to the planet's surface, thus increasing the temperature.

As water vapor accumulates in the atmosphere, the amount of infrared radiation trapped increases to the point where the surface temperature begins to rise appreciably. When the water vapor pressure reaches the point when condensation or freezing occurs, accumulation of water vapor and increase in temperature are halted. By contrast, on Venus the atmosphere contains a high proportion of carbon dioxide and water vapor. Proximity to the Sun may have started to evaporate its oceans, creating more water vapor. The resulting runaway greenhouse effect completed the evaporation.

The early greenhouse effect may explain the faint Sun paradox. As nuclear fusion has increased the mean molecular weight of the Sun, its luminosity and, therefore, the solar energy input to the Earth has increased. Some 4·5 billion years ago, the luminosity is believed to have been about 75 percent of today's value. Other things being equal, the reduced energy input during the early history of the planet should have resulted in an ice-covered Earth although there is clear evidence that this was not the case. Did the greenhouse effect counteract the influence of the faint Sun?

such as argon, krypton and xenon, are of particular interest as they are not susceptible to loss from the atmosphere, condensing at very low temperatures and forming few compounds. Their abundance can be considered to be a constant feature of the atmosphere throughout its history. Nevertheless, they are depleted in the atmosphere relative to their abundance in the Solar System and so some must have been lost from the Earth soon after, or during, its formation.

As the Earth solidified, outgassing from the hot interior occurred beginning the formation of the present-day atmosphere, dominated by nitrogen, oxygen, argon and carbon dioxide. Volcanic eruptions were the major source of atmospheric gases until the origin of life. As the degree of volcanic activity grew, water, carbon dioxide, hydrogen sulfide and nitrogen (the secondary atmosphere) became increasingly important. Some water vapor was dissociated in the high atmosphere by ultraviolet radiation producing oxygen and hydrogen. This may have been the main source of oxygen in the early atmosphere. Photochemical conversion of oxygen to ozone resulted in the formation of a "screen", protecting the lower atmosphere and the Earth's surface from ultraviolet light.

Eventually, the saturation point of water vapor was reached. Any further water subsequently released into the atmosphere could not remain as water vapor but began to condense to form oceans. Carbon dioxide was then dissolved in the oceans and equilibrium was ultimately reached. At this stage it is reasonable to assume that the water vapor and carbon dioxide levels in the atmosphere were not too different from today.

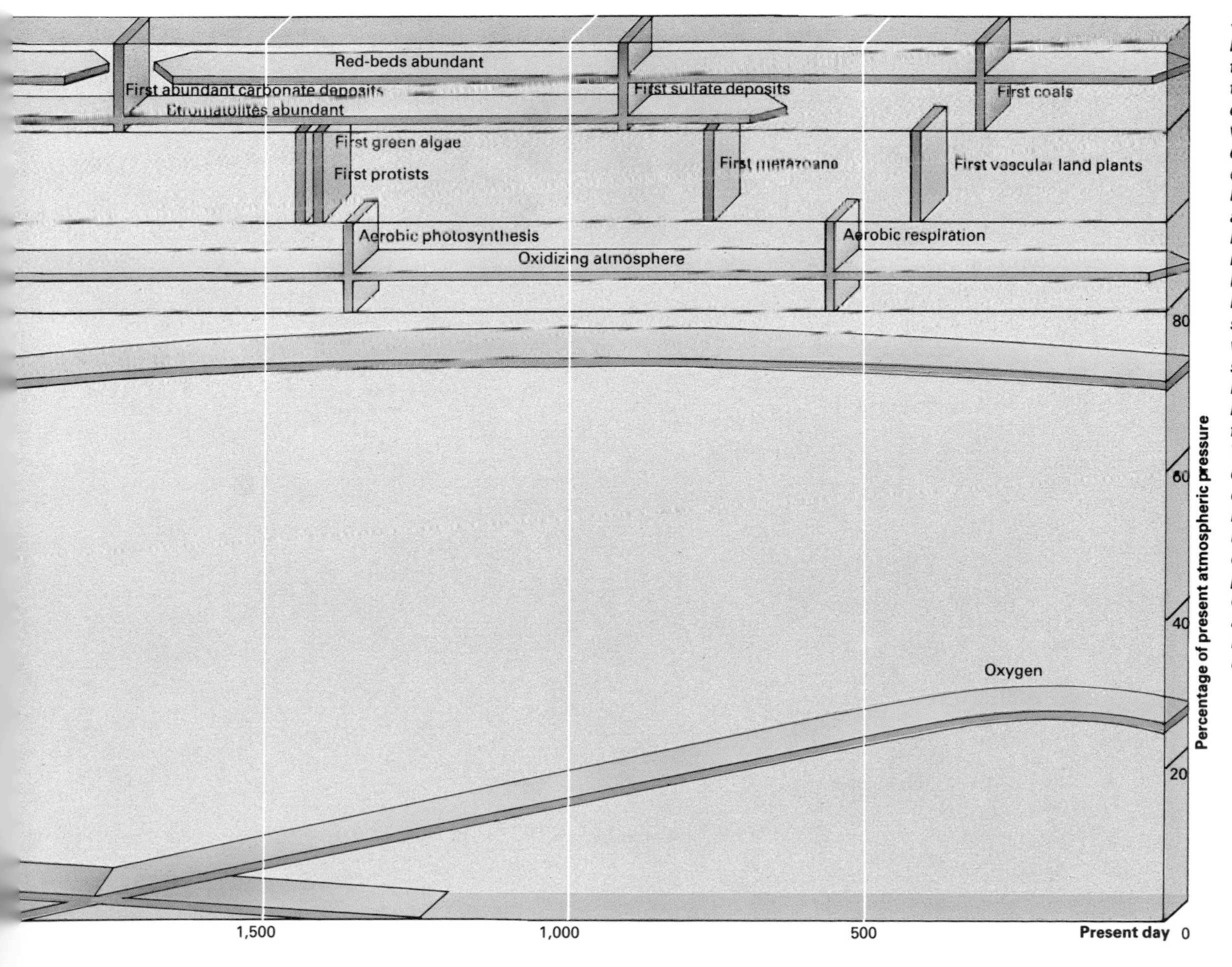

◀ The Earth's atmosphere has not always been as it is today. The proportions of the different gases have changed since the earth was formed. Carbon dioxide and hydrogen were once important but these have now largely died away. In their place there has been a build-up of nitrogen and oxygen. Hydrogen, being a very light gas, was lost into space. Carbon dioxide was incorporated into sediments, particularly limestone. The nitrogen built up by outgassing from the Earth's interior. The most important component, from our point of view, is oxygen. This did not enter the atmosphere until photosynthetic plants evolved. The chemical processes of life depend on the kinds of gases available. Early life-forms had quite different modes of respiration from today's.

At first, the vast majority of the volatiles that formed the early atmosphere were held in the solid Earth. Gases were injected by volcanic activity and the atmosphere grew. There was, and still is, a continuous recycling of the volatiles in these reservoirs. Tectonic activity injects volatiles into the atmosphere, sedimentary rocks take them up, and they are then returned to the mantle in subduction zones.

Ammonia, which photodissociates readily, has a relatively short atmospheric lifetime. Early sources include volcanoes and photochemical reactions. Once life had developed, fermentation would have produced ammonia, although it may have been required earlier for the development of amino acids.

One of the major differences between the composition of the early atmosphere and that of the present day was the lack of free oxygen. It is likely that the early atmosphere was strongly reducing. This is a condition for the development of life. The transition from a reducing to an oxidizing atmosphere may have depended on the balance between the sources and sinks of oxygen and hydrogen. Today, the volcanic injection of hydrogen into the atmosphere is roughly equal to its loss to space. In the early atmosphere, the volcanic source may well have been much greater and there were no major sources of oxygen.

Volcanic gases

Present-day volcanic fumes contain water, hydrogen, carbon dioxide, and small quantities of sulfur gases, nitrogen, chlorine and, possibly, argon and ammonia. Water and, to a lesser extent, carbon dioxide are the dominant components; free oxygen is rarely released from magma. It is reasonable to suppose that the early volcanoes injected similar gases into the atmosphere. Although these gases formed the basis of the early atmosphere, the actual composition was complicated by processes such as photochemical reactions, condensation and precipitation, and interaction with the surface reservoir. For example, the temperature of the Earth's surface was sufficiently low to permit condensation of water vapor and the formation of the Earth's oceans. Carbon dioxide was dissolved in surface water and formed carbonate minerals.

▼ Much of the early atmosphere was formed from gases emanating from below the Earth's surface through volcanoes and fumaroles. This is still happening today in volcanic areas such as here, in Hawaii.

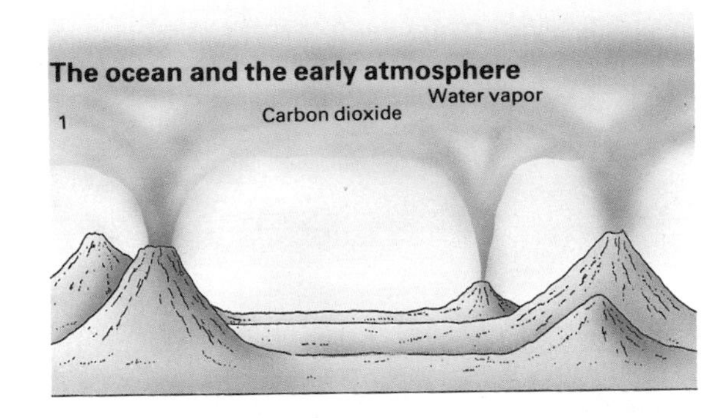

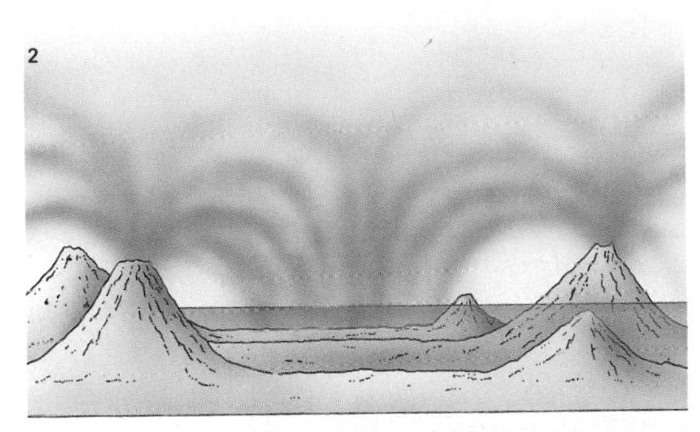

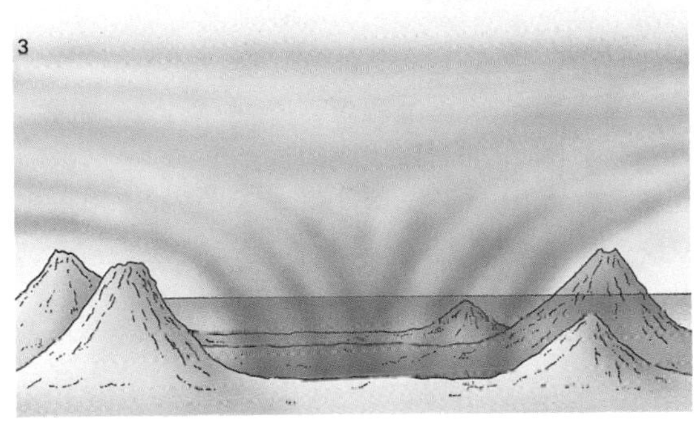

▲ The important gases placed in the early atmosphere by volcanic activity were carbon dioxide and water vapor (1). The vapor condensed into ocean water (2) and the carbon dioxide was dissolved in it and turned into carbonates (3).

Once life formed, a major source of oxygen developed, and the balance was tipped in favor of an oxidizing atmosphere.

At different stages in the Earth's history, the atmospheric composition has been dominated by different processes. Following the initial formation of the atmosphere, physical and chemical processes became of paramount importance then, as life originated, biological factors increased in influence. The development of lifeforms using different energy sources produced a sequence of changes in atmospheric composition. The hydrogen and, to a lesser extent, carbon dioxide content of the atmosphere were affected. The major change occurred when the mechanism of photosynthesis, and the resulting release of oxygen, evolved. Oxygen in the atmosphere may be indicated by "red beds" of oxidized iron some 2 billion years ago. If so, this marks the transition from a reducing to an oxidizing atmosphere.

By the end of the Precambrian biological and geological processes had taken over as the major agent of change in atmospheric composition. Climate and continental geography played an important part. Evidence, drawn largely from the fossil record, suggests that the composition of the atmosphere has been approximately as at present for the past 500 million years.

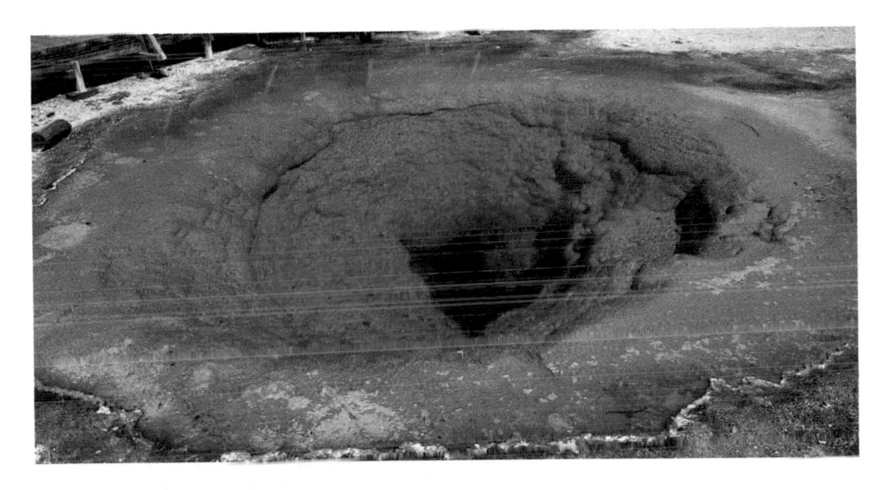

◄ ▲ The types of algae (above) and bacteria (left) that existed in oxygen-starved environments of the early Earth can still be seen today. The hot springs of Yellowstone Park support a colorful and spectacular collection of anaerobic life forms existing on the chemicals of the hot waters.

See also
Origin and Evolution 69-72
Development of the Oceans 121-8
Climate and Climatic Change 129-40

More recent atmospheric changes

Carbon dioxide levels may have varied by as much as a factor of four during the Phanerozoic. This is of interest at present because of its bearing on the study of the climatic effects of pollution. There is some evidence from Antarctic ice cores of reduced levels during the last major glaciation, 14,000 to 22,000 years ago. This is most probably due to increased storage in the colder oceans or the biosphere. Possibly the warmth of the period that followed the last glaciation may have been caused by increased atmospheric carbon dioxide content.

During the past 100 million years, the oxygen content of the atmosphere has been affected by a number of geological processes – sea-level changes, shoreline length and temperature, for example. The amount of oceanic phosphate may also have affected oxygen levels. It is, therefore, difficult to speculate about the net effect.

During this period, the major influences on oxygen content have been geological. For example, as temperatures have fallen during periods of glaciation, the solubility of oxygen in the oceans has risen. The oxygen content of the deep ocean water would have increased and fossilization of organic carbon would have decreased as less ocean floor was underlain by anaerobic sediments. The net result would have been a decrease in atmospheric oxygen. The breakup of the supercontinents that existed 100 million years ago resulted in an increase in shoreline. And it is in near-shoreline areas and in shallow shelf seas that most organic carbon is buried. The increased rate of burial of carbon could well have resulted in an increase in atmospheric oxygen, as would the spread of water across the land at times of rising sea-level. Decreasing ocean phosphate levels, likely under such conditions may, however, have counteracted this effect.

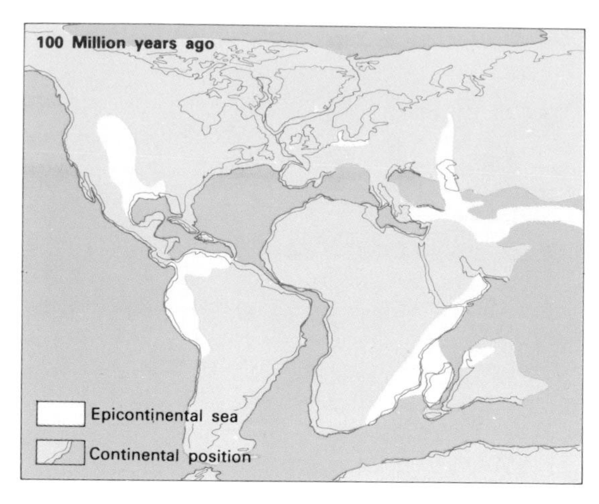

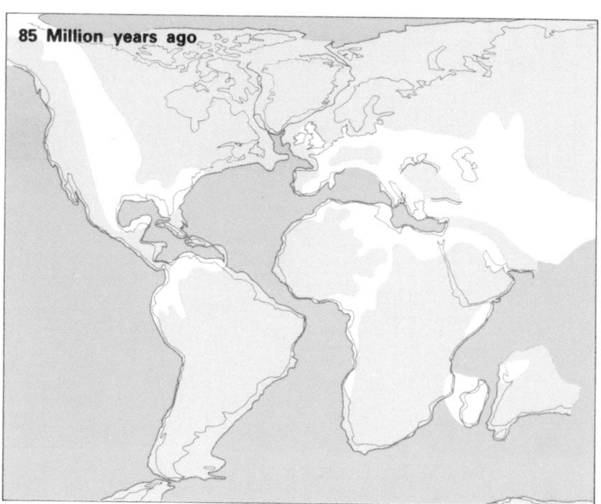

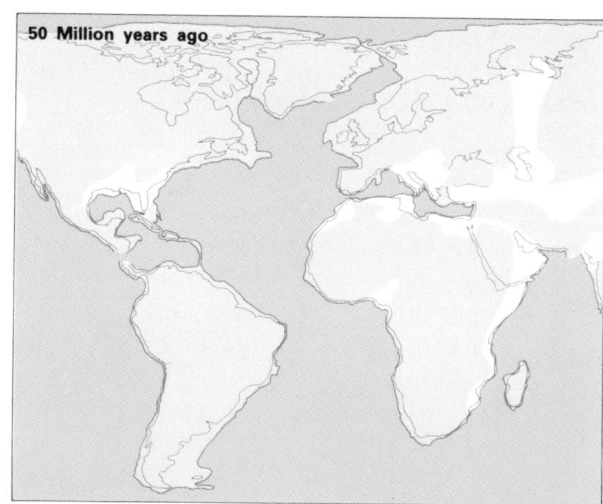

◄ *In the last 100 million years the shape and distribution of the continents has changed significantly. Pangea has broken up and the pieces drifted apart. This has greatly increased the amount of coastline, and hence the area of epicontinental seas. Even in the last few thousand years the sea level has risen and fallen, shown by the number of raised beaches that exist.*

▲ *The significance of the varying amounts of epicontinental seas is that such seas present a large area of water to the atmosphere. Some gases can be absorbed from the atmosphere while others can be given up from the water. The net effect would be to increase the oxygen content of the atmosphere with increasing areas of epicontinental seas, particularly during the last 100 million years.*

Development of the Oceans

Ocean water – not as unchanging as once thought... Salinity – the saltiness of the ocean, and how this has changed...Circulation – movement of seawater across the globe, driven by density and temperature... PERSPECTIVE...The largest surface feature on Earth... Mixing of the sea layers and its effect on fertility... Ancient circulation and oil formation...Hot springs from ocean ridges

Until recently geologists have regarded the sea as constant; unchanging in the composition and flow of its waters, a dependable setting for the evolution of life. Its main interest has been the record of earth history left on the edges of the continents by its unceasing rise and fall. However, acceptance that the continents have shifted endlessly with time, meant that the ocean circulation and the climates of the past must have changed also. The discovery of hot springs on the deep-sea floor has called into question the constancy of the composition of seawater. Suddenly, the role of the ocean in the history of the Earth looms larger, and has become more interesting.

The Earth is a water planet. Land today occupies less than one third of its surface, and even less at many times in the past. This huge volume of water makes our world special; without it, life as we know it would not have existed.

The ocean water is far from pure. Its salinity – the total amount of dissolved salts – is 33 to 38 parts per thousand, containing virtually all elements, though most occur in minute amounts. Sulfate, chloride, sodium, and magnesium dominate, followed at some distance by calcium and potassium. The proportions of most dissolved elements are remarkably constant – it is possible to calculate with precision the concentrations of such elements if the salinity is known. This implies that, except near shores or ice, seawater is the result of varying dilutions of a standard solution.

Not all dissolved substances are constant, however. Some are gases, mainly oxygen and carbon dioxide, their concentrations determined by exchange with the atmosphere and by biological activity. Others are nutrients essential to life in the sea, such as phosphorus or nitrogen, or are, like calcium carbonate or silica, used by organisms in the construction of their shells and skeletons.

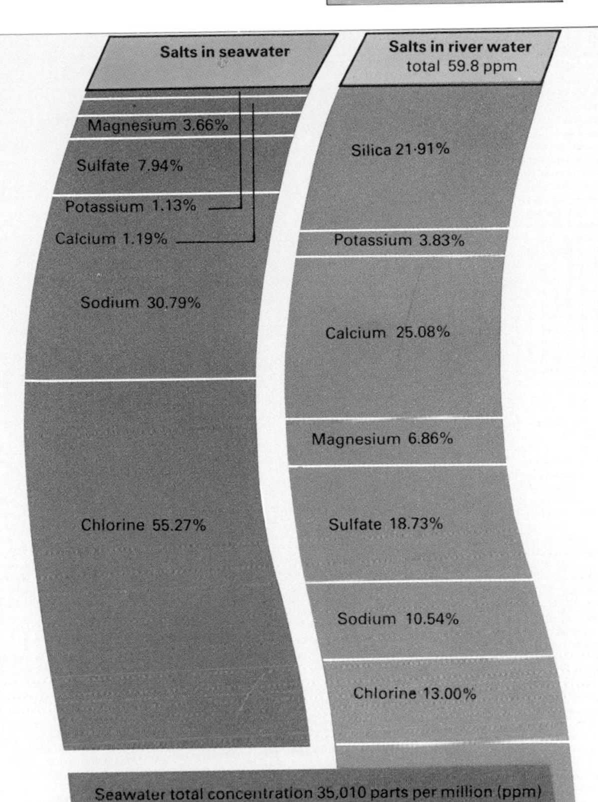

▲ Seawater contains dissolved minerals at a concentration of between 33 and 38 parts per thousand. It has in solution nearly all the elements that exist. Sodium and chloride together form common salt and this constitutes more than 85 percent of the substances in solution, making seawater quite unsuitable for drinking or for irrigation. Some parts of the seas are saltier than others, but the proportions of the dissolved elements are quite constant. In contrast the average river water has only about 0·2 percent of the salinity of seawater, and this varies greatly.

▲ Land and sea are distributed very unevenly over the globe, most water being in the Pacific.

The largest feature on Earth

When seen from space, our planet has a predominantly blue color. This is due not only to the haze of atmosphere, but also the vast area of water beneath.

With a surface area of 362 million km^2 and a mean depth of 3·8km, the ocean contains 1,350 million km^3 of water. The amount of dissolved substances is equally large; the sea holds enough salt to cover Europe to a depth of 5km, and enough organic matter, alive and dead, for several meters of slime on North America.

The vastness of the ocean warns us not to make quick judgements regarding human impact. Many have reported that oil pollution occurs everywhere on the ocean surface. A simple calculation shows that a layer only 0·01mm thick would require 120 percent of the 1982 world oil production. Moreover, it would last only days or weeks before being destroyed by oxidation, mixing or bacterial activity. In the early 1970s, high mercury levels in fish livers also raised the cry of pollution. The current mercury level in the ocean is only 0·00006mg/l. To double this, with no assured ill effect, 12,000 years of the 1982 world production would be required.

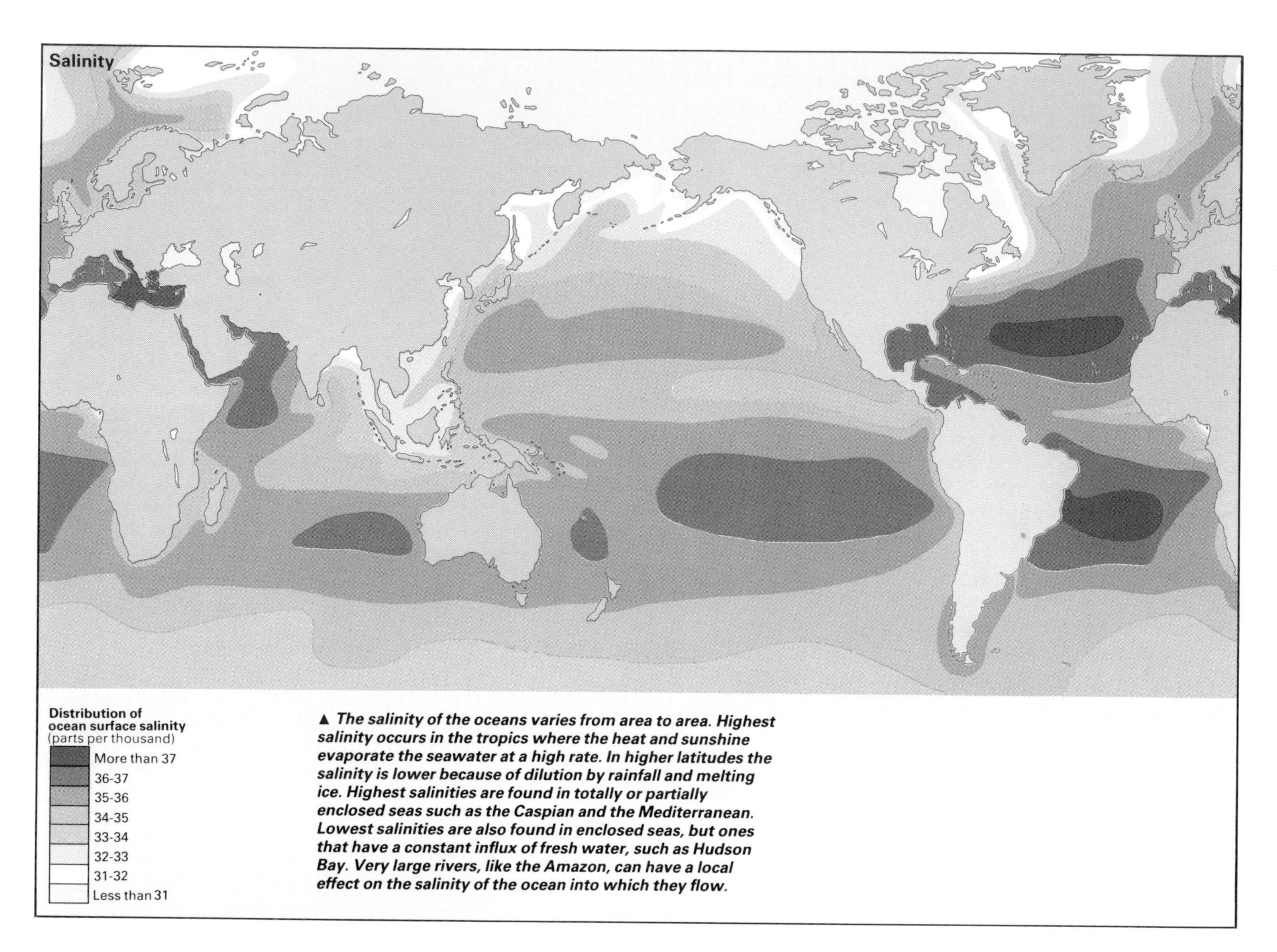

▲ ***The salinity of the oceans varies from area to area. Highest salinity occurs in the tropics where the heat and sunshine evaporate the seawater at a high rate. In higher latitudes the salinity is lower because of dilution by rainfall and melting ice. Highest salinities are found in totally or partially enclosed seas such as the Caspian and the Mediterranean. Lowest salinities are also found in enclosed seas, but ones that have a constant influx of fresh water, such as Hudson Bay. Very large rivers, like the Amazon, can have a local effect on the salinity of the ocean into which they flow.***

▲ ***There is plenty of geological evidence for the salinity of the sea in past times. Thick beds of salt are laid down where inland seas dry up. More local effects can be seen where rocks are formed from mud on a beach. A pool of water cut off from the sea may dry out in the sun, and the salt precipitate out as large crystals. These crystals may later dissolve away leaving depressions in the mud of their exact shape. The next layer of mud will fill these hollows and, when the whole sequence is turned to stone, the crystal shapes wil be preserved as "pseudomorphs".***

It is generally accepted that the oceans are at least as old as the continents (◀ page 109). Where did the water come from, and how and when? Water is believed to have originated in the mantle. More than 4 billion years ago, the outer layer melted, leaving the present mantle as a residue. A light primitive crust formed on top, with water and an atmosphere above. It is possible that the ocean filled gradually with water issuing as steam from volcanoes. It is now known, however, that steam from present-day volcanoes is mainly recycled groundwater or seawater, and that very little "new" water enters the ocean. The ocean must have been full quite early. This is confirmed by the sea-level record; as long as 600 or 700 million years ago the sea stood even higher than it does now. The gradual, perhaps intermittent, conversion of primitive crust into continents may have released some more, especially between 3·5 and 1 billion years ago, but no one knows how, nor how much, nor when. The primordial ocean may therefore have been shallower than the present one, but it was probably as large.

Establishing the history of the ocean's salinity

What about the history of salinity? Nearly all dissolved salts in the sea come from weathering on the continents and are brought to the ocean by rivers. They arrive at different rates; were one to start with an ocean of pure water, it would take 260 million years to supply all the sodium, but only a century for the dissolved aluminum. The time

◀ *A large proportion of the gas belched out by volcanoes is water vapor. This is probably how most of the water reached the surface in the first place. However, nearly all the water erupting from modern volcanoes is actually recycled seawater.*

Circulation of seawater through volcanoes

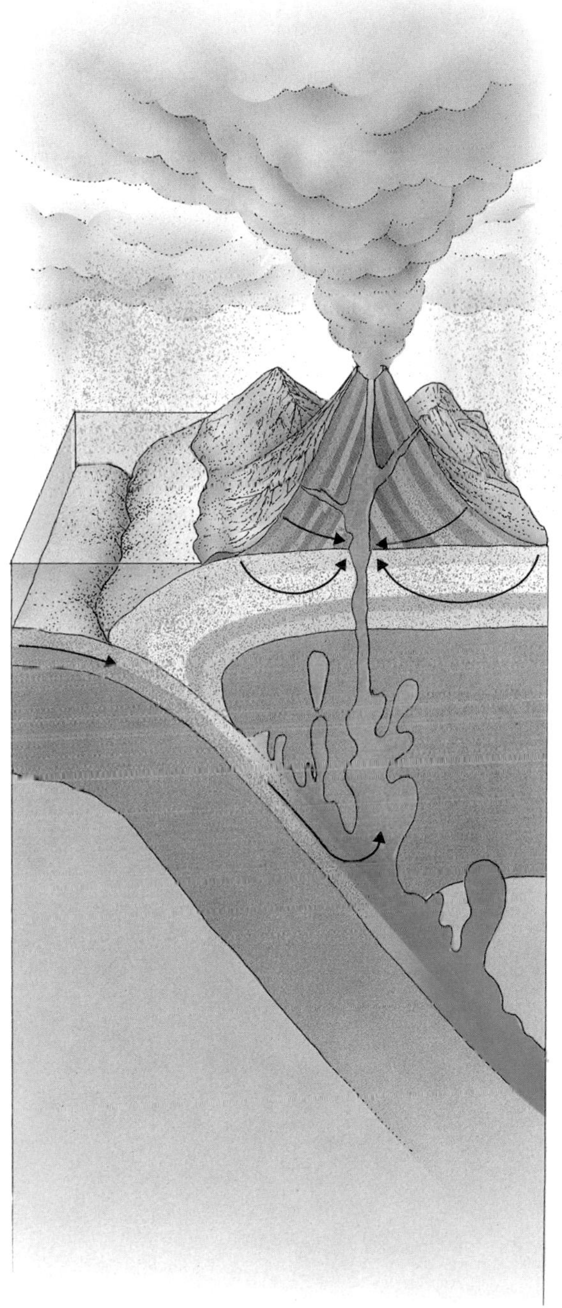

▲ *When sediments are laid down on the floor of the ocean, they trap a great deal of seawater in them. Eventually these sediments, in great thicknesses, find their way to subduction zones, where they are swallowed up along with the crust of the ocean plate. At depths in the subduction zone, where the pressures and the temperatures begin to break up the downgoing plate, the water is sweated out of the sediments and rises to the surface along with the newly formed magma. There it erupts with the magma and eventually falls as rain on the ocean again.*

needed to replace each element in the sea is called its residence time. If the salinity is to remain constant, elements must leave the ocean at the end of their residence time, no sooner and no later.

Whether the salinity has remained constant with time is hard to prove, because minerals or sediments recording past salinity are rare and generally unstable. The assumption is that it has, and it rests mainly on lack of evidence to the contrary, and on one simple idea. The salt concentration of the body fluids in organisms is the same as that of seawater. If this reflects the salinity of the sea when life began there 3 billion years ago, a constant salinity since then is plausible.

Alas, it also seems impossible. The continents did not approach their present size for several billion years, and so the salt supply must have increased slowly; a reasonable estimate is that the salinity remained below its present value until more than a billion years after life began. Since then, the salinity can have remained constant only if salt were eliminated at the same rate as it was being supplied.

It is not clear how excess salt leaves the sea. The deposition of rock salt, or evaporites, in the Permian for example, does not suffice to eliminate billions of years of input. The Swedish geochemist L.G. Sillen has proposed that the excess is precipitated as clay-like minerals, but the process has not been confirmed on the right scale, and the expected vast deposits of such minerals are not there. There is a major quandary here, one which has not received the attention it deserves.

Animals and plants of the sea are all dependent on the circulation of seawater

The recent discovery of hydrothermal springs on ocean ridges (➧ page 128) partly answers the question of the history of the ocean's salinity. Water passing through the hot new crust dissolves and supplies to the sea virtually all of the iron, manganese, lithium and barium in seawater. It also provides calcium and silica equivalent to one fourth of the river input, not to mention much carbon dioxide. It loses all of its magnesium to the rock, and probably large quantities of sodium and chlorine as well. Estimated from the helium-3 in seawater (an isotope of mantle origin) the volume of water involved globally is so large that the springs must be a major source, or sink, for many elements. Clearly, the traditional geochemical budget of the sea as it stands needs substantial revision. Because the hydrothermal process depends on the spreading rate, and is not closely tied to the rate of weathering and river supply on continents, the salt content of the sea is certain to have varied over time, though it is not yet known how.

Nutrients, carbon dioxide, calcium carbonate and silica are not constant, and obey different rules. In the middle of the Mesozoic Era, large scale use of carbonate and silica for shells and tests by plankton started the era of calcareous and siliceous oceanic oozes. Since then, these substances have been controlled mainly by life.

The carbon dioxide content of the oceans depends on exchange with the atmosphere; add carbon dioxide to the air, and somewhat less than half enters the sea. There, it exists in equilibrium with carbonic acid and carbonate ions. The surface waters are saturated with carbonate and organisms can utilize it without fear of depletion. At depth, however, the sea is undersaturated – the calcareous shells of dead organisms dissolve as they sink and rest on the bottom. Below a level known as the lysocline, the rate of solution becomes significant, and increases downward. At a depth of about 5km the rate of deposition of carbonate equals the rate of solution. Below this "calcite compensation level", the sediments are carbonate-free. All of this causes the

Fertility of the sea

The microscopic plants at the base of the oceanic food chain require, besides water, carbon dioxide, and sunlight, nutrients such as phosphorus and nitrogen. The nutrients near the sunlit surface are soon exhausted, limiting the productivity. In deeper water, decay of sinking organic debris regenerates nutrients, but the stable ocean stratification makes it difficult to bring them to the surface. Consequently, most of the ocean is no more productive than a desert. Ocean currents stir the water near the thermocline, making them somewhat fertile, but the main agent in the recycling of nutrients is a process called upwelling. In this, currents or winds from the land drive the surface water seaward, so that water from 1,000 to a few hundred meters depth brings nutrients to the surface, making upwelling zones the most fertile in the oceans. More than 99 percent of the productivity comes from less than 10 percent of the ocean's area. There, off Peru, Japan, California, or West Africa, occur most of the great fisheries. Since this process is dependent on the prevailing winds, any change in the wind pattern could have a grave effect on these industries.

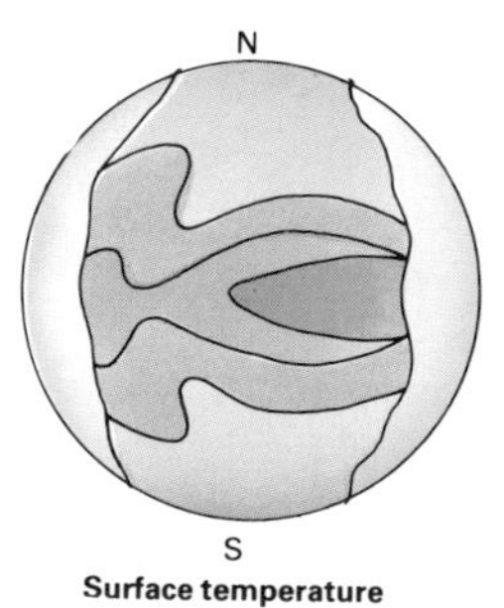

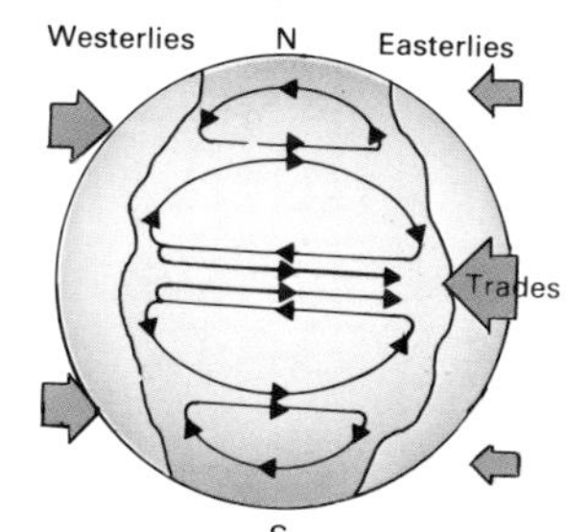

Seawater circulation in hot springs

Cold inflow
Altered zone
Hot

▲ ▶ The temperature of the ocean is much greater on the surface than it is at depth. The world-wide distribution of the surface temperatures of the ocean is influenced by the large scale circulation pattern of ocean currents. This, in turn, is caused by the prevailing winds.

Temperature
Thermocline
Depth (km)

◀ Ocean currents exist in the rocks of the ocean floor as well. The hot mineral-rich springs at constructive plate margins are caused by the recirculation of seawater. This cools the newly formed plates and adds dissolved minerals from the new crust to the ocean water.

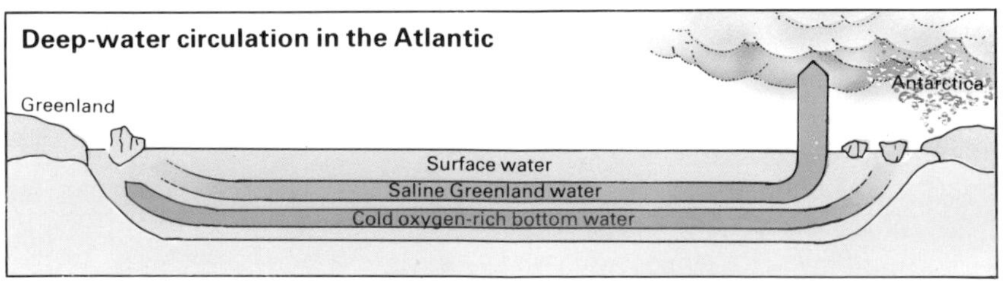

◀ The Atlantic Ocean bottom currents are generated by the freezing of seawater in the north and south. As the water freezes it leaves its salt behind in concentrated solutions. These heavy waters sink and travel over the ocean floor beneath the temperate and tropical surface layers.

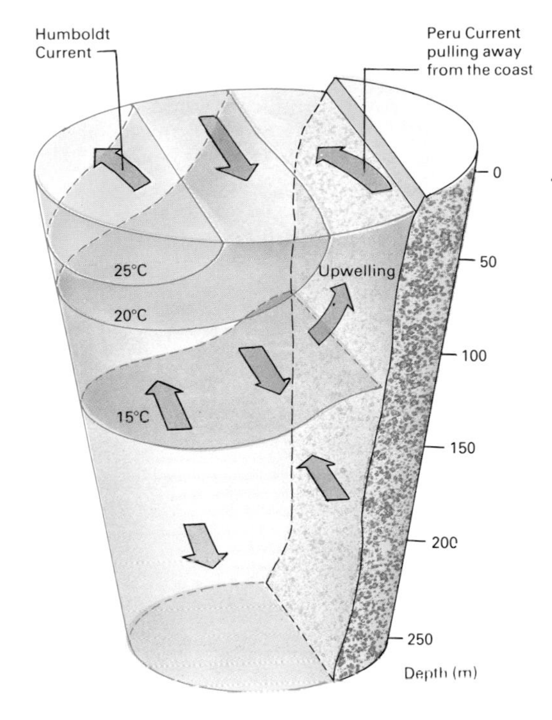

▲ *Vertical currents are found off Peru where the prevailing Trade Wind system causes an ocean current to pull away from the coast. Cold water from the depths of the Pacific then wells up up to take its place. This cold water is rich in nutrients and is the basis of the extensive anchovy fishery industry of the area.*

▼ *Only the sunlit surface waters of the ocean can support plant life. The floating plant life – the phytoplankton – determines the distribution of all animal life that feeds on it, and hence all animal life of the sea. Its distribution is determined by water temperature, current systems and nutrient-rich upwellings.*

vast area of the seafloor to resemble a mountainscape in winter, with snowcapped hills where the carbonate sediments lie, and brown muddy valleys where no carbonates are deposited.

The depth of compensation depends on the production rate at the surface, on the depth of the lysocline, and on the rate at which solution increases below it. If the surface production is increased, while keeping the other factors the same, the compensation depth will increase and a larger area will be covered with calcareous ooze. On the other hand, if the carbon dioxide is increased, for example by the burning of fossil fuel, the equilibrium in the water will shift, and it will be more corrosive. The compensation depth then rises, and carbonate sediments dissolve. Calcareous ooze thus functions as a buffer, minimizing the effect of changes in carbon dioxide and carbonate in the water. The buffering, however, depends on the rate of circulation and mixing in the deep ocean, and is slow compared to the rate at which carbon dioxide is being added to the atmosphere at present.

The oceans' circulation

Circulation of the ocean currents is driven by the differences in the density of seawater. These differences are related to salinity and temperature. The density of seawater increases with decreasing temperature but, unlike freshwater, it is densest at freezing point. Ice, on the other hand, is always lighter than water. The latter has the larger influence. Cold, saline waters are denser than warmer, less saline ones.

The surface waters of the ocean are heated by the Sun, and stirred by wind and waves to a depth of 100 or 200m. There a sharp temperature drop, the thermocline, separates the surface water mass from the much colder deep water, which has a temperature of 2 to 4°C. The ocean is thus stratified with a thin, light warm layer on top of a much larger mass of rather uniform deep water, making it difficult to bring dense water from the deep to the surface.

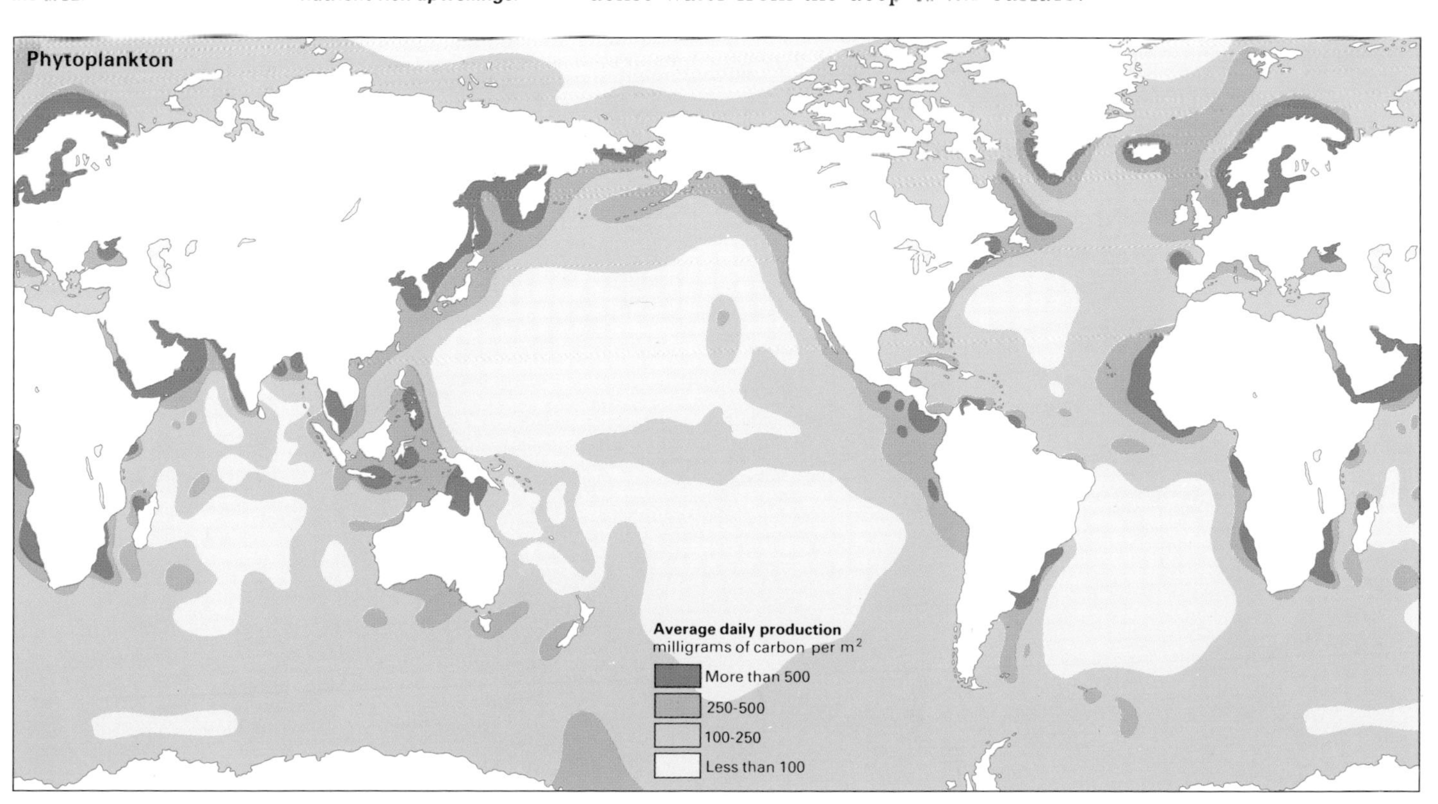

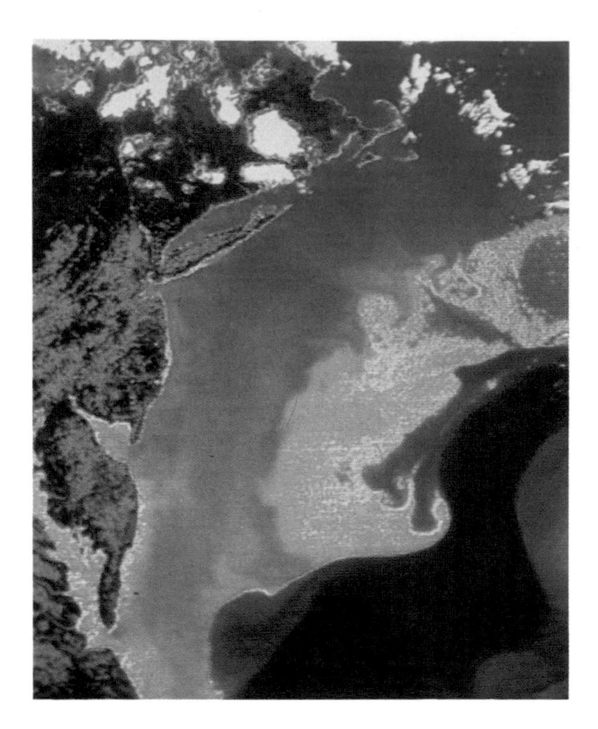

▲ *An important surface current is the Gulf Stream or North Atlantic Drift. From the Gulf of Mexico it skirts the eastern United States (seen here in false color image) and crosses the north Atlantic to bring warm water to Scotland and Norway.*

Above the thermocline, the surface circulation is driven by the main wind systems, the subtropical trades and the temperate westerlies. In a simple ocean surrounded by continents, the trades would drive equatorial currents westward. These, upon reaching the opposite shore, would return as an equatorial countercurrent and in two large gyres in the Northern and Southern Hemispheres. The surface temperature reflects this pattern. The water is warmest in the eastern equatorial zone where it has traveled a long distance in a warm climate. Cool currents flow equatorward along the eastern shores of the ocean.

The circulation of the real oceans, although more complicated, is analogous, and paleoceanographers can infer the general features of past oceans. An important role is played by seaways between oceans. The free passage around Antarctica permits the water to flow east around the Earth, thereby remaining long in a cold zone. It isolates the southern continent from the warmer waters to the north.

The deep circulation of the present ocean is driven by temperature and salinity differences created in the southern oceans by the freezing and melting of Antarctic ice. The freezing ice excludes salt, and a cold water mass of high salinity forms underneath. It sinks, travels north along the bottom of all three main oceans, and finally returns to the surface in the far north. Cold, saline water is also formed near Greenland. It travels south at intermediate depth, and rises near Antarctica to replace the sinking abyssal water. Being slightly warmer, it evaporates more and provides snow to the southern continent. The abyssal water completes its travel in 500 to 1,000 years, in contrast to the surface circulation which needs years or at most decades for its travel. Consequently, any change introduced in the deep water takes millennia to reach equilibrium, against decades for alterations of the temperature or composition of the surface water.

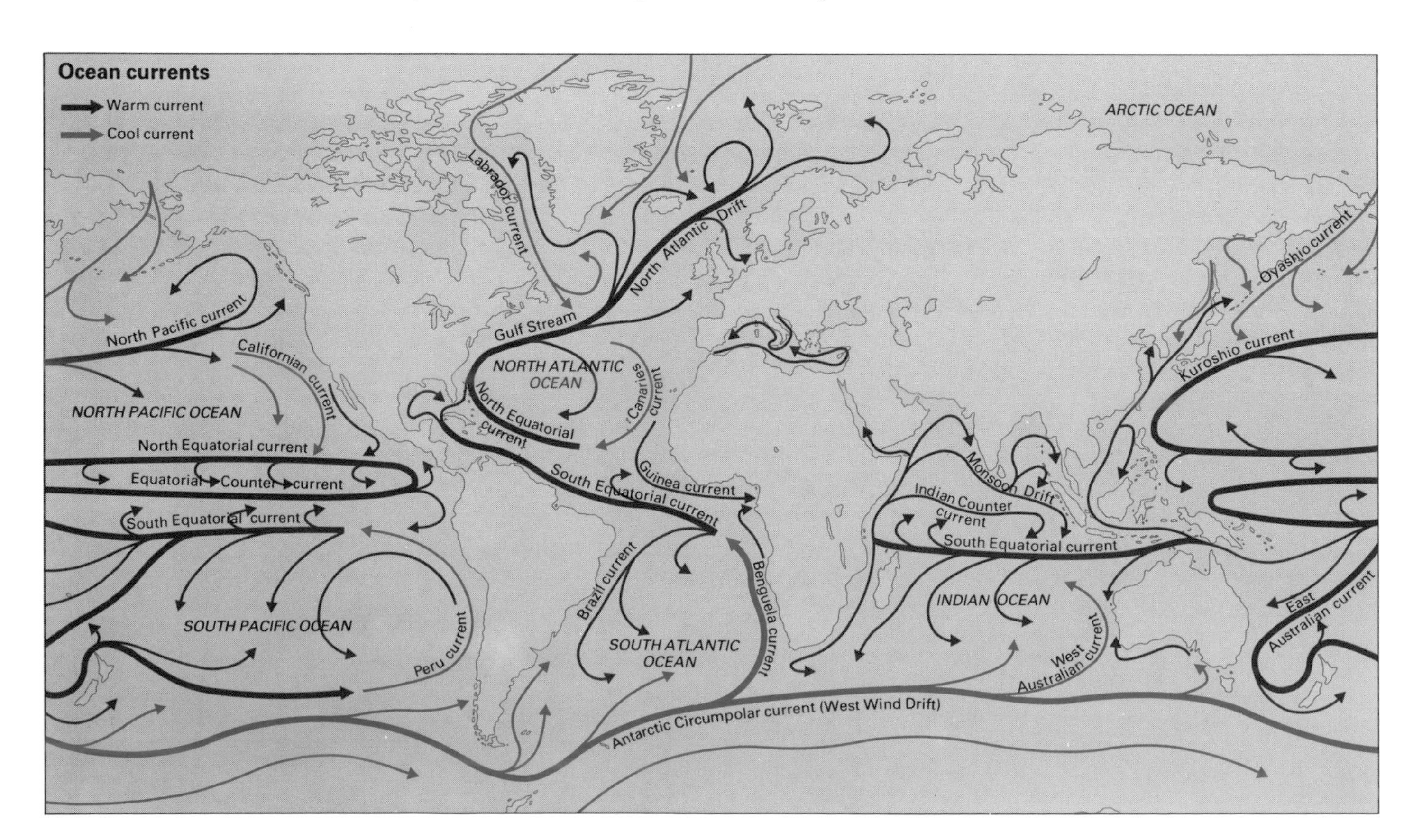

▼ *The generalized circulation pattern determined by the prevailing winds (◆ page 124) is severely disrupted by the presence of continents. The pattern in each ocean is based on the generation of large-scale gyres.*

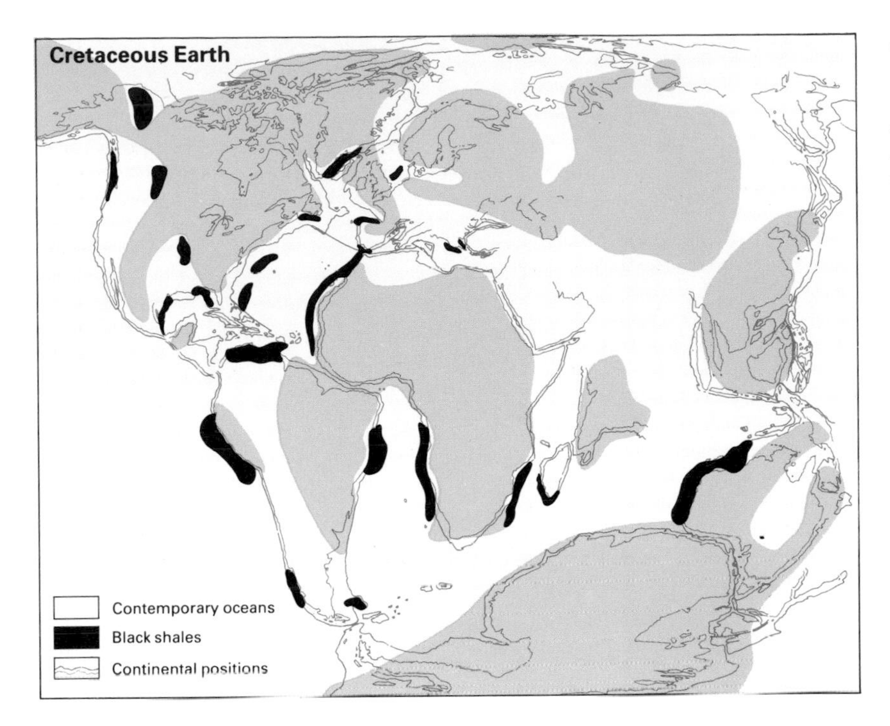

Cretaceous deep water – a source of oil

The temperature of Cretaceous deep water was 10-15°C instead of the present 2°, pointing to warm seas at high latitude. Evidently, the ocean stratification was less stable, and the deep circulation, driven by so small a density difference, must have been sluggish. This deprived the abyss of oxygen, limited organic decay, and caused intermittent deposition of black shales. These shales are the source beds for many of the world's richest oil fields. Ironically, the Cretaceous oceans, rich in potential oil, cannot have been fertile because the lack of organic decay limited recycling of nutrients. The abyssal flow was complicated further because, with such a small temperature range, saline water formed by evaporation in warm shallow seas may have been a major local source of deep water. The Mediterranean is an example. Here evaporation exceeds the influx of freshwater, so furnishing a brine of such density that, notwithstanding its 15° temperature, it spreads across most of the central Atlantic Ocean just above the abyssal water. An intricate pattern of abyssal environments may therefore have existed during the Cretaceous.

Evaporation
Atlantic
Mediterranean
Surface water
Bottom water

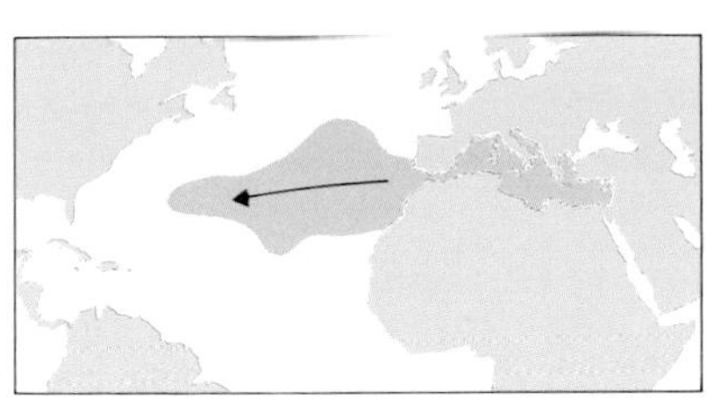

▲ ◄ In Cretaceous times the breakup of Pangea was under way, but the oceans had not developed to anything like their present sizes. There were no cold depths, and hence few convection currents. This meant that the deepest parts of the oceans were poor in oxygen and any organic remains deposited in them did not decay. Nowadays oil is sought in rocks that were deposited in the ocean depths in those times. The main currents at the time may have been caused by heavy salty waters flowing out of enclosed seas, as happens close to the Mediterranean today.

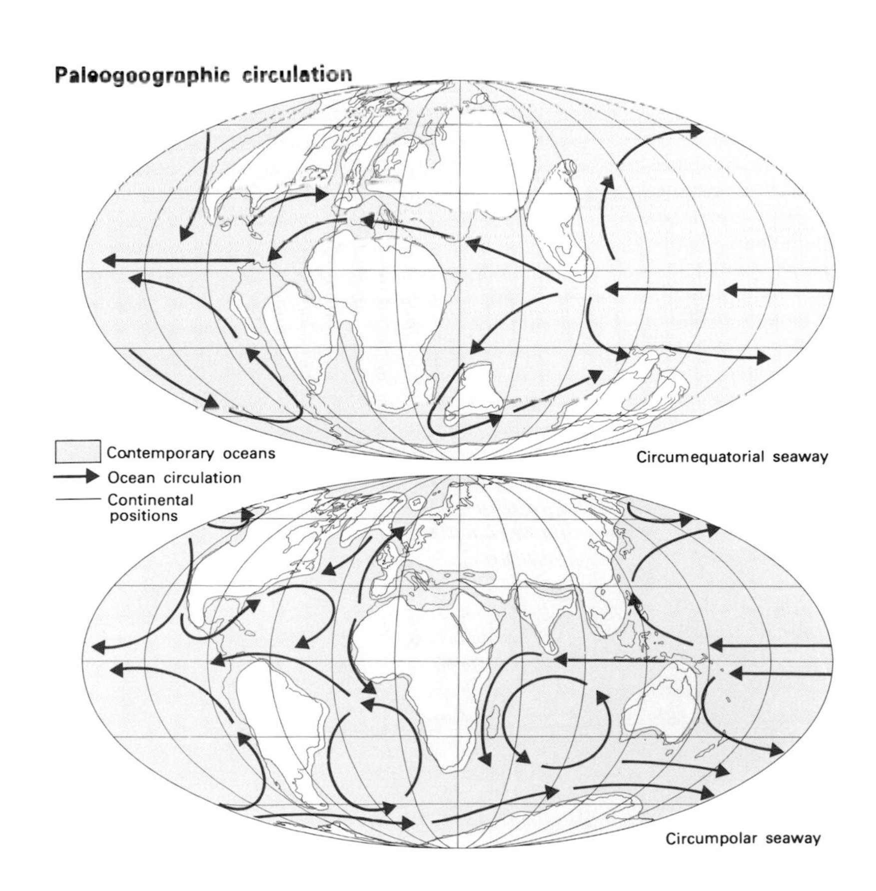

Oceans of the past

Continental drift continuously alters geography, closing and opening seaways, and its impact on climate is large although as yet only partly understood. With an open circum-equatorial seaway, the equatorial water will be warmer before being deflected north and south than is the case today. This reduces the temperature contrast between equator and poles, and the high latitudes will be temperate. A free path around the polar regions, on the other hand, creates a cold current which isolates the polar region from the warmer world. During the last 200 million years, the Earth has gone from a single continent to the current state of dispersed lands. Throughout the Mesozoic an equatorial seaway dominated the circulation and the higher latitudes were warm. In the middle Cenozoic, this seaway was closed in several places, and a southern circum-polar one opened. The result was a cold, soon ice-capped Antarctica, a steep latitudinal temperature gradient, and eventually the late Cenozoic ice age.

◄ The global circulation of seawater changed profoundly as the continents separated. As the northern part of Pangea – Laurasia – separated from the southern part – Gondwana – an equatorial seaway opened up around the world in the Mesozoic. This meant that warm water could spread round the globe, warming even the northern and southern oceans. By the middle Cenozoic several barriers to this flow had developed and, instead, a circum-polar current developed in the south. This had the effect of separating the waters of the cold pole from those of the warm equator.

See also
The Dynamic Earth 9-24
An Expanding Earth? 109-12
Development of the Atmosphere 113-20
Sedimentary Environments 157-72

▲ The first of these chemical oases was found in 1977 by the submersible ALVIN on the spreading ridge near the Galapagos Islands.

Hotsprings on the deep-sea floor

The new crust on ocean ridges dissipates much heat, mostly by conduction. Heatflow measurements, however, indicate an annual shortfall of 10^{16} kilocalories. The crust might also be cooled, analogous to an automobile engine, by circulating cold ocean water through fissures. The hot water would return to the surface as hydrothermal springs. The existence of such springs is implied by the presence in seawater of helium-3, an isotope of mantle origin. In 1977, the submersible ALVIN, guided by a minuscule temperature anomaly, discovered hydrothermal springs on a ridge crest east of the Galapagos Islands, and so confirmed their existence. Since then, the springs have been found to be common on ocean ridges. Their combined flow accounts for the entire shortfall of heat dissipation.

Satisfying as the confirmation of a hypothesis is, it was eclipsed by the unexpected discovery of a rich and complex spring fauna, unlike anything known from the deep-sea floor. The animals, ranging from primitive limpets to giant clams and even larger tubeworms, belonged to many new species, with special adaptations to filter-feeding and life in oxygen-poor water. Their food supply turned out to be based on the hydrogen sulfide abundant in the spring water. Bacteria, utilizing this compound as a source of energy, are the base of a food chain which includes filter-feeders, scavengers and rare predators – a fascinating, unique exception to the dominant role of photosynthesis. The tubeworms even possess tissue housing sulfide-reducing bacteria, an internal food supply analogous to the chloroplasts of green plants.

▼ Around the hot springs on the sunless ocean ridges, the dissolved chemicals feed bacteria. These feed other creatures giving rise to a rich fauna.

Climate and Climatic Change 13

Atmospheric circulation – the climate generator... Complicating factors...Climates of the past – ice ages...Reasons for climatic change – atmospheric dust and an inconstant Sun...Human influence – air pollution...PERSPECTIVE...Seasons...Why climates change...Viking voyages...Dust veil index...Nuclear winter

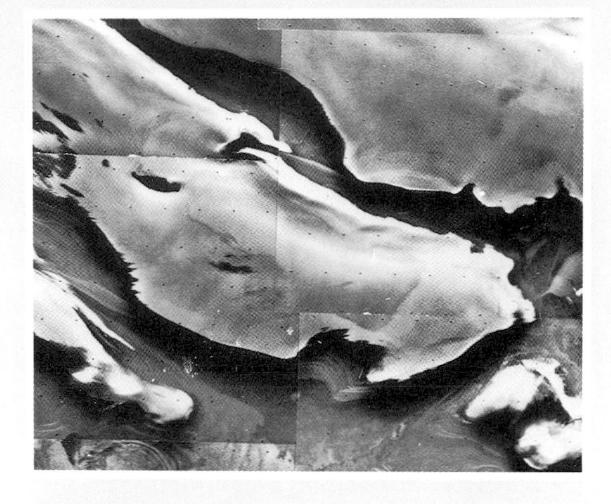

▲ *The Martian weather can be seen by the changing icecaps, and the winds that raise duststorms.*

▼ *The climate of Venus is hard to imagine, pressure and heat being so different from Earth.*

Why is our planet Earth different from its nearest planetary neighbors in space, Venus and Mars? The answer is obvious – Earth carries life, while Venus and Mars are sterile. And why does it carry life? It does so because of the presence of large quantities of free water, the oceans that are such a distinctive feature of our planet.

Venus, Earth and Mars all have atmospheres, and their atmospheres may well have originated in the same way, possibly by outgassing from volcanoes (◀ page 118). In each case, the original atmosphere must have consisted of a mixture rich in carbon dioxide and water vapor. On Venus, closer to the Sun than we are, and hotter, the water stayed as vapor, and, together with the carbon dioxide, trapped heat from the Sun (the "greenhouse effect"), sending surface temperatures soaring still farther. On Mars, a little farther from the Sun than Earth is, the water has frozen, leaving a thin, lifeless carbon dioxide atmosphere.

On Earth, however, the water could form oceans, streams and lakes, in which life could emerge, and life in its turn transformed the atmosphere, converting carbon dioxide into oxygen and paving the way for life to progress from sea to land. Those same oceans which provided the initial home for life are the key to an understanding of the climate of the Earth. They dominate the weather machine. They provide the moisture which forms clouds and falls as rain and snow, and they carry heat from the equator to the poles.

► *The Earth's climate is very variable. The continents heat up more quickly than the oceans, and different parts of the continents absorb and lose heat at different rates. Air moves constantly from areas of high pressure to areas of low. Moisture in the atmosphere varies from place to place, and can be transported around by the air currents. All these factors produce the weather, and it is the average of all the weather conditions over a long period of time that defines the climate of any area.*

▲ Polar climates are characterized by constant cold weather. At the poles there tend to be areas of high pressure, where dry air descends and spreads outwards. The high latitude means that the Sun has little warming effect, each ray having to travel far through the atmosphere.

◄ Temperate climates are, as their name suggests, neither searingly hot nor bitingly cold. The change of seasons means that in the winter the main climatic influence is from the polar cold area, while in the summer it is the tropical air masses that affect the weather.

► Desert climates are dry, and usually hot. They occur in latitudes where global circulation brings dry air down from the upper reaches of the troposphere, in continental areas far from the sea, or where the winds have dropped their moisture on intervening mountain flanks.

►► Equatorial climates are hot and wet. Constant sunshine over the tropical continents causes the warm air there to rise, forming a low pressure area and sucking in wet winds from the north and south. The relentless heat and rainfall produce the luxuriant tropical forests.

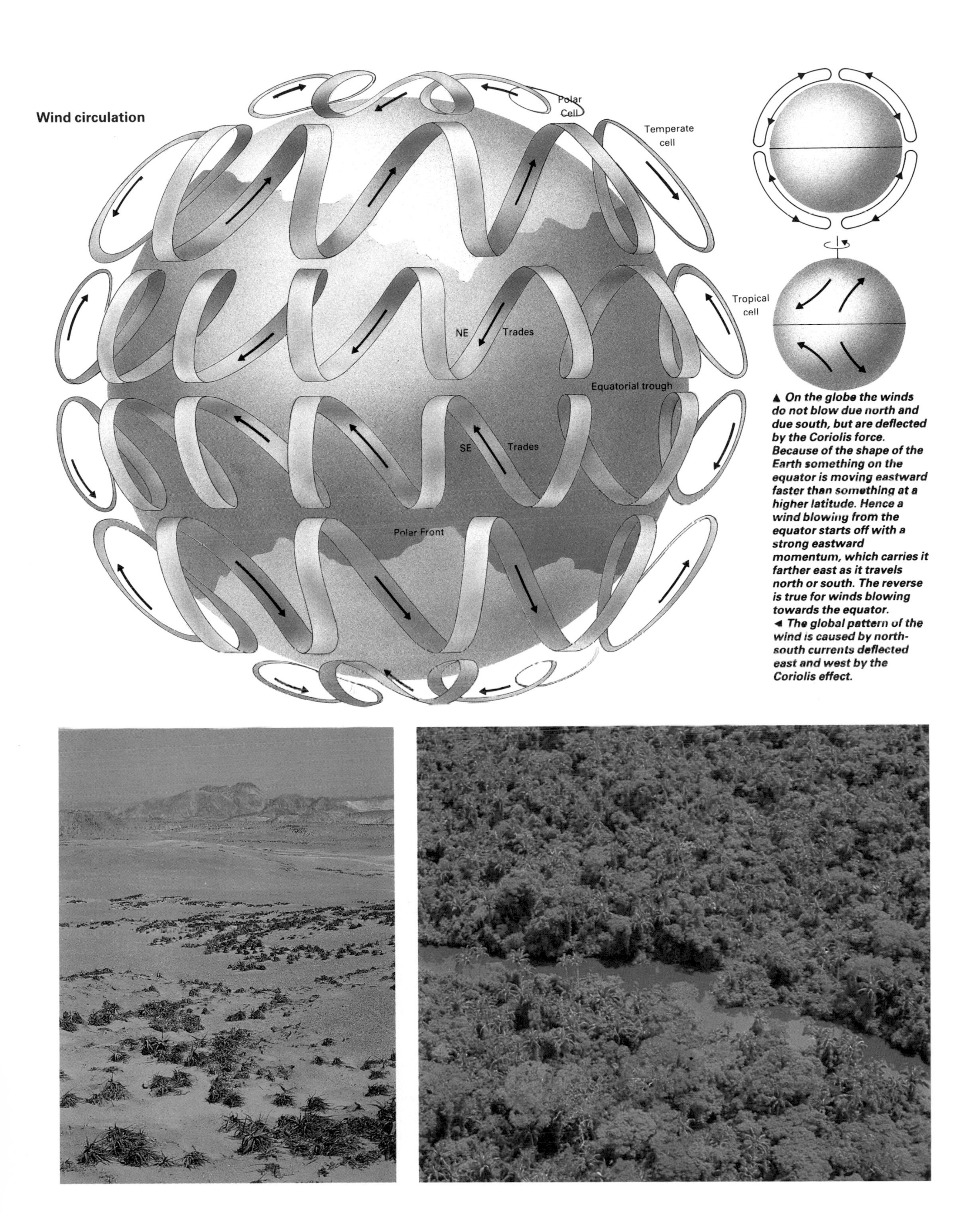

▲ *On the globe the winds do not blow due north and due south, but are deflected by the Coriolis force. Because of the shape of the Earth something on the equator is moving eastward faster than something at a higher latitude. Hence a wind blowing from the equator starts off with a strong eastward momentum, which carries it farther east as it travels north or south. The reverse is true for winds blowing towards the equator.*

◄ *The global pattern of the wind is caused by north-south currents deflected east and west by the Coriolis effect.*

The seasons

Imagine a line joining the center of the Earth to the center of the Sun, and another line through the Earth from pole to pole. The angle between these two lines is not a right angle, 90°, but about 23·5° less. At one end of the Earth's orbit around the sun, the North Pole is, as a result, tilted this much towards the sun, and the Northern Hemisphere experiences summer. But as the Earth moves in its orbit, the direction of this tilt of the Earth stays fixed in space. Six months later, the Northern Hemisphere is tilted away from the Sun and it is midwinter; in the south, however, it is now midsummer. In between these extreme states, during fall and spring, the tilt is neither towards nor away from the Sun, and all regions of the globe get the same amount of light and heat.

Because of these seasonal changes, the climatic boundary between the Northern and Southern Hemispheres shifts during the course of the year. This meteorological equator, the Intertropical Convergence Zone, is the region where tradewinds from north and south of the equator meet. As it shifts, all of the climate zones in one hemisphere are squeezed, while those in the other expand.

▼ The Sun shining on the Earth determines the overall wind circulation. As the tilt of the Earth varies from one part of the year to another, the Earth appears to move up and down beneath this circulation pattern and presents a slightly different face to the Sun each season. Hence any part of the Earth's surface comes under the influence of different wind systems at different times of the year.

The climate of any particular region of the Earth is a measure of the "average weather" you are likely to experience there. Is it likely to rain in May or in September? Will the summer be hot and dry? Will the winter be frosty or mild? If the same kind of weather repeats, most years, at about the same time of year, then that is part of the region's climatic characteristics. Climate itself may change – during the latest Ice Age, the climate of much of Britain, Europe and North America was much colder, year in and year out, than it is today. A snapshot view of present day climate is revealed by what happens to the air set into motion in the equatorial region as a result of solar heating.

The circulation of air

The circulation patterns that define weather and climate zones on the surface of the Earth take place almost entirely within the bottom 11km of the atmosphere, the troposphere, or weather layer (◆ pages 114-5). Convection is limited to this layer of atmosphere because higher up there is a layer, the stratosphere, in which the atmosphere is hotter than the upper troposphere, because oxygen in the stratosphere absorbs ultraviolet energy from the Sun. Hot air only rises if the air above is colder.

A great deal of the Sun's heat near the equator goes not just into warming the air at ground level, but to evaporating water from the oceans, so the hot, rising tropical air is also moist. As the air rises it cools, and deposits this burden of moisture as rain. So the tropics are wet and covered with vegetation. This climate zone covers a band from about 5° south to 10° north, the difference being due to the uneven distribution of land between the two hemispheres. The air

The climate of a hypothetical continent

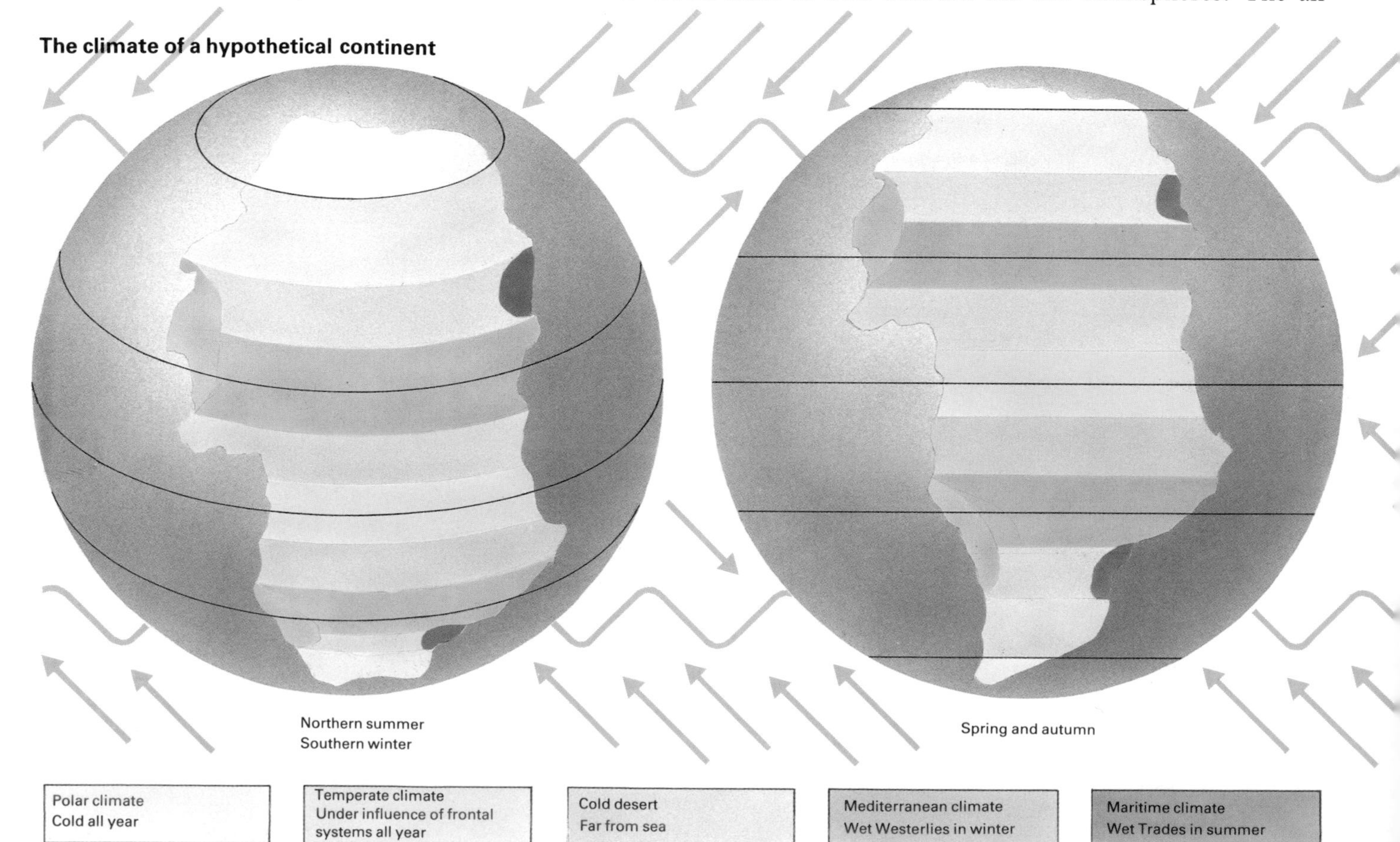

lifted from the equatorial regions by convection is pushed sideways by more rising air, and then returns to the surface of the Earth in a band of latitudes beyond about 20° on either side of the equator. That air then returns to the equator at surface level, completing one cycle as the northeast and southeast tradewinds, which blow into the tropical zone from about 20° north or south of the equator.

Beyond the belt of the trade winds, the weather systems are dominated by the descending air that has risen from the tropics, dropped its moisture as it cooled, and been pushed sideways by more rising air. When this air descends, it is already dry. As it falls, it warms up because pressure is increasing – the process is exactly the same as the way a bicycle pump gets hot when you use it – and this increases its ability to store moisture. So on the other side of the tradewind zone from the equator each hemisphere is dominated by hot, dry descending air which forms the subtropical high pressure systems. This zone is the region of our planet dominated by deserts the Sahara, the central Australian desert, and others.

In the tropics, weather is predictable and reliable – hot and wet. In the desert region, it is equally predictable – hot and dry. But in between there are variations, because the whole pattern of climate zones shifts north and south with the seasons. So there are regions, such as in northwest India, where the rainfall comes with tropical strength, but only for part of the year. These seasonal rains are called the monsoon. Relatively small changes in climate can shift a fertile monsoon region into the desert region, with disastrous consequences for its inhabitants. This seems to be happening today in the Sahel region of Africa, south of the Sahara desert.

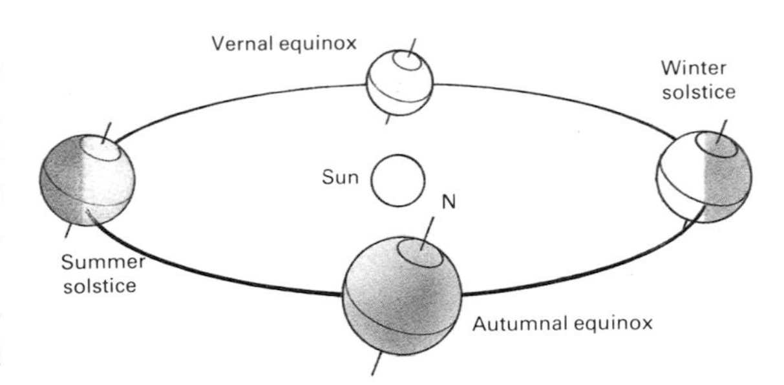

▲ The Earth has a constant tilt of 23·5°. As the Earth orbits the Sun, the North Pole is tilted towards the Sun by this angle at summer solstice, and away from it at winter solstice.

▼ As a result different parts of the Earth see the Sun take different paths around the sky. At the poles the path is parallel to the horizon, while at the equator it is vertical.

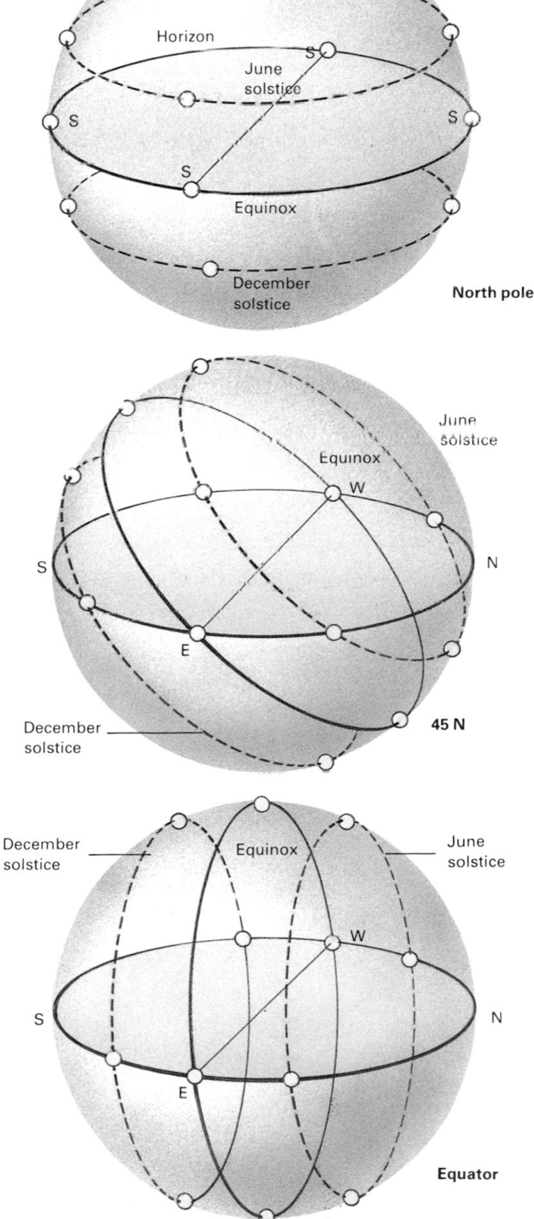

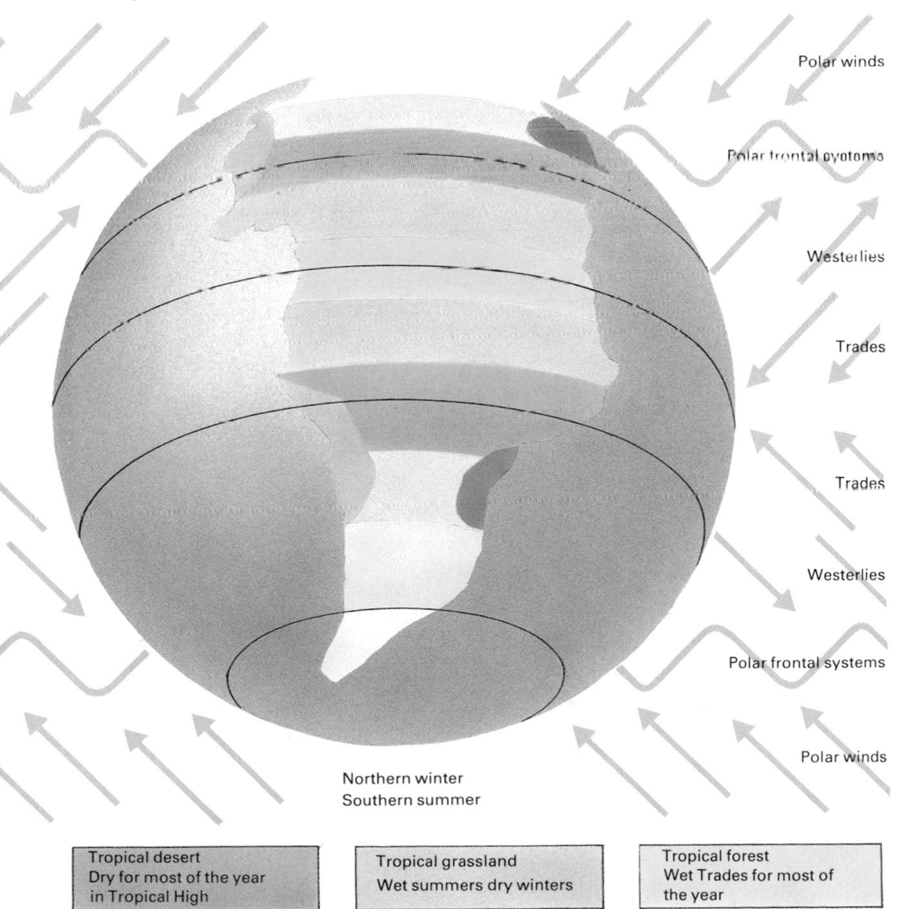

Westerly winds dominate the weather pattern in the first zone of climate poleward from the hot desert zone. This is called the Mediterranean zone, because the Mediterranean itself is the archetype for this climate. Such regions lie between 30 and 40° latitude, and they are dominated by hot, dry summers and warm, wet westerlies in winter. A Mediterranean climate only occurs on the west of a continent, because the winds are too dry to produce much rain by the time they reach the east of the landmass. Apart from the Mediterranean itself, the other classic "Mediterranean" climate is found in northern California. The best wines tend to come from regions with this weather pattern.

Farther to the north or to the south, the difference between western seaboards and continental interiors is equally marked. The zone from about 40 to 60° in the Northern Hemisphere and 35 to 55° in the Southern is called "temperate" because the name was coined by west Europeans. But the continental heartlands in this latitude band are now described as having "continental" climates, with very cold winters and very hot summers, even though they may be in the temperate zone. Siberia is the classic example. In the temperate regions proper (western Europe, New Zealand) prevailing winds off the ocean keep temperatures from falling too low in winter or rising too high in summer. This makes for an ideal climate for agriculture.

At higher latitudes still, there are the sub-polar regions, dominated by snow and ice. Like the dry, hot deserts nearer the equator, these are regions dominated by descending dry air forming part of another convective cell system. The temperate zones, between the polar cold and the more equable Mediterranean climates, are in an analogous position to the monsoon regions, caught between two climate zones. A small shift can bring either sub-polar cold or sub-tropical warmth to the temperate regions, which is why they have such changeable weather patterns.

Complicating factors

The pattern is further complicated by the rotation of the Earth, so that instead of blowing straight from north to south, or south to north, the prevailing surface winds are given a sideways twist. At the latitude of North America and Europe, this makes the prevailing winds blow from the west, bringing moist air off the oceans and onto the British Isles and Oregon, for example, while leaving the heart of North America relatively dry in the lee of the Rocky Mountains.

Local weather and climate patterns are also affected by geography. Where the winds blow across a mountain range, they will be forced to rise and cool, dumping their burden of moisture as rain or snow. So regions in the lee of mountains, such as the North American plains, tend to be drier than their latitude would suggest. Britain, dominated by westerlies, is much wetter than Siberia, reached by those same westerlies only after they have traveled a long way over land.

With all these factors to consider, the definition of climate as the "average weather" that a particular region is likely to experience, over a long time, is one fraught with uncertainties. Does a particularly severe winter, or a prolonged summer drought, signify an abnormal fluctuation of the weather, or is it just one of those things that ought to be expected from time to time? Whenever any extreme weather conditions occur, anywhere in the world, some experts claim them as evidence of a changing climate, while others will say that they are normal fluctuations, part of the established long term pattern.

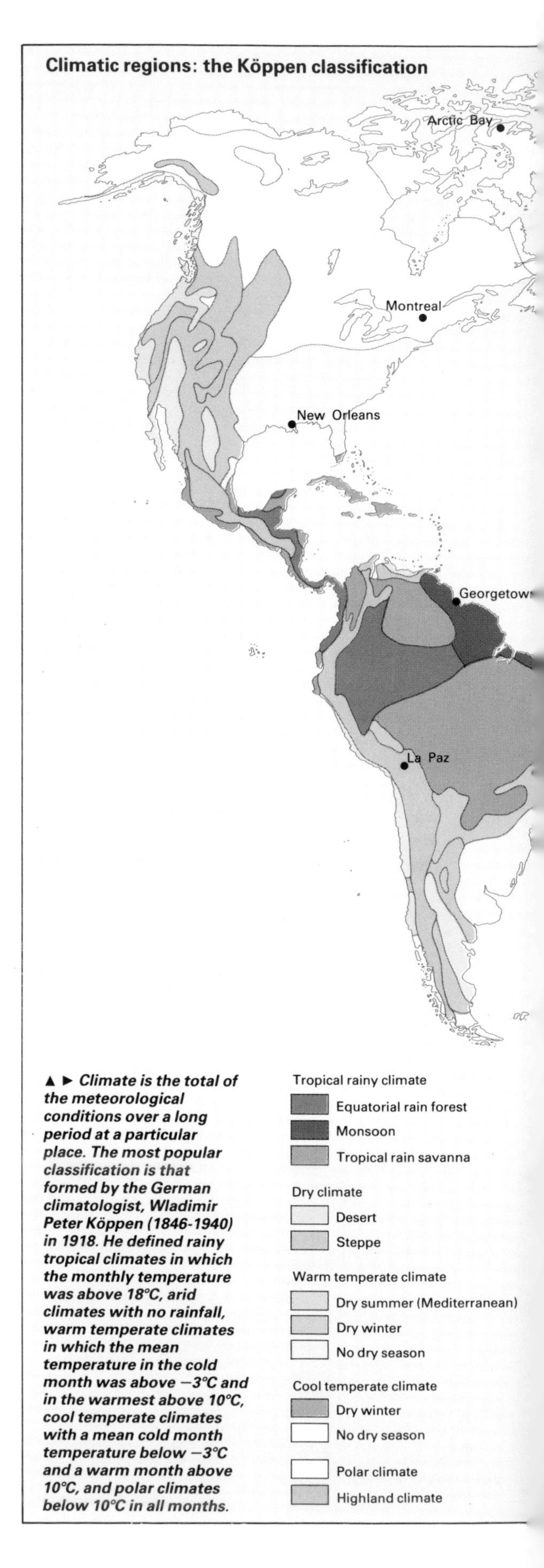

▲ ► Climate is the total of the meteorological conditions over a long period at a particular place. The most popular classification is that formed by the German climatologist, Wladimir Peter Köppen (1846-1940) in 1918. He defined rainy tropical climates in which the monthly temperature was above 18°C, arid climates with no rainfall, warm temperate climates in which the mean temperature in the cold month was above −3°C and in the warmest above 10°C, cool temperate climates with a mean cold month temperature below −3°C and a warm month above 10°C, and polar climates below 10°C in all months.

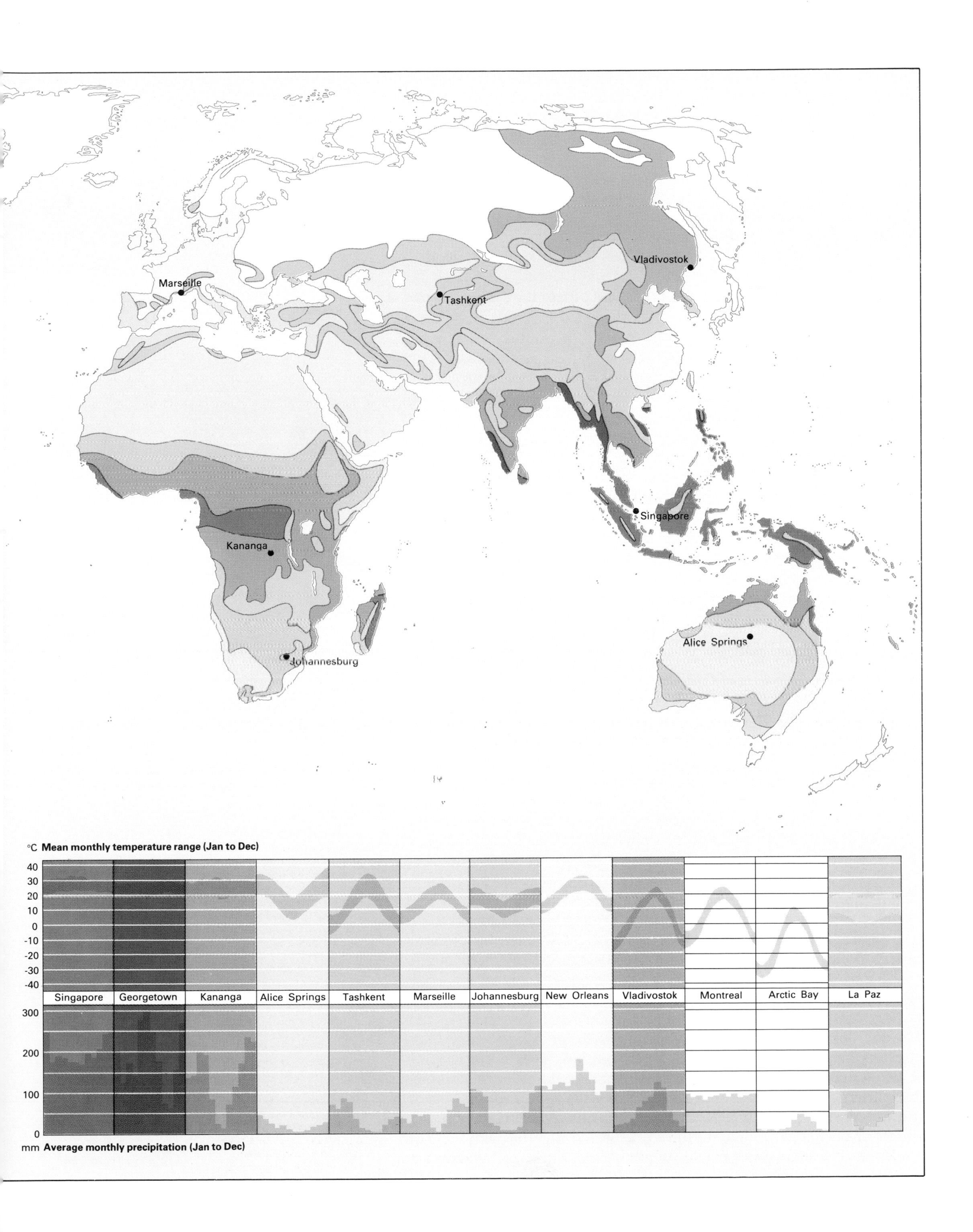
Marseille
Tashkent
Vladivostok
Kananga
Singapore
Johannesburg
Alice Springs
°C Mean monthly temperature range (Jan to Dec)
40
30
20
10
0
-10
-20
-30
-40
Singapore
Georgetown
Kananga
Alice Springs
Tashkent
Marseille
Johannesburg
New Orleans
Vladivostok
Montreal
Arctic Bay
La Paz
300
200
100
0
mm Average monthly precipitation (Jan to Dec)

Climates of the past – the ice ages

Moving away from the kind of fluctuations that occur in a human lifetime, there are genuine, global changes in the climate of our planet, climatic changes on which all of the experts agree.

Taking the longest perspective of all, looking back 4·6 billion years to the formation of the Solar System, the earliest climatic patterns must have been very different from today. The Sun itself was cooler than it is now, and the atmosphere of our planet was very different from its present composition, much richer in carbon dioxide and perhaps carrying a significant trace of ammonia. Both of these gases contribute to the greenhouse effect, which prevents heat from the ground escaping into space. Together, they may have counteracted the lower solar temperature to keep the waters of the Earth from freezing. But really very little is known about the climate of the Earth during the first 80 percent of its life. Geological records only begin to provide detailed information for the past 500 million years.

The evidence comes from the remains of animals and plants, and the nature of sediments and erosion patterns from different times. For example, during the Jurassic Period from 213 to 144 million years ago, many parts of the world experienced a cool, wet climate during which the great dinosaurs flourished. At other times the Earth's surface has been scoured by the glaciers of ice ages. The first great puzzle of climatic change is why such ice ages should occur, and although the details remain to be filled in, the broad outline of the picture emerged in the 1960s and 1970s, with the establishment of the theory of plate tectonics.

The geography of the globe is constantly changing, and the broad pattern that we think of as normal – the present day distribution of the continents – is, in fact, highly unusual. We think it is normal to have polar ice caps, and yet, throughout the past 4·6 billion years, it has been rare for even one of the poles to be covered by ice. The reason comes back to that dominant feature of our planet, its large oceans of water. Water stores and transports heat much more effectively than air does, and an ocean current can carry warmth to high latitudes far more effectively than the winds can. Normally the warm water from the tropics easily penetrates up to the poles, and keeps them free from ice. But from time to time a large land mass drifts across one of the poles and blocks the flow of warm water. When that happens the situation is similar to Antarctica today. The frozen continent is locked in the grip of ice until continental drift carries it away to lower latitudes once again and lets water in to defrost the pole.

This is, almost certainly, the way in which all the great ice ages of the past were produced. The ice did *not* cover the whole Earth, freezing continents even at low latitudes. Instead, the continents themselves moved to the coldest parts of the globe, and froze accordingly.

But there is another way to freeze a polar region, a much more rare geographical pattern which has existed for the past few million years in the Northern Hemisphere, but may never have occurred on Earth before. The Arctic Ocean is itself almost surrounded by land masses. Although there is still a polar sea, there is no way for the warm water from the oceans farther south to get in to the Arctic basin, and as a result the ocean is so cold that it carries a thin layer of ice. The shiny white surface of the ice reflects away heat from the Sun, and helps to keep the pole frigid, while the cold influence of the icecap extends on to the surrounding land masses, bringing Arctic winters to Siberia, Alaska and Canada.

► The sequence of cold and warm spells during the Pleistocene can be detected in the sediments of the time. Some microscopic foraminifera coil to the left in cold periods, but coil to the right in the warm. The proportions of the two forms in ocean sediments indicate contemporary temperatures. Proportions of oxygen isotopes (below) show about 18 cold periods during the last 1·6 million years.

Left-coiling *Globorotalia truncatulinoides* from cold water

Right-coiling *Globorotalia truncatulinoides* from warm water

Warmer

Colder

Age (Mya)

0 0.4 0.8 1.2 1.6 2.0

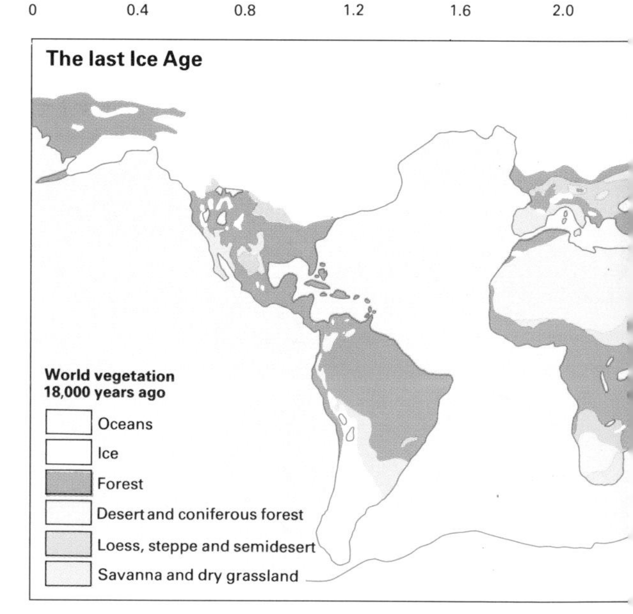

A model for climatic change

The whole suite of climate changes can be explained by the Milankovitch Model, named after the Yugoslav scientist who developed the first complete version of the theory between the two World Wars. His calculations – and modern refinements of them – depend both on the geometry of the Earth's orbit and the tilt of its axis. An imaginary line drawn through the Earth from pole to pole does not make a right angle with a line from the equator to the Sun. That is why we have seasons – when one hemisphere is tilted towards the Sun it receives more warmth, and it is summer; when it is tilted away, the high latitudes are cold and dark, it is winter.

But the amount of this tilt is not constant; the Earth wobbles as it orbits the Sun, rather like a child's spinning top wobbling on the floor, but in a much more stately fashion.

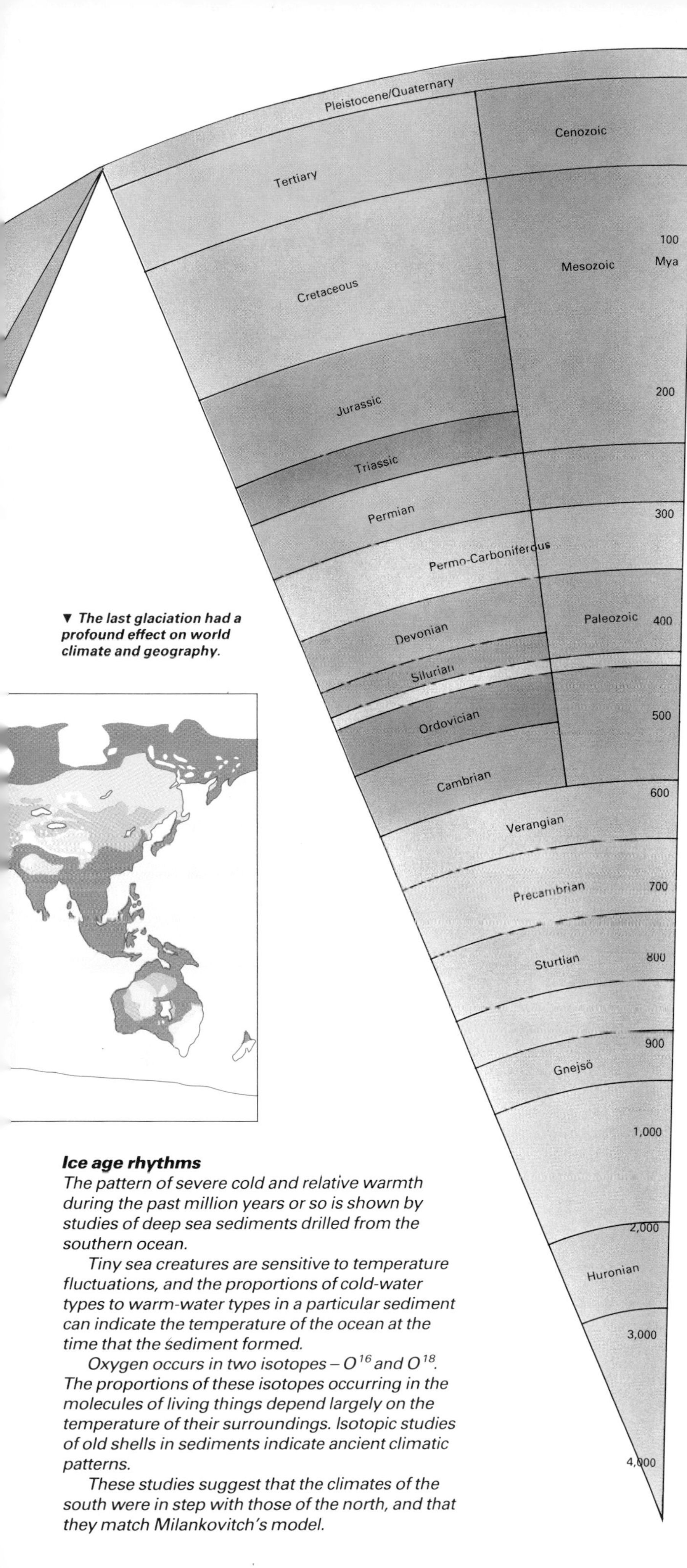

▼ ***The last glaciation had a profound effect on world climate and geography.***

Ice age rhythms

The pattern of severe cold and relative warmth during the past million years or so is shown by studies of deep sea sediments drilled from the southern ocean.

Tiny sea creatures are sensitive to temperature fluctuations, and the proportions of cold-water types to warm-water types in a particular sediment can indicate the temperature of the ocean at the time that the sediment formed.

Oxygen occurs in two isotopes – O^{16} and O^{18}. The proportions of these isotopes occurring in the molecules of living things depend largely on the temperature of their surroundings. Isotopic studies of old shells in sediments indicate ancient climatic patterns.

These studies suggest that the climates of the south were in step with those of the north, and that they match Milankovitch's model.

▲ ***James Croll, who saw the astronomical connection***

Pioneer of ice age theory

The Scot James Croll, born in 1821, was the pioneer who first developed a detailed astronomical theory of ice ages, the forerunner of the Milankovitch model. Too poor to attend university, he worked as a millwright and as a carpenter, devoting much of his effort to reading rather than to his work. A failure also in business, after some time as an insurance salesman and a period of unemployment, in 1859 he found the ideal occupation, as janitor in the Andersonian College and Museum in Glasgow. The job left him plenty of time to read his beloved books. It also gave him access to one of the finest libraries in the land, and he later commented, "I have never been in any place as congenial".

Studying more intensively than ever, he began to publish scientific papers about ice ages in 1864. This eventually led to an academic appointment, and a book "Climate and Time", in 1876. Election as a Fellow of the Royal Society duly followed.

The heart of Croll's work is the realisation that changes in the geometry of the Earth's orbit alter the balance of heat between the seasons. In the 19th century there were no adding machines to aid his work, which involved laborious calculations of the way in which the changes in the Earth's wobble through space alter the balance of heat throughout the year. It involved paper and pencil calculations of the forces acting not just between the Earth, Moon and Sun, but also the influence on the Earth's orbit the gravity of the other planets. Croll's Royal Society Fellowship was hard earned and merited.

◄ ***The Pleistocene ice age, occurring in the last two million years or so, is often thought of as the only ice age of any significance. In fact ice ages have occurred throughout geological time. There were at least four during the Precambrian, one in the Ordovician and a particularly important one in the Pennsylvanian and Permian. These have left their traces in the rocks of the time in the form of tillites – fossilized beds of glacier-transported material – and rocks that have been scratched and polished by the passage of glaciers across them.***

In 1683 the ground in southern England was frozen to a depth of more than a meter, and the English Channel had belts of pack ice 5 kilometers wide

▲ There are many eye-witness accounts of the changes in climate. The "Little Ice Age" in Europe, which began in the 13th century, was particularly severe in the 17th. Paintings of the Alps made about this time show the glaciers extending much farther down the valleys than they do today. The freezing of the River Thames was a regular event, and the ice was so thick that fairs could be held upon it.

Fifty million years from now, the steady widening of the Atlantic Ocean at 2cm per year will have "opened the gate" for warm water to penetrate the Arctic basin and defrost the North Pole. Meantime, however, we are still subject to the unusual geography which has made for this repeating pattern. The key factor is the presence of a frozen pole surrounded by plenty of land on which snow, if it does fall, can settle and build up white sheets to reflect heat away into space. Under such conditions, even small changes in the amount of heat reaching the ground can have a profound effect on climate, and it is these particular changes that drive the ice age rhythm.

Reasons for climatic change

Why should the climate change on these timescales of decades and centuries? There are two main possibilities, and both may be involved. First, there is no doubt that large volcanic eruptions contribute to short-term cooling of the globe. They throw dust and sulfur dioxide gas high into the stratosphere, where it forms a fine haze blocking some of the heat from the Sun and preventing it from reaching the ground. In 1815, a huge eruption in Indonesia was responsible for so much veiling effect that a year later there was virtually no summer in New England and western Europe. In parts of England, temperatures that summer were 3°C below normal. In New England snow fell in

Viking voyages

Oxygen isotopes from the layers of the Greenland glacier reveal the fine climate fluctuations over the past two thousand years, and explains the fate of Viking voyagers around the north Atlantic.

The first recorded attempt to settle Iceland failed, in AD 865, just at a time of cold fluctuations revealed by the ice core, and this failure gave the country its name. Just nine years later, as the climate improved, a successful colony was established. In AD 985 it sent its own settlers west to Greenland, and named it in their first flush of enthusiasm during a minor climatic optimum.

But it was not to last. In the 13th century the pattern changed, and the cold of the Little Ice Age began to set in. By the 14th century the Greenland colony had collapsed; the Iceland colony itself survived the following centuries only because of its value as a base to whalers.

The dust veil index
Professor Hubert Lamb of the University of East Anglia in Norwich, England, has compiled a formula, called the Dust Veil Index, that relates the climatic influence of volcanoes through the centuries to the amount of material that their eruptions blow into the stratosphere. He divides volcanoes into two categories – those that send dust to the lower stratosphere, 20-27km above the Earth's surface, and those that send dust to much greater altitudes, up to heights of 50km. The more important group, from the point of view of the Earth, is the first. Very great amounts of material can be sent to these altitudes by huge eruptions like those of Krakatau (sometimes spelled Krakatoa) in the East Indies in 1883 and El Chichon in Mexico in 1982. These produce a dense and long-lasting veil of both dust and fine droplets of sulfuric acid, the latter formed when sulfur dioxide gas reacts with the water of the atmosphere.

Professor Lamb's studies show that after two major eruptions in 1783 the average temperature over the Northern Hemisphere fell by nearly 1·5°C; the famous eruption of Krakatoa is taken as the benchmark for his Index, however, being given an arbitrary rating of 1,000 on the Dust Veil Index. All other eruptions can be measured relative to this. The index shows that the relative warmth of the early and middle 20th century was at least partly due to a clearing of the atmosphere during a period of volcanic quiet. The indications are that this has now come to an end, and that average temperatures may fall.

◄ Dust blasted into the sky by volcanic eruptions may be responsible for short-term climatic changes. The eruptions from the crater of El Chichon in the Yucatan in 1982 spread 16 million tonnes of pulverized rock into the atmosphere. The veil of dust so formed absorbed some of the Sun's warmth and led to a lowering of temperatures.

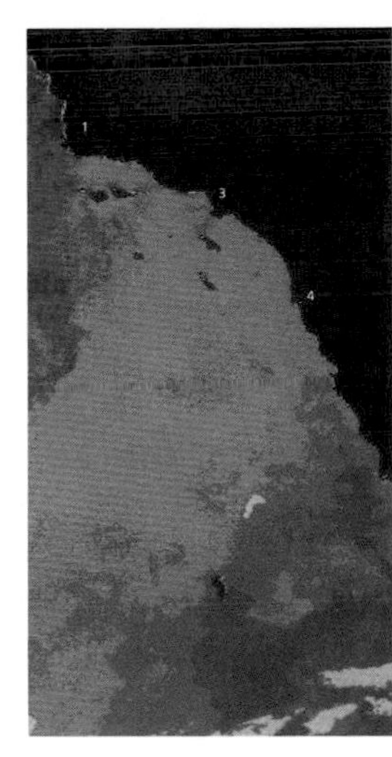

▲ In 1982 El Chichon's eruption disrupted the trade winds and halted the upwelling of cold water along the west coasts of the Americas (***◄ page 124******), bringing abnormal weather conditions to these areas. The false color images show the normal cold water (green) off southern California in January 1982 (left). In January 1983 (right) the waters are abnormally warm, as shown by the red color.***

June and there were frosts every month of the year.

It takes either one very large volcanic eruption, or a succession of lesser ones, to have such an effect on weather worldwide. The British climatologist, Hubert Lamb, of the University of East Anglia, has made a lifetime study of these links between volcanic activity and climate, showing that much of the worst weather of the Little Ice Age (1645–1715) can indeed be linked to the changing volcanic dust veil. And, indeed, the relative warmth of the 20th century did "coincide" with a period of volcanic calm, worldwide. The volcanoes started erupting again in the 1960s, and have kept active since coinciding with a downturn in the weather.

The other possibility is the fluctuation of the Sun itself. A change of only one percent in the amount of heat reaching the Earth from the Sun could change average temperatures by a full degree (Celsius), and astronomers have recently been forced, reluctantly, to take this possibility seriously. There is evidence, from studies of old astronomical records, that the Sun actually changes in size, breathing in and out by a tiny amount with a rhythm of about 180 years, a rhythm that is also found in the climatic records. Such studies are at the cutting edge of current research, and cannot be taken yet as proof. But, for what it is worth, such evidence as there is points, once again, to a cooling of the globe as we enter the 21st century.

See also
Earthquakes and Volcanoes 25-32
Development of the Atmosphere 113-20
Development of the Oceans 121-8
Landscape Features 181-96
Changing the Landscape 213-28

The human effect on climate

For the first time, human activities may be beginning to alter the natural climatic state of our planet. The problem is related to a build-up of carbon dioxide gas in the atmosphere, a direct result of burning fossil fuel and destroying the tropical forests. Carbon dioxide is a trace gas about 0·03 percent, or 300 parts per million (ppm). Since the industrial revolution of the 19th century, however, this proportion has risen from about 280ppm to nearly 350ppm. This is important because carbon dioxide is very good at trapping infrared radiation. When the Sun's rays (which are *not* absorbed by carbon dioxide) warm the surface of the Earth, that surface (land or sea) radiates heat outwards at infrared wavelengths. Any that is trapped by carbon dioxide keeps the surface of the Earth warmer than it would otherwise be – hence the term "greenhouse effect". There is no doubt that this greenhouse effect must be occurring as the build-up of carbon dioxide proceeds, although there is considerable doubt about how big the effect will be and how soon we may notice it. Most climatologists agree, however, that by the time the natural proportion of carbon dioxide in the air has doubled, which might happen late in the 21st century, the world may have warmed, on average, by about 2°C.

That does not sound much, but this average warming may disguise a tiny change in the tropics and an increase of as much as 8-10°C near the poles. That is the basis for newspaper scare stories about the "threat" of the greenhouse effect, melting ice caps, flooded coastal regions and the rest. But there are important caveats.

First, any pronounced greenhouse effect lies a generation or more ahead, giving us time to take action to avert the damage or ameliorate it. Second, in view of all the evidence that the Little Ice Age may be about to return in full force, perhaps a little greenhouse effect may be no bad thing. And, finally, the farmers and agriculturalists have pointed out that carbon dioxide is actually good for crops, encouraging them to grow and making them more efficient in their use of water. The study of changing climate shows that it is indeed, *always* changing, on many timescales, and that the best mankind can hope to do is to learn to predict the changes and bend with them.

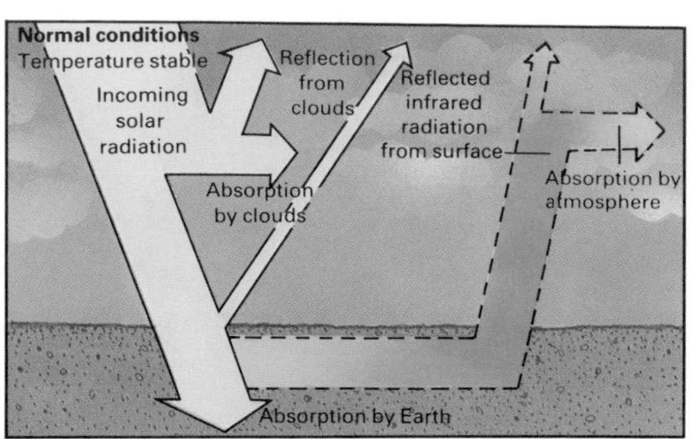

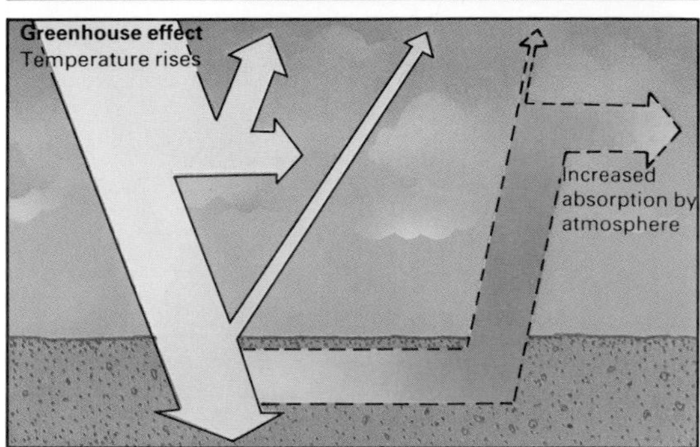

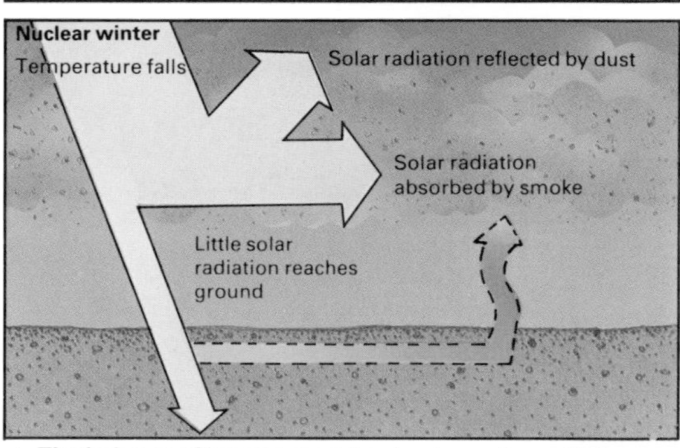

▲ ***The balance of incoming and outgoing solar radiation may be artificially disturbed by added carbon dioxide producing a hot "greenhouse effect", or by dust and smoke from nuclear warfare producing a cold "nuclear winter".***

▲ ***Pollutants are constantly entering the atmosphere.***

Nuclear winter

A particularly dramatic alteration of the Earth's climate by human interference, could be the result of a nuclear war. Following an exchange of nuclear missiles between the superpowers the pall of smoke and dust from the direct effects could cool the Earth's surface and produce what scientists have come to call a "nuclear winter". The pall would consist of dust raised by the initial explosions, and of smoke generated by burning cities and forests. The dust, being a light color, would scatter back the rays of the Sun, the smoke and soot would absorb much of the Sun's heat. The result would be a lowering of temperatures beneath the cloud by between 10°C and 20°C for many months. Coastal areas would remain relatively warm because of the effect of the water, and fogs and intense storms would result from their interaction with the bitter cold of the twilit continents. The resulting cold and darkness which would last for some considerable time would kill off many growing crops and the effect on human populations would be far greater than the direct effect of the war itself.

Rocks and the Rock Cycle

Rocks – the crust's raw materials...The rock cycle – their destruction and regeneration...Weathering – rock decay...Lithification – sediments into sedimentary rocks...PERSPECTIVE...Magma... Igneous rock types and textures...Metamorphic rocks and their origin...Sedimentary rocks and their raw materials

Modern geologists see the Earth's surface layer as part of a dynamic system, with drifting continents, spreading ocean floors and interacting lithospheric plates. But long before these phenomena were discovered – as long ago as the early 19th century, in fact – geologists came to recognize that the material at the Earth's surface is part of a dynamic system of a different sort. For rocks are continually being created, altered, broken down and recycled. The surface environment is perpetually changing, and throughout the vast span of geological time it has never looked precisely the same for two years, or even two days, in succession.

Geologists define rock as any naturally formed mass of mineral matter found in the Earth's crust. Under this definition loose material such as sand and mud can be regarded as rock. More commonly, however, rocks are thought of as the solid stony masses made up of a number of minerals fused or cemented together.

Magma – molten rock

Magma is the molten rock that resides in parts of the Earth's upper mantle and lower crust. Since this is the origin of the material that erupts to the surface in volcanoes, and solidifies to form igneous rocks (▶ pages 142-4), it is tempting to think of magma as the liquid phase of volcanic rocks. In fact this is not so, since magma goes through all sorts of composition changes in the process of cooling and solidifying. Though molten, however, few magmas are entirely liquid. For one thing, they contain up to about 14 percent by volume of dissolved gases, chiefly water vapor and carbon dioxide. Many also contain solid material. This is possible because magma is not a pure substance with a precise melting point. It is a mixture of silicates and oxides with different melting points, and all constituents will only be completely liquid if the temperature is higher than that of the highest melting point. Magmas are thus complex mixtures of solid, liquid and gases, having different compositions, temperatures and mobilities.

Two main types – basaltic . . .

Basaltic magma is derived by partial melting of the peridotite (▲ page 63) that constitutes the asthenosphere and is exemplified by that which rises at oceanic ridges to form new oceanic lithosphere. The asthenospheric material, which is partially molten already, begins to rise along oceanic ridge zones and, as it does so, further partial melting takes place because of the decrease in pressure. The first minerals to melt in this new phase (the ones with the lowest melting points) are those that constitute basalt. And because basaltic magma is less dense than the peridotite from which it is derived, it rises more rapidly and in preference. Basaltic magmas contain about 50 percent silica (SiO_2) and have temperatures in the range 900°C-1,200°C.

. . . and granitic

Granitic magma, by contrast, is typically generated in plate collision regions, beneath the mountain ranges produced in continent-continent collisions and in the subduction zones at ocean-continent collision boundaries. In the latter case, as the oceanic lithosphere with its thin covering of water-rich sediment descends into the Earth's interior, it begins to melt along the upper edge. Some of the magma thus produced rises to the surface where it results in volcanic activity. Some, however, fails to reach the surface but cools within the crust to form granitic masses. Magmatic activity in subduction zones is much more complex than that at oceanic ridges and is far from perfectly understood. Nevertheless, it is clear that granitic magmas are much richer in SiO_2 (60-70 percent) and have a lower temperature (below 800°C) than their basaltic counterparts. For both these reasons, granitic magmas are less mobile.

◀ The Earth's crust is made of rock – aggregations of minerals. The slow processes of rock formation are difficult to record – the solidification from a melt, the compaction of masses of sediment, or the high-pressure alteration of minerals deep inside a mountain. However, some igneous rocks form in a matter of minutes when a volcano erupts.

Igneous Rocks

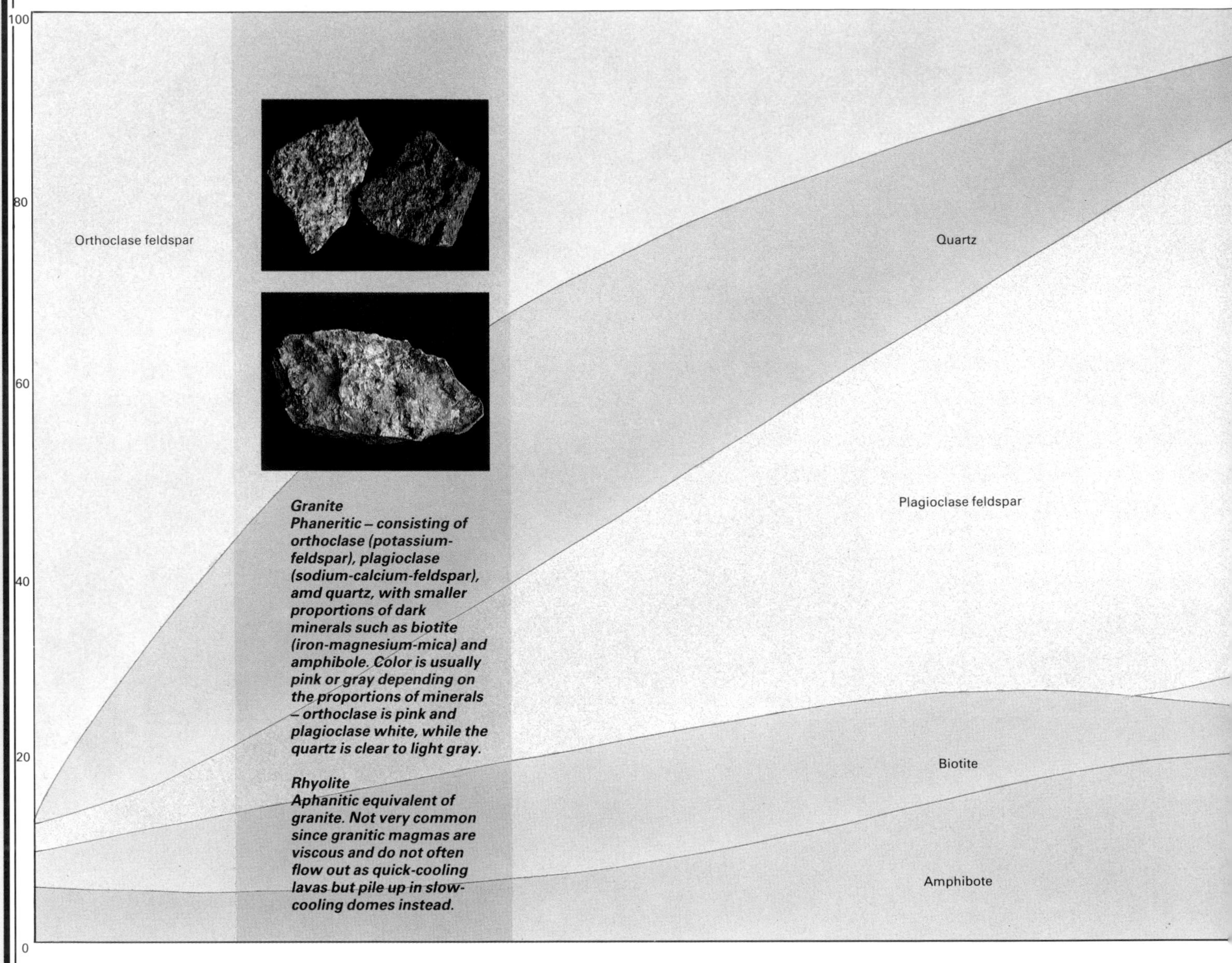

Granite
Phaneritic – consisting of orthoclase (potassium-feldspar), plagioclase (sodium-calcium-feldspar), amd quartz, with smaller proportions of dark minerals such as biotite (iron-magnesium-mica) and amphibole. Color is usually pink or gray depending on the proportions of minerals – orthoclase is pink and plagioclase white, while the quartz is clear to light gray.

Rhyolite
Aphanitic equivalent of granite. Not very common since granitic magmas are viscous and do not often flow out as quick-cooling lavas but pile up in slow-cooling domes instead.

▲ Igneous rocks are classified both on their texture – coarse or fine – and their composition – the proportions of their minerals. Most contain varying proportions of the common minerals feldspar, quartz, mica, amphibole, pyroxene and, possibly, olivine. These proportions can be plotted on a chart as shown, and a particular rock's composition can be defined by its position along the horizontal axis of the chart. For example, the position of granite and rhyolite shows them to contain about 30 percent orthoclase feldspar, 28 percent quartz, 20 percent plagioclase feldspar, and some biotite mica and amphibole. Gabbro and basalt, however, have mostly pyroxene with some plagioclase and olivine.

Igneous rock types

Igneous rocks seldom, if ever, have the same compositions as the magmas from which they are ultimately derived. Because these source materials are mixtures of silicates and oxides with different melting points, a rise in temperature will cause some of the constituents to melt before others. The magma that finally rises towards the Earth's surface will thus have a composition which is different from that of its source material and which depends on the temperature at which it leaves behind the remainder of the source material.

Further changes in composition may take place as the magma cools again. As the temperature falls, the higher melting point constituents will crystallize out first. Then some of the molten material may drain away before the whole crystallization is complete. The minerals that crystallize out of the magma early may even react with liquid remaining to form minerals of quite different composition.

It follows that a magma of a single composition may give rise to igneous rocks of numerous different compositions. Add to that the fact that not all magmas have the same composition in the first place and the number of potentially different igneous rock compositions is huge. Then there is the question of texture. Igneous rocks of the same composition can have quite different textures depending on their cooling histories.

The five basic textures in igneous rock

Texture means the size, shape and characteristic arrangement of the rock's constituent minerals. A rock with glassy texture has no crystals at all, for the rock cooled so rapidly that crystals had no time to grow. The atoms that were randomly distributed in the magma remain "frozen" in their disordered state in the final rock. Glassy textures are formed, for example, in the surface layers of lavas that cool and solidify very rapidly under water and in small volcanic fragments that are shot out in a molten state into the much cooler atmosphere.

A magma that cools rather more slowly gives rise to rock with an aphanitic, or fine-grained, texture. Crystals have time to grow but they are

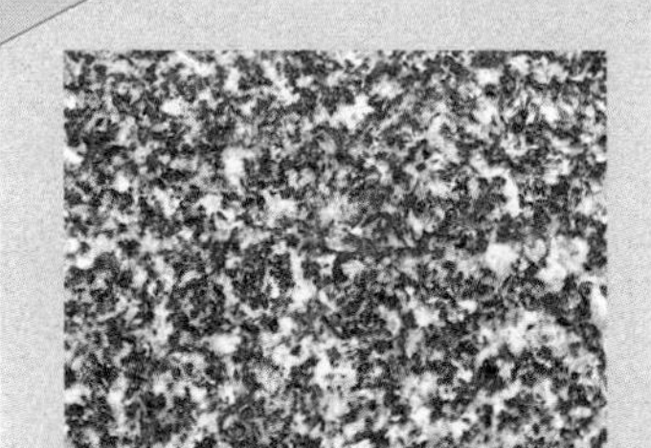

Diorite
Phaneritic – similar to gray granite but consisting largely of plagioclase and amphibole, with smaller amounts of biotite, quartz and orthoclase.

Andesite
Aphanitic and extrusive form of diorite. Found at destructive plate boundaries, probably formed by partial melting in subduction zones. The name comes from the Andes, where rocks of this composition are common.

Gabbro
Phaneritic – consisting of pyroxene and calcium-rich plagioclase feldspar, with smaller amounts of olivine and amphibole. The dark minerals give it a dark green to black color. Rare at the Earth's surface but may be the main constituent of the lowest layer of oceanic crust, and possibly of continental crust.

Basalt
Aphanitic extrusive equivalent of gabbro. The most common extrusive rock, forming most lava flows.

Peridotite
Phaneritic – consisting almost entirely of olivine and pyroxene. Rare in the Earth's crust but thought to be the main constituent of the mantle.

Pyroxene

Olivine

very small and may be detected only under a microscope (not by eye in a hand sample).

Magma that cools even more slowly than that produces a rock with a phaneritic, or coarse-grained, texture in which the crystals are visible to the naked eye in a hand sample. Phaneritic textures are characteristic of large intrusives – rocks that have cooled very slowly underground, where they are surrounded by insulating rock. Aphanitic textures, by contrast, are characteristic of extrusives which, exposed to the atmosphere, cooled rather more quickly.

In aphanitic and phaneritic rocks the interlocking crystals in any particular sample are usually about equal in size, suggesting that cooling took place fairly uniformly. In some rocks, on the other hand, there are two distinct sizes of crystals – large ones, known as phenocrysts, set in a matrix, or groundmass, of smaller ones. This texture is called porphyritic and the rocks involved, porphyries. A porphyritic texture indicates a two-stage cooling – a first stage of slow cooling, possibly intrusive, during which the phenocrysts formed, followed by a second stage of more rapid cooling, possibly extrusive, which produced the surrounding matrix.

Fifthly, there are pyroclastics, or rocks of pyroclastic texture. These are rocks ejected from volcanoes (◆ pages 25-32). They do not have a structure of interlocking crystals but comprise fragments, often broken and deformed, that were either welded together when hot or cemented together later when cold.

Texture is important because it provides geologists with information about a rock's cooling history. It is thus a necessary part of igneous rock classification. But it is not sufficient by itself, because a rock of a particular composition may take a number of different textures depending on the circumstances in which it was formed. A complete classification must therefore also involve composition expressed in terms of the mineral constituents. Using both texture and composition, geologists have set up a dual classification system that reveals the relationships between the most important igneous rock types.

Rocks are not a permanent feature of the Earth's surface

The rock cycle

Igneous rocks are formed when magma from the Earth's interior rises, cools and solidifies. Some magma moves up through volcanic conduits or fissures and is extruded onto the Earth's surface where it forms extrusive igneous rocks such as basalt and andesite (◀ pages 142-3). No sooner do extrusives appear than they begin to weather under the action of rain, ice, chemicals and plants, by which agencies they are broken down into smaller and smaller fragments.

Some magma never reaches the surface, however, rising only part way through the crust where it comes to rest to form intrusive igneous rocks such as granite and gabbro (◀ pages 142-3). As long as they remain unexposed, intrusives cannot be weathered by climatic phenomena (although they can be altered chemically by percolating fluids). But if, as frequently happens, the overlying rocks are eroded away, intrusives become no less susceptible to erosion than extrusives and begin to break down.

The end product of weathering of igneous rocks is small particles of sediment, many of which are transported by rivers and streams and finally deposited in the oceans. There they get buried under succeeding sediments and gradually become compacted and cemented, or lithified, into hard sedimentary rock such as sandstone and shale (▶ pages 150-1).

Some of this sedimentary rock will get very deeply buried and, by intense heat and pressure, be converted into a third type of rock known as metamorphic rock, such as slate and schist (▶ pages 146-7). If the rock is then subjected to even greater heat and pressure, it may well melt to produce magma which is then available to form new igneous rock. What began as an igneous rock thus ends up as igneous rock, having passed through two other rock types in the process.

This change from igneous to sedimentary to metamorphic to igneous rock is known as the rock cycle. But not all rock material goes through the full cycle. Igneous rocks can become buried and thus get metamorphosed without passing through the sedimentary stage. In the form of oceanic lithosphere they may also, with or without metamorphism, be returned to the ultimate source of most magma, the asthenosphere. Sedimentary and metamorphic rocks may (for example, when they are thrown up into mountains) be weathered directly back into new sedimentary rocks. On the other hand, some of the sedimentary and metamorphic, as well as igneous rocks, thrown up into mountains may become metamorphosed or remetamorphosed in the process and only then begin to weather into sediments. In short, there may be many different routes that rock can follow.

In principle, the rock cycle could take place even if the continents, ocean floors and lithospheric plates were perfectly stationary. In practice, however – and this is something the 19th century geologists did not know – the global tectonic processes set the conditions under which the rock cycle can operate. Thus for the most part it is the existence of the plate boundaries, the zones of weakness in the lithosphere, that allows magma to reach the surface in the first place. Likewise, the mountain heights in which the greatest erosion occurs are the result of colliding plates. Moreover, it was largely the displacement of water by new spreading oceanic ridges that enabled the continents to be inundated at various times and thus become covered in sediment. In short, were it not for the continual upheavals provided by global tectonics, the Earth's continents, if they existed at all, would have long ago been eroded to an inactive sea-level surface.

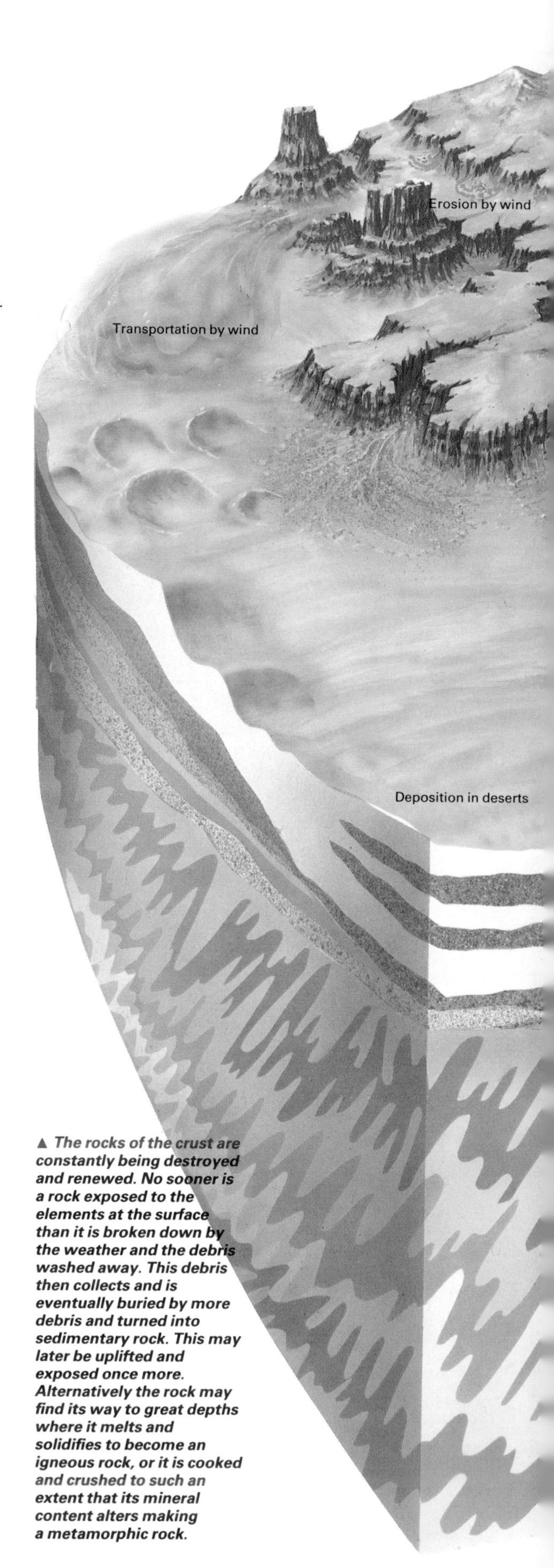

▲ The rocks of the crust are constantly being destroyed and renewed. No sooner is a rock exposed to the elements at the surface than it is broken down by the weather and the debris washed away. This debris then collects and is eventually buried by more debris and turned into sedimentary rock. This may later be uplifted and exposed once more. Alternatively the rock may find its way to great depths where it melts and solidifies to become an igneous rock, or it is cooked and crushed to such an extent that its mineral content alters making a metamorphic rock.

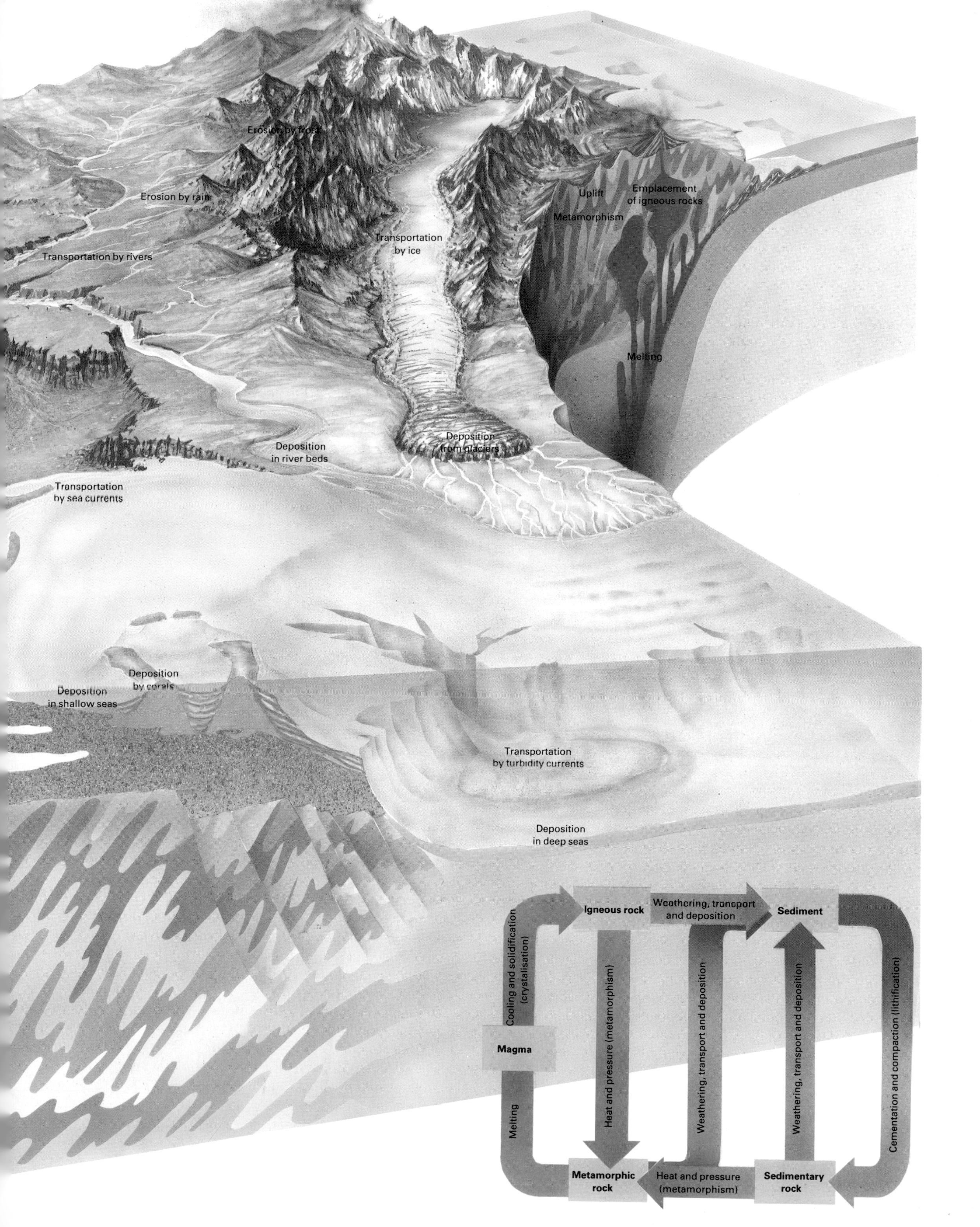

Erosion by frost
Erosion by rain
Transportation by rivers
Transportation by ice
Uplift
Emplacement of igneous rocks
Metamorphism
Melting
Deposition in river beds
Deposition from glaciers
Transportation by sea currents
Deposition by corals
Deposition in shallow seas
Transportation by turbidity currents
Deposition in deep seas

Igneous rock
Weathering, transport and deposition
Sediment
Cooling and solidification (crystalisation)
Heat and pressure (metamorphism)
Weathering, transport and deposition
Weathering, transport and deposition
Cementation and compaction (lithification)
Magma
Melting
Metamorphic rock
Heat and pressure (metamorphism)
Sedimentary rock

Metamorphic Rocks

Mountain range

Some directional pressure

Slate
Produced by low-grade regional metamorphism of shale. Consisting of mica, chlorite and quartz. The chlorite and mica are platy minerals and align themselves parallel to one another giving the characteristic cleavage along which the rock splits easily. The original bedding of the rock may still be visible, but this has nothing to do with the final cleavage – determined by the pressure direction.

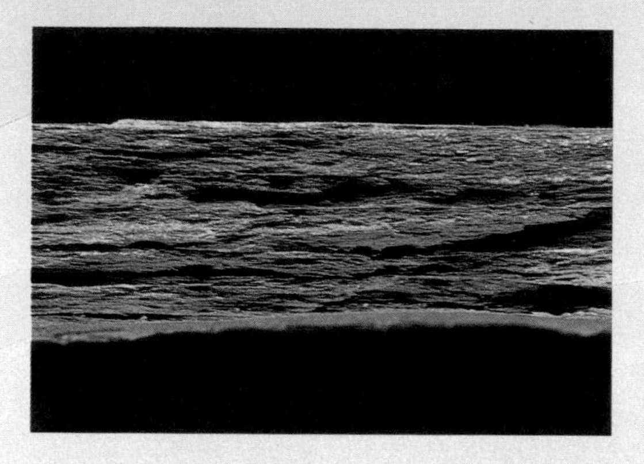

Quartzite
Produced by regional metamorphism of sandstone. Where the original sandstone was pure and consisted entirely of quartz, then the quartz will merely recrystallize in the quartzite to form a tightly-packed mosaic. This produces a white sugary-looking brittle rock. If there were impurities in the original rock the quartzite will have a streaked or mottled appearance.

Non-directional pressure

Schist
Produced by high grade metamorphism of fine-grained rocks. The mineral composition depends on the nature of the original rock and the grade of metamorphism. Under very high pressures minerals such as garnet, staurolite and kyanite will form. The rock is strongly foliated, with flat platy minerals lying along a particular direction determined by the direction of the applied stress.

Very great directional pressure

Very great non-directional pressure

Gneiss
Produced by high grade regional metamorphism of many kinds of rocks. A coarse-grained irregularly-banded rock in which bands of quartz and feldspar alternate with bands containing mica. The terms paragneiss and orthogneiss refer to gneisses derived from sedimentary and igneous rocks respectively. However, many gneisses are so altered that it is impossible to identify the original rocks.

Hornfels
Produced by thermal metamorphism of rocks in contact with a high temperature igneous body. A hard flinty-looking rock. The minerals are present as a fine mosaic in which it is both difficult to see and identify individual grains. Hornfels consists of common minerals, for example, andalusite and cordierite, formed only by thermal metamorphism.

Metamorphic rocks

Metamorphic rocks are rocks that have had their textures, structures and even mineral composition changed by intense heat and/or pressure. They comprise large parts of the vast continental shield areas and of the basement beneath the stable platforms, accounting for about 85 percent of at least the upper 20km of the whole crust. They can be seen to be highly deformed.

Most metamorphic rocks are formed in the roots of ancient mountain belts under the intense forces generated where tectonic plates collide (regional metamorphism). Others are produced by the heat of igneous intrusions (contact metamorphism) or in deep fault zones where the chief influence is pressure (dynamic metamorphism). Sometimes, in low-grade metamorphism, the original rocks may do little more than become more compact. In high-grade metamorphism, by contrast, bedding planes, fossils, vesicles and similar phenomena are completely obliterated, and the source rock cannot even be identified. Metamorphism of whatever type is usually considered to take place at temperatures greater that 300°C.

The most conspicuous property of many metamorphic rocks is a layered or banded structure, known as foliation, which was produced by the alignment of minerals in the plane at right angles to the direction of the pressure. In fine-grained low-grade metamorphic rock, the foliation may not be visible to the naked eye but it is obvious by the way in which the rock breaks. The classic example is slate, which formed by the metamorphism of shale. Slate splits easily into thin sheets along closely spaced internal surfaces. This type of foliation is known as slaty cleavage.

Under more intense metamorphic conditions, however, the foliation becomes visible because of the generation of larger crystals by recrystallization along the direction of foliation. This type of foliation is known as schistosity, and the resulting medium-grained to coarse-grained rocks are known as schists. Schists also break easily along cleavage planes. They form from a variety of rock types the most common of which is shale.

During high-grade metamorphism minerals can segregate, forming alternating layers of light and dark minerals. This type of foliation is known as gneissic layering and the resulting metamorphic rock as gneiss. This foliation is not related to splitting of the rock but results from layers of different mineralogical composition. Gneiss is usually formed from granite or diorite, but it can also come from other rocks.

Metamorphism may not produce foliation when the original rock is largely a single mineral. Small crystals simply grow larger and develop an interlocking coarse-grained texture known as granular, or just unfoliated. Examples are marble, produced by the metamorphism of limestone, and quartzite, which is derived from sandstone.

◄ Metamorphic rocks are classed as thermal metamorphic rocks if heat was the major force in their formation, or regional metamorphic rocks if pressure is the main influence. The former may have little obvious structure in them, while the latter often show twisted bands and shear planes, indicating the great stresses under which they formed.

◄ ▼ The raw material of sedimentary rocks is usually debris worn away from rocks that have already been formed. The weather does this work. Some minerals are unstable when they are exposed to the elements, and may alter or break down. This weakens the rock that contains them and the more robust mineral constituents fall out and are washed away by the rain. Granite (left) is a case in point. The feldspars decay to soft clay minerals, and the quartz and mica grains fall loose, eventually to be transported to the sea where they settle as sands. These may eventually become sandstones. The calcite that makes up almost the entire mass of limestone (below) dissolves in the carbonic acid in rainwater and is washed away. It may eventually precipitate out of solution elsewhere and form new beds of limestone.

Weathering

All rocks exposed to the atmosphere undergo weathering that gradually breaks them down into rubble and, ultimately, fine sediment. The two types of weathering are chemical and mechanical.

The chief agent of chemical weathering is water, both pure water and water with chemicals dissolved in it. Many minerals are soluble in water to some extent, although some are more soluble than others. Rock salt (sodium chloride), for example, is completely soluble and so only exists at the Earth's surface in very arid regions. Although few rocks are soluble completely, many contain soluble components that can be removed selectively, a process known as leaching. Unpolluted rain is almost always pure, but each year rivers take down to the oceans about 4 million tonnes of dissolved matter (➧ page 157).

Although not all the minerals in rocks are soluble, impure water can increase the number of soluble minerals by chemical reaction. Water containing dissolved carbon dioxide (from the atmosphere and vegetation) is in effect a weak acid, carbonic acid, which reacts with many minerals. Minerals high in iron (such as olivine and pyroxene) are particularly vulnerable to attack by water containing dissolved oxygen from the atmosphere. Such minerals rust, which means that the iron combines with oxygen to form the reddish-brown iron oxide hematite or in some cases the yellowish iron hydroxide limonite.

Mechanical weathering is the disintegration of rock by purely physical processes which usually involve attacking the rock along its natural zones of weakness. The chief mechanisms are the freeze-thaw cycle of ice, root growth in plants and sheeting (the removal of pressure causing expansion and fracturing). The most conspicuous result of the early stages of physical weathering is piles of angular rock fragments, known as talus, at the bases of rocky slopes.

▲ ▼ *Physical, as well as chemical, forces have their part to play in the formation of sedimentary material. In dry areas such as deserts (above), fine particles of sand caught in the harsh wind sandblast exposed rock surfaces wearing them down. Where there is a great variation between daytime and nighttime temperatures the constant expansion and contraction of the surface layer of a rock may weaken it (below). These processes rarely work on their own, but usually a combination of physical and chemical effects conspire to reduce a rock to rubble and dust.*

One mechanism is more contentious. Where the day-to-night temperature fluctuations are particularly large (greater than 30°C), one would expect that the rapid alternation of rock expansion and contraction causes the rock to fragment. Yet recent attempts to simulate the effect in the laboratory have failed, even though shattered pebbles observed in desert areas show signs of thermal disintegration.

Weathering, whether mechanical or chemical, is the first stage in the process of erosion, by which topographic high spots are gradually worn down and topographic low spots are filled with the resulting sediments. Erosion involves not just weathering but also the transport of the weathered material. During that transport, however, further weathering may take place. For example, rock fragments that enter and get carried along by a river or stream will grind against other fragments and hence wear away into smaller particles. Rock fragments carried along in the base of a glacier will likewise be ground against the hard rock beneath. Erosion and weathering go hand in hand to achieve the final result – minute particles of sediment.

The rates at which rocks are weathered depend on a number of factors, including rock type, climate and altitude. Because mechanical weathering takes place primarily at fractures, highly fractured rocks weather most rapidly. Likewise rocks with a high proportion of soluble minerals will weather quickly. Climate has a number of influences. It dictates the amount of water available, which affects the rates of chemical weathering and the freeze-thaw ice cycle. Second, it dictates temperature. Warm water is a far more effective weathering agent than is cold water, and temperature fluctuations are essential in the ice cycle. Third, climate dictates the nature and quantity of vegetation. Plants work away in rock fractures, while rotting vegetation provides carbon dioxide for the acidification of the water.

Sedimentary Rocks

Sedimentary rocks
*The term sedimentary rocks applies to those that have been built up from layers of material that have been deposited by natural processes. These processes usually involve currents, and give rise to accumulations of sediment such as sandbanks or mudflats (**➧ pages 157-72**). Unlike igneous or metamorphic rocks, heat has played no part in their formation.*

There are three main processes that eventually give rise to sedimentary rocks, and sedimentary rocks are classified according to the processes that produced them.

The first class of rocks are the clastic sedimentary rocks. These are made of pieces of older rocks that have been broken down. Their fragments have been transported away from their original location and have been deposited elsewhere. Sometimes, in coarse clastic rocks such as conglomerate and breccia, the chunks of the original rock are quite obvious. More usually the sediment consists of merely the original mineral grains separated from their parent mass, as in sandstone. The chemical composition of the original minerals may even have been altered in the process as in shale and clay. A special type of clastic rock, known as pyroclastic, is made up of the dust and fragments ejected from volcanoes, but this is usually regarded as an igneous rock.

The second class are the chemical sedimentary rocks. The material for these rocks was not suspended in the moving water, as in the clastic rocks, but actually dissolved in it. When this water evaporated, in a warm shallow sea or a playa lake, the concentration of the dissolved substance became so great that it precipitated from solution and deposited as a bed at the bottom. Such beds are sometimes called evaporites. Rock salt and various forms of limestone were formed in this way. The name chemical is something of a misnomer, since physicists and chemists regard the solution and precipitation of soluble substances as physical changes rather than chemical processes.

The third type of sedimentary rock is the biogenic sedimentary rock. This is essentially made from the remains of living creatures. They are usually identifiable by the obvious presence of the fossils of the creatures that went into making it. This is certainly the case with shelly limestone, which is made up of masses of shells cemented together. Sometimes, as in the case of chalk, the individual remains are far too small to be seen with the naked eye. In other examples, such as coal, the structure of the original living matter is obscure, having been altered during rock formation.

All these rocks tend to be instantly recognizable in outcrop, since they usually form beds – flat-lying layers which reflect the layered nature of the original sediment and the currents and depositional processes that accumulated them.

► The three types of sedimentary rocks are clastic, chemical and biogenic. The first are formed from fragments and particles broken off other rocks. The second form from chemicals precipitated from the waters in which they were dissolved. The third are produced from the hard remains of living things.

Mountains (scree slope)

Desert

Playa lake

Rock salt
Chemical sedimentary rock formed when seawater evaporates depositing the chemicals dissolved in it. This usually happens when an arm of the sea is cut off and dries up. Sometimes there remains a connection with the open ocean, giving a constant supply of salt water and very thick beds.

Conglomerate
Clastic sedimentary rock formed from large rounded pieces of rock cemented together. Its occurrence represents a fossilized shingle beach and usually overlies an unconformity. The composition varies depending on the original rock broken down to form the shingle.

Shingle beach

Sandy beach

Ocean

Deep ocean basins

Breccia
Coarse-grained clastic sedimentary rock, like conglomerate, but formed form irregular jagged fragments. Derived from the collapse, rather than the erosion, of preexisting rocks. May represent a fossilized scree slope or a caved-in cavern.

Desert sandstone
Medium-grained clastic sedimentary rock formed from consolidated desert sand. Individual fragments tend to be rounded and of an even size. Usually a red color due to iron oxide produced in desert conditions.

River sandstone
Clastic sedimentary rock formed from the sand piled up in river beds, sandbars and deltas. Lacking the red iron staining of desert sandstone, it may contain structures that show it to have been deposited by a river current (▸page 164), and may be interbedded with shale.

Shelly limestone
Biogenic sedimentary rock formed as shells of dead sea creatures accumulated on the sea bottom and were buried. The shells may be broken and cemented together with inorganic calcite. Often the shells are all of one sort.

Shale
Fine-grained clastic sedimentary rock formed from consolidated mud. May be derived from river sediments or from the deep sea. Often contains much organic material and fossils. Shale splits easily along the bedding planes.

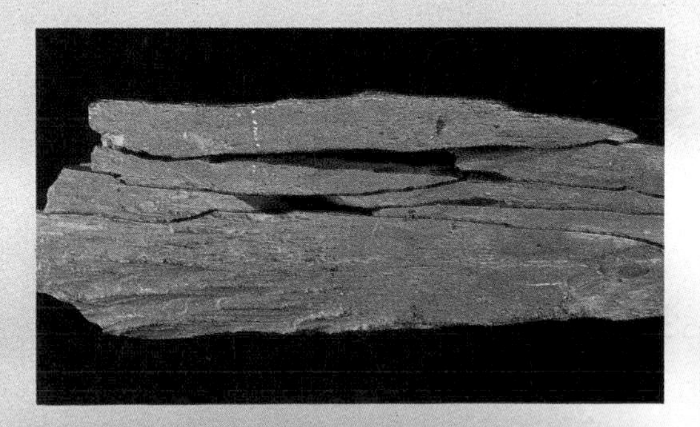

See also
Earthquakes and Volcanoes 25-32
Internal Composition 57-68
Dating the Rocks 73-80
Sedimentary Environments 157-72
Landscape Features 181-96

From sediment to sedimentary rock

A sediment is not a sedimentary rock. Sediments include gravel, sand, silt and mud. Sedimentary rocks derived from these are conglomerate, sandstone, siltstone and shale respectively. Changes must take place in the one before it can become the other.

The first process involved is compaction. Once a sediment is laid down, as a sandbank or the mud of a seabed or whatever, it is then followed by others in a sequence. More material is later laid down upon it, and this continues until the phase of sedimentation is finished. By this time the great weight of the upper sediments bearing down on the lower, compresses those at the bottom, squeezing out air and water from between the individual particles, and distorting and crushing the particles themselves until they flatten out and interlock with each other. This compressional effect is added to by subsequent Earth movements which may throw these beds into folds, shear them with faults, and otherwise exert all sorts of pressures on them.

The next process is that of cementation. Groundwater seeping through whatever spaces are left in the sediment deposits the minerals it carries in solution. The most common mineral deposited is calcite. It forms tiny crystals growing on each grain of sediment, and they grow together, uniting the whole into a solid mass. It is exactly the same process as the manufacture of concrete, where hard-core rubble and gravel is cemented together by lime to form a tough, durable building material. In fact, worn masses of concrete on a beach are often mistaken for pieces of conglomerate, even by experienced geologists!

Occasionally the individual mineral grains will recrystallize or alter their chemical composition – calcium in calcite may be replaced by magnesium to form dolomite. The minerals in beds of limy mud may separate out and, when lithified, form alternating beds of limestone and shale. This whole process of the formation of sedimentary from a sediment is known as lithification, or diagenesis.

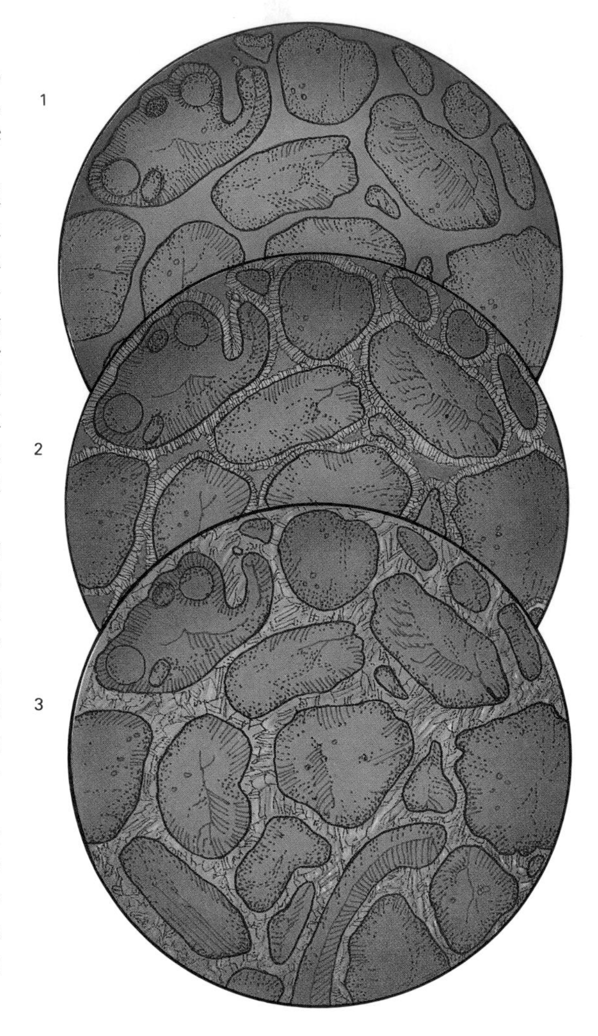

*◀ ▲ **A sediment does not become a sedimentary rock, for example the sand in the foreground (left) does not become sandstone like that in the background, until it is consolidated. Two processes are involved (above). The sediment is compressed (1), interlocking the particles together and eliminating most of the air spaces. Then minerals such as calcite are deposited (2) from percolating ground water in whatever spaces are left and these cement (3) the sediment particles into one solid mass.***

Faulting and Folding

Deformation of the Earth's crust – breaking and bending of the layers...Fold mountains – the tortured result of intense geological pressures... PERSPECTIVE...Faulting and fault types...Folding and fold types

Rift valleys and block mountains

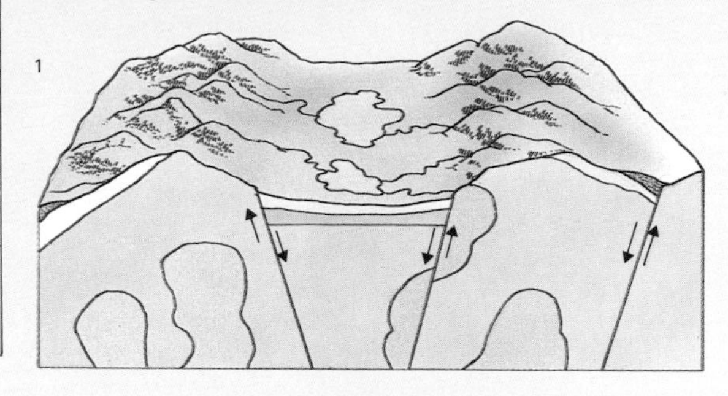

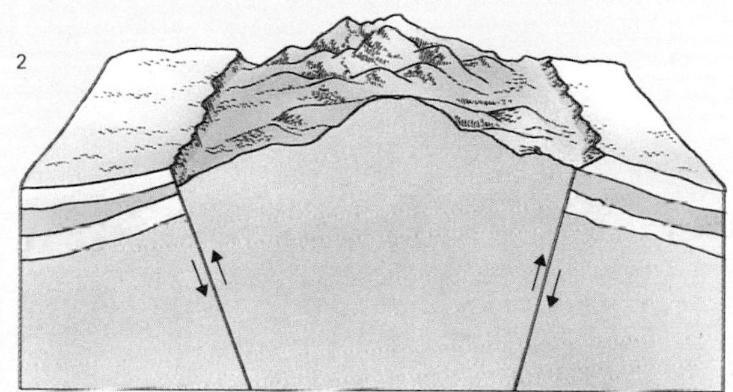

Deformation of the crust in the form of faulting (▶ page 154) and folding (▶ page 155) take place on all scales from centimeters upwards, but by far the most important is that represented by the world's mountain ranges. Not all mountains are primary deformation structures. For example, deformation is minimal in isolated volcanic cones and highlands formed form plateau basalts (◀ page 93). Of those that are, there are three main types – fault-block mountains, upwarped mountains and fold mountains.

Fault-block mountains are the result of a tensional stress, producing a series of normal faults. One of the best known examples is the Basin and Range province of the USA, from California to New Mexico. Upwarped mountains, by contrast, are the result of compression which creates an uplift of part of the crust. Examples include the southern Rocky Mountains and the Black Hills of Dakota.

Fold mountains, the largest and most complex of all mountain belts, are also compressional, involving folding, faulting, metamorphism and igneous activity. They are due to the collisions of lithospheric plates – ocean-ocean, ocean-continent or continent-continent. The Himalayas are a spectacular example of ones still forming, the Urals an example of ones which ceased to rise long ago.

▲ ▼ ***Geological structures, such as folds and faults, are important in determining landscape. However, the surface of the land rarely follows the structures exactly, erosion shaping the final appearance. A block subsiding between two faults produces a structure called a graben (1). This may show as a rift valley on the surface, but erosion of the high areas and sedimentation in the low even out the topography so that the landscape feature is insignificant compared with the depth of the geological feature. The same is true of a horst block, uplifted between two faults, which may form block mountains (2). Those in Oregon (below) are typical.***

Folds and faults show past Earth movements, and indicate the direction of the pressures that produced them

Faulting and fault types

A fault is a fracture in the Earth's crust along which displacement has taken, or is taking, place. The face of the fault, the surface over which the slippage occurs, is called the fault plane. The angle that the fault plane makes with the horizontal is known as the dip; so that a perfectly vertical fault, for example, has a dip of 90°. Except for vertical faults, there will be rock above the fault plane, known as the hanging wall, and rock below the fault plane, known as the footwall. The strike of the fault is the direction in which the break at the surface (the fault trace) lies or, if the fault is entirely underground, the direction in which the upper edge of the fault plane lies. Strike is measured with respect to normal geographical coordinates – for example, east-west, northwest-southeast.

With this terminology it is possible to classify faults according to the nature of the relative displacement between the two crustal blocks involved. A fault in which the movement is up or down the dip is called a "dip-slip fault". If the hanging wall moves downwards with respect to the footwall, the fault is a "normal fault"; if the hanging wall moves upward relatively, the fault is a "reverse fault". Most normal faults are steeply inclined, typically having dips in the range 65°–90°. But reverse faults are less restricted in their dips. A reverse fault in which the dip is less than about 45° (i.e. one in which the fault plane is tending towards the horizontal) is known as a "thrust fault". In the largest thrust faults, in mountain belts such as the Alps and Himalayas, one block may be thrust over the other by as much as 50km. Thrust faults in particular and reverse faults in general result from compressional forces, whereas normal faults are a consequence of tensional stresses trying to pull the crust apart.

A fault in which the movement is directed horizontally (along the strike) is known as a "strike-slip fault". If an observer on one side of the fault sees the opposite block to have moved to the right, the fault is said to be right-lateral, or dextral. If the opposite block were moving to the left, the fault would be left-lateral, or sinistral. Strike-slip faults, which usually have a high dip, can have lengths up to many hundreds of kilometers. Strike-slip faults generally give rise to little relief because they have no motion in the vertical direction. Dip-slip faults do give rise to relief, especially if there is more than one in close association. Where there are two parallel normal faults, for example, the crustal block between them may either fall to produce a rift valley, or graben, or rise relative to the adjacent blocks to form a horst.

Of course, in nature not all faults are either completely dip-slip or completely strike-slip. A fault combining vertical and horizontal motion is known as an "oblique-slip fault". Nor do real faults have perfectly plane fault planes or perfectly straight fault traces. If they did, it would be much more difficult for a fault to get "locked" by frictional forces, slippage would be much less jerky, and there would be far fewer very large fault-induced earthquakes. In practice, however, fault planes are usually very rough surfaces and fault traces very wriggly lines.

Faults

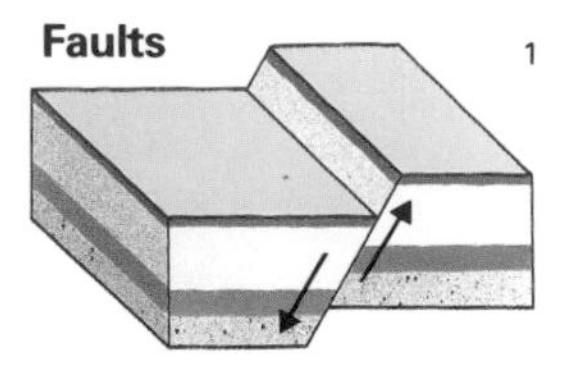

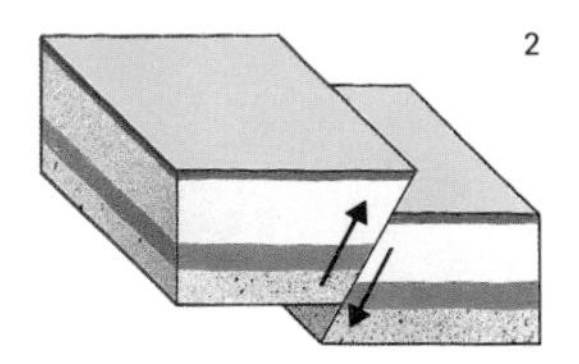

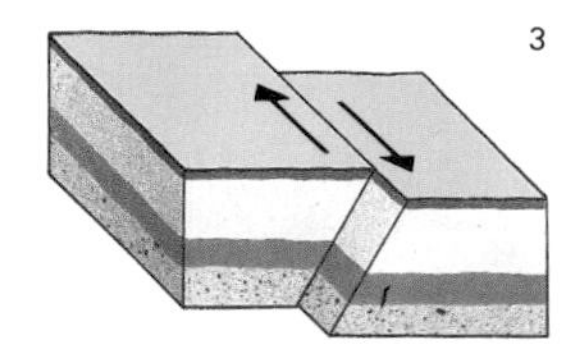

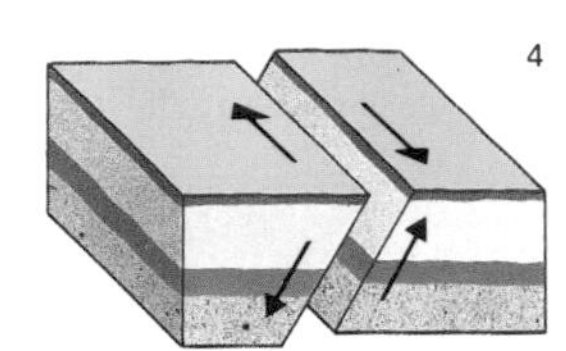

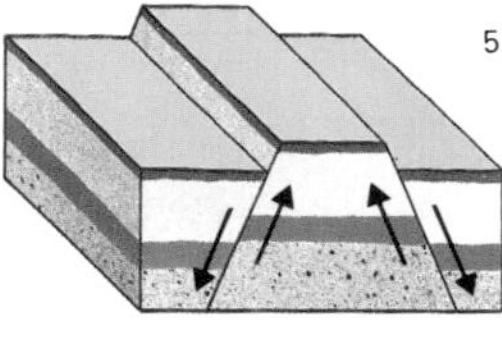

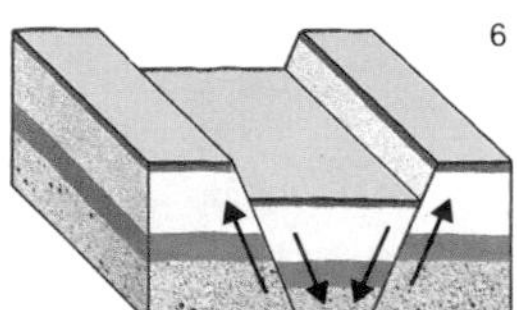

◄ There are many types of fault. When one block has slid down the fault face in relation to the other it produces a normal fault (1). When it appears to have moved up it is a reverse fault (2). A strike-slip fault (3) has moved the blocks sideways. An oblique fault (4) has both vertical and horizontal movements. Horsts (5) and grabens (6) result from blocks moving between faults.

▲ When a fault is exposed in a cliff face or an excavation it is often possible to see at a glance what type of fault it is. This example in a road cutting in Utah consists of a series of parallel normal faults. The amount and direction of movement is obvious by the displacement of the prominent dark bed. Often the broken ends of the beds are curled over in the direction in which they have been dragged.

▲ **Folds that arch upwards are called anticlines, and those that sag downwards are known as synclines. They are rarely found separately but usually form continuous sequences such as here, in Wales.**

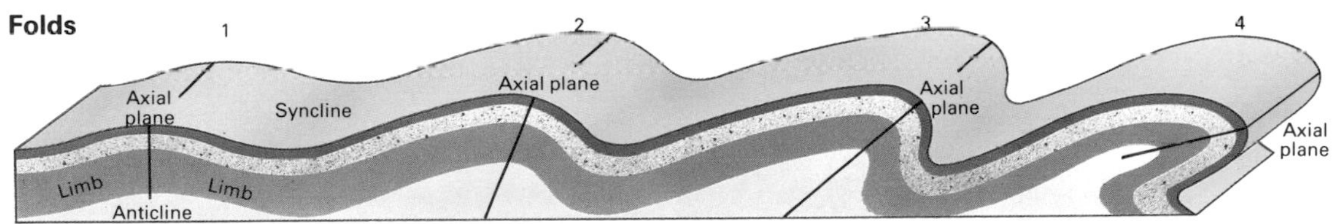

▲ ***Folds occur when beds of compressed rock bend rather than break. The axial plane is the theoretical surface that passes through the "hinge line" in each successive bed. A vertical axial plane gives a symmetrical fold (1), a sloping one an inclined fold (2). An overturned fold (3) has both limbs sloping the same way. A fold pushed so far over that it has collapsed is a recumbent fold (4).***

◄ ***Chevron folds are sharply angular crinkles, usually between less deformed beds of rocks.***

Folding and fold types

A fold is a bend or warp in the strata of the Earth's crust in response to pressure. The two sides of a fold are called the limbs and the angle between them the interlimb angle. The surface bisecting the interlimb angle is known as the axial plane, and the line along which the axial plane cuts the folded rocks is the axis of the fold. The axis is usually at right angles to the direction of the pressure that caused the folding. If the axis is not horizontal it is said to plunge, the angle of the plunge being the angle at which the axis lies below the horizontal. If the strata in the fold are arched upwards, the resulting structure is known as an "anticline", but if the fold goes downwards it is a "syncline".

In a symmetrical, or upright, fold both limbs are inclined at the same angle and the axial plane is vertical. If the limbs are inclined at different angles, however, the axial plane will not be vertical and the fold is said to be asymmetrical, or inclined. A fold which is very asymmetrical may even be overturned so that its limbs dip in the same direction, albeit by different amounts. And if a fold is so much overturned that its axial plane is near to horizontal, it is said to be recumbent.

Linear folds do not go on for ever; they fade out at each end rather like the ruckles in a piece of cloth. However, not all folds are linear. A "dome" is a mild upwarping of the crust, or an anticlinal fold, which plunges away from a point radially in all directions. In its synclinal opposite, the basin, the strata plunge radially inwards towards a point. Domes and basins have a very wide range up to diameters of hundreds of kilometers.

See also
The Dynamic Earth 9-24
Earthquakes and Volcanoes 25-32
Dating the Rocks 73-80
The Growth of Continents 101-8
Landscape Features 181-96

Fold mountains – the three styles of mountain building

When two oceanic plates converge, the less vigorous is subducted beneath the other. The top of the descending slab partly melts, the resulting magma rises, and volcanoes form at the surface, gradually building up from the ocean floor as an arc of volcanic islands. As time goes by, the island arc becomes more substantial because of repeated volcanism, erosion begins to generate sediment, and more sediment is scraped from the top of the descending lithosphere. The end result is a mature arc – a mountain range of islands surrounded by water. An example is the Aleutian arc in the northern Pacific.

Though surrounded by ocean, island arcs usually lie quite close to continents, being separated from them by narrow back-arc basins. There are situations in which subduction zone, arc volcanism and continental edge are in close proximity. The position here is more complex than in an island arc, partly because there is much extra material present in the form of both continental crust and sediment eroded from it onto the continental shelf. The result is a complicated mountain range which is folded, faulted, metamorphosed and intruded by igneous activity. The best known example is probably the Andes along the edge of South America.

The final and most complex stage of the process occurs when the spreading oceanic lithosphere forces a second continent into the subduction zone. As neither continent can be subducted, one overrides the other. Subduction ceases, but the continents continue to be forced against each other. Perhaps the best example is the way that India is still being pressed firmly into the Asian mainland by spreading in the Indian Ocean, resulting in the Himalayas.

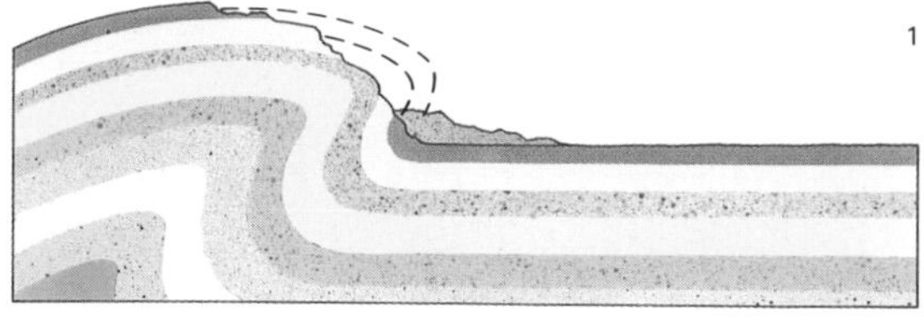

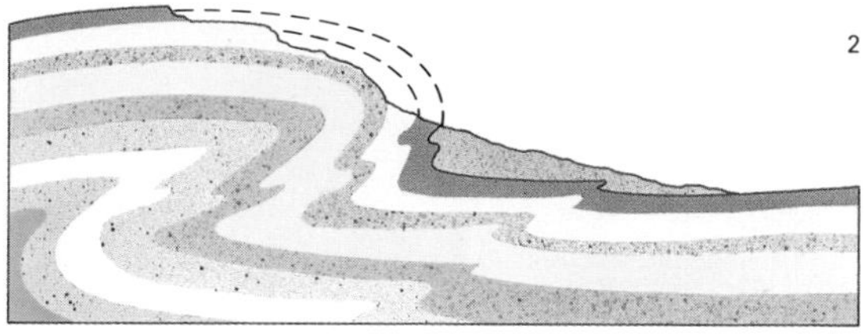

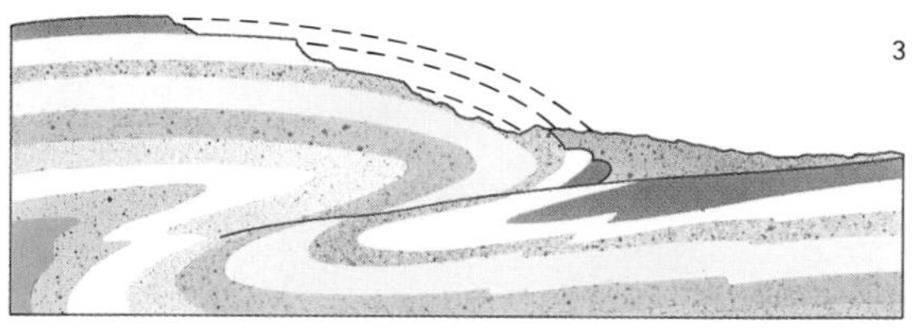

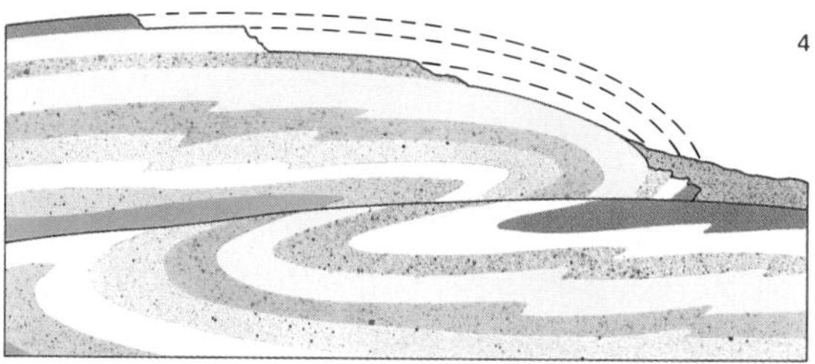

▲ *Overthrusts are extreme developments of the recumbent type of fold. Sustained pressure from one direction folds the beds and pushes them over. Continued pressure then pushes the uppermost arm of the fold over the lower, shearing it away along a thrust plane. This does not form gently rounded topographic features on the surface, however, since erosion is working on the higher portions at every stage of the process.*

◄ *A fold mountain chain such as the Rockies is such a mass of folds that it is next to impossible to work out the sequence of events that produced the deformation. Beds are folded, and these folds are themselves folded, and then folded again, making fold mountains the most complex geological features on Earth.*

Sedimentary Environments

16

Erosion and deposition...Deserts – where dry sediments lie...Rivers – active sediment movers... Shorelines and deltas – the sea's edge...Reefs – rocks from life...Ocean sediments – clays and oozes...PERSPECTIVE...Sediment distribution...Early theorists...Great Barrier Reef...Explanations... Ocean drilling

The face of the Earth is marked by a constantly changing pattern of erosion and deposition of material. Sediment originates at the Earth's surface by the disintegration and decomposition of older rocks. The products of disintegration and decomposition gradually make their way into some form of transport system for sediment and are eventually deposited. These materials are termed "clastic" sediments, meaning broken or fragmented. To these must be added sediments which form by precipitation from lake or ocean waters or by the accumulation of the skeletons of dead organisms (◆ page 150).

Processes of erosion are perpetually transferring material from land areas to the sea, but there are very marked differences in rates of erosion over the face of the Earth. Tropical areas of intense chemical weathering are eroded most rapidly, while in temperate and cold regions the rate of erosion is relatively low. The amount of material eroded from the land is a complex function of rainfall, relief and vegetation.

Distribution of sediment

The global patterns of sediment yield can be matched with the amount of material delivered to the oceans, but it is clear that a considerable mass of sediment never reaches the sea. It is estimated that 33 percent of the sediment of the Yellow River in China is deposited within the river valley and another 43 percent is retained in the delta region. It should be constantly borne in mind that only a small proportion of the modern river sediment brought to the coastlines of the world accumulates on the floor of the deep sea.

The distribution of sediment yields reveals some startling contrasts. The Eurasian Arctic is the biggest drainage basin in the world with several extremely large rivers, yet this huge area contributes only a small amount (84 million tonnes) of sediment yearly to the Arctic Ocean. On the other hand, Chinese rivers and those draining the Himalayas carry an enormous annual sediment load of over 3 billion tonnes. Expressed slightly differently, the high yield rivers between Korea and Pakistan, including the Mekong, the Irrawaddy and the Ganges, contribute about one half of the total world input of sediment to the ocean.

▼ Eroded material is constantly being transported from the decaying highlands of the Earth and deposited in the lowlands and the seas. However, the area of a river basin is no guide to the volume of its sediment. The vast basins of northern Asia bring little sediment to the Arctic Ocean, while those carrying the greatest volume of sediment are those that drain from the rapidly-eroding Himalayas.

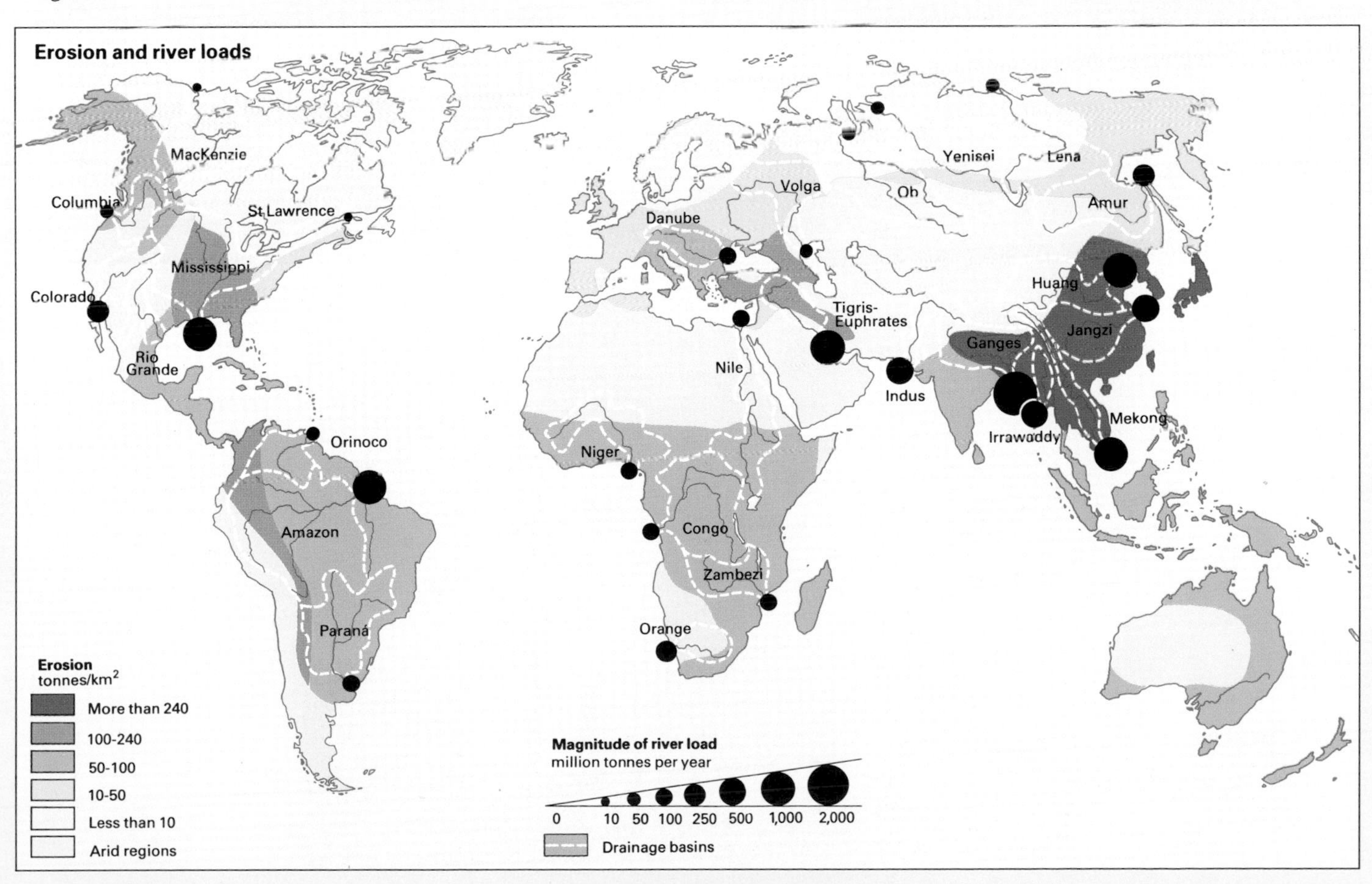

Only a small proportion of the Earth's desert surface consists of sand

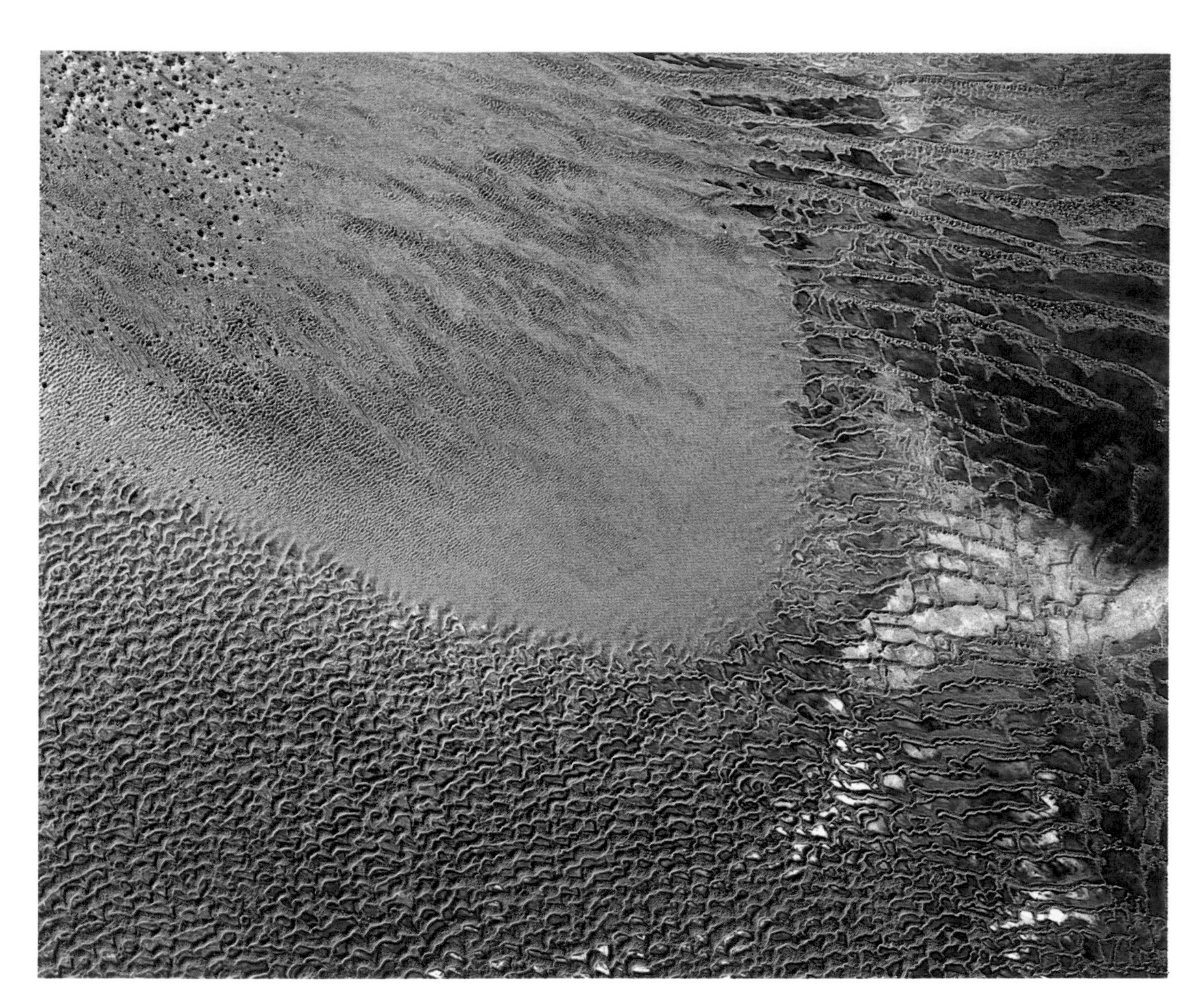

▲ ***Even from space the largest sand dunes of the dry sandy desert areas of the Earth can be seen. In this Landsat photograph of Liwa Hollows in the United Arab Emirates the ripples represent strings of crescentic barchan type dunes, while the streaks are elongated dunes of the seif type. The prevailing wind is from the top left.***

▼ ***Sand dunes are always on the move, shifted by the wind. The air currents blow individual grains along the ground. They roll up the windward faces of the dunes then roll down and collect in the sheltered lee side. In this way the dune advances across the desert, continuously being eroded on one side and built up on the other.***

The world's deserts

Deserts are a particularly dramatic feature of the Earth's surface. They occur in both Arctic areas, where sediments are derived from wasting glaciers, and tropical zones where vast sand seas punctuated by scattered oases are found. Areas of low rainfall occur as two discontinuous belts around the latitudes of 20–30° and are associated with persistently high atmospheric pressures. Deserts also occur in the centers of large continental masses.

The great sand seas or "ergs", sometimes as large as 500,000 square kilometers, are the main sites of accumulation of wind-blown sand. In the large ergs of the Sahara and Arabia, the wind sculpts the mobile sand into distinct configurations termed "bedforms". The largest of these, called draas, have superimposed on them smaller bedforms (dunes) and still smaller features (ripples). In some deserts where there is less sandy sediment, as in Australia, draas are absent.

Most sand set in motion by the wind moves as small rippled bedforms. These are termed impact or "ballistic ripples", because the sand grains bounce into one another, and their crests are oriented transversely to the wind direction. Another kind of ripple forms on

▲ ► Small scale sand ripples are superimposed on large scale dunes in very sandy areas such as Algeria. Different wind patterns give different sand dune types. Barchans (1,2) are the crescentic type. Star-shaped dunes (3) result from irregular winds. Seif dunes (4) develop parallel to the wind direction on bare rock surfaces.

◄ Desert sandstones in characteristic dune bedding structures are the marks of ancient deserts. This Permian desert sandstone is in southwest England.

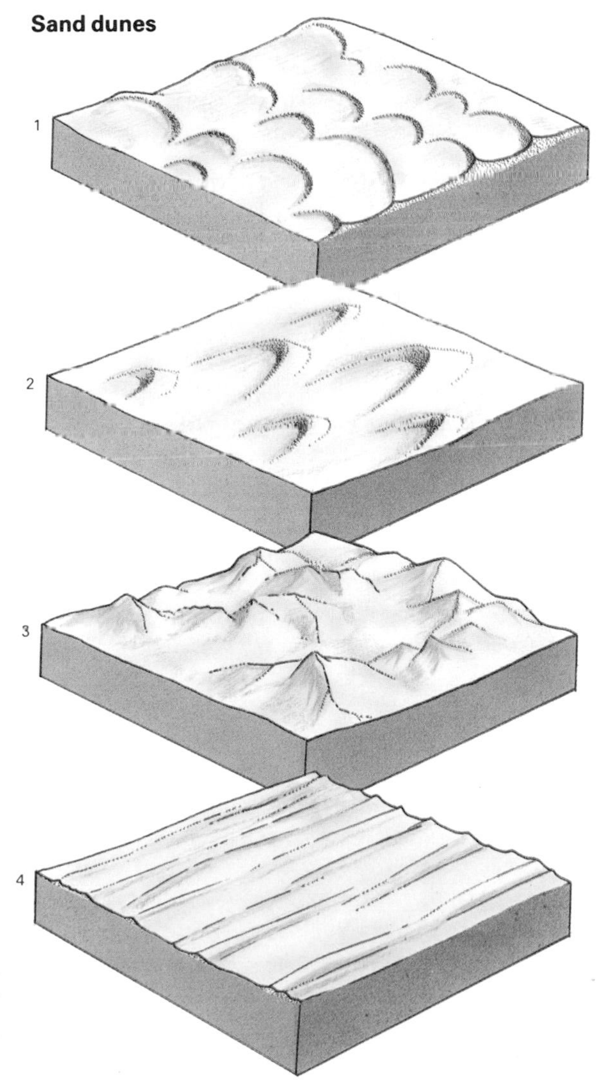

damp sediment. Moving sand grains adhere to the sediment surface in rather irregular three-dimensional patterns. During periods of high winds, the desert may be scraped down to the level of the water table, producing extensive platforms covered with these "adhesion ripples".

Ripples are superimposed on larger scale bedforms such as dunes and draas. The rate of movement of these large bedforms is related to their size, with the smaller forms able to move more quickly. Draas may move at about 1cm per year, but may take over 10,000 years to grow in the first instance. Dunes and draas take on a variety of forms ranging from long-crested transverse features with steep avalanche faces on the downwind side, to crescentic, parabolic and longitudinal or "seif" forms, elongated parallel to the wind direction. In prevailing winds these bedforms retain a clearly recognizable form, but where winds are variable, complex three-dimensional dunes and interference patterns result.

In modern wind-blown dunes and draas internal discontinuities representing old slip faces can be studied and matched successfully with ancient deposits.The internal structure of wind-blown dunes reflects the successive positions of slip faces.

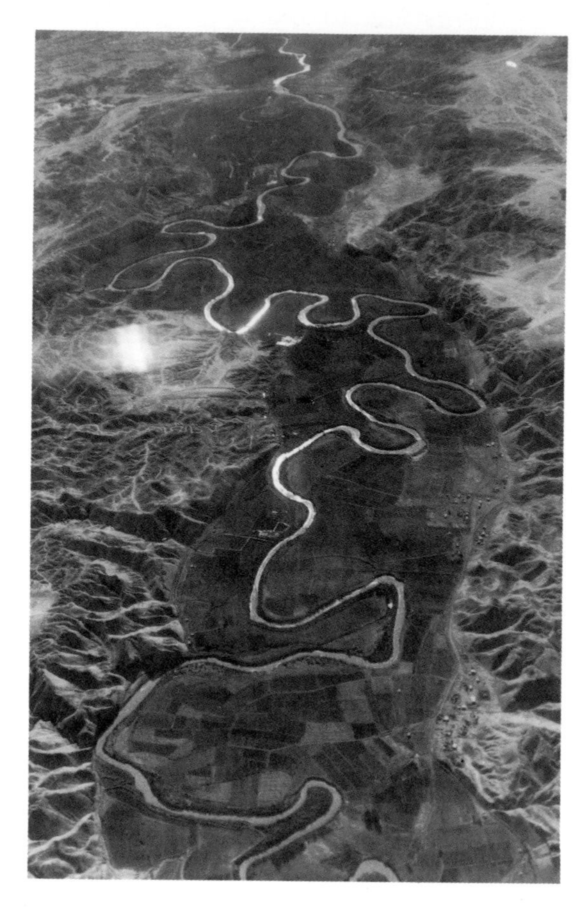

▲ *A meandering river is always changing its course, shifting to a different part of its flood plain in a matter of centuries. It spreads sediment over the plain at times of flood.*

▼ *Levees, banks bordering a meandering river, are built up at times of flood. The floodwater over the levees is slacker than in the channel and the silt is deposited there.*

Levees

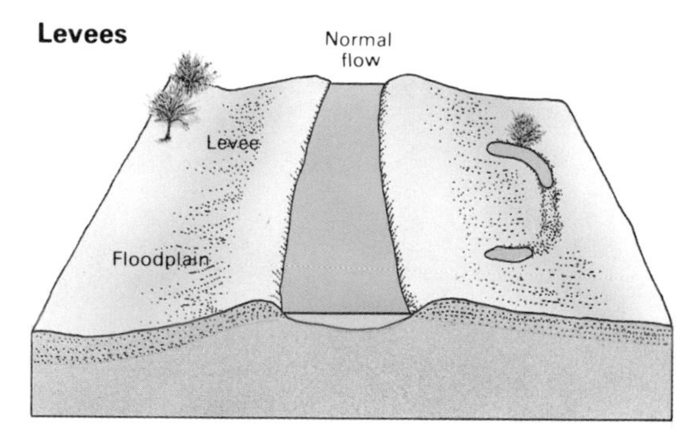

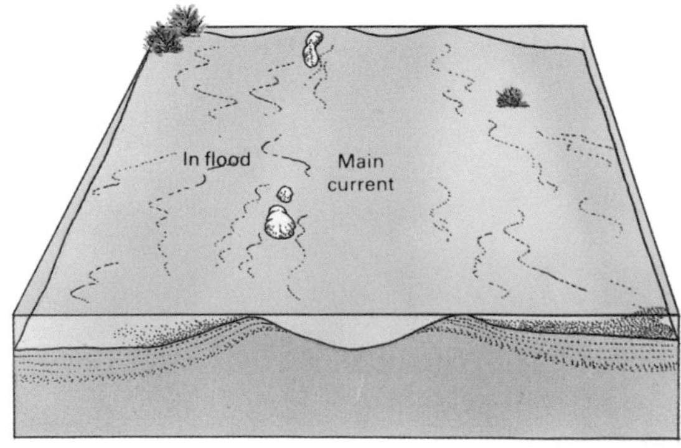

The world's rivers – transporters and depositers of sediment

Rivers are not only major "transporters" of sediment, but are also important "depositers". River types vary enormously according to climate, relief and vegetation, but can be broadly classified into highly sinuous forms which migrate relatively rapidly across their flood plains and less sinuous forms which have several streams with constantly shifting banks. The deposition from these two types of river differ in detail. However, sites for deposition can also be subdivided into "channels" which confine normal flows, and areas covered only by floodwaters termed "overbank areas" or flood plains.

When highly sinuous streams meander they do so by deposition of sediment, usually sand, on their inside banks and by cutting away at the outside banks of river bends (➧ page 186). In this way a body of sandy sediment is built up called a "point-bar". This is the major site of deposition in a highly sinuous stream. Sometimes the point-bar deposits possess an internal structure with surfaces dipping at low angles towards the deepest part of the channel. These represent old point-bar surfaces preserved during progressive changes in the channel. Because river valleys are generally areas of subsidence, in time the channel and point-bar deposits become buried like sandy building blocks. The distribution of these sandy building blocks within the sedimentary deposits of the river valley gives rise to a characteristic fluviatile "architecture". The precise interpretation of such architectural patterns in terms of river processes and basin development is a subject of considerable concern to sedimentologists.

Rivers of low sinuosity tend not to deposit point-bars because they do not meander. Instead, their numerous channels divide and rejoin, producing a complex mosaic of variously shaped bars. Rivers of this type commonly do not carry large loads of suspended fine-grained sediment, overbank deposits are generally thin and the banks of the river very unstable. As a result, they tend to "comb" their flood plains leaving behind a sheet of sand and gravel. Ancient rocks of this type show many intersecting channeled surfaces incised into low bars, and their highly interconnected sheet–like nature is quite unlike the architecture of ancient highly sinuous stream deposits.

River channel patterns may change in space and with time. During the last Ice Age (late Pleistocene), when thick ice caps covered much of the continents and sea level was considerably lower than today, the Mississippi River became deeply entrenched into its valley. As the ice melted and sea level rose, the Mississippi deposited a sheet of coarse grained sediment from low sinuosity streams. As sea level continued to rise, a much greater amount of suspended material was carried by the river, providing cohesive banks, and the present day highly sinuous river pattern was established.

Deposition from a braided stream

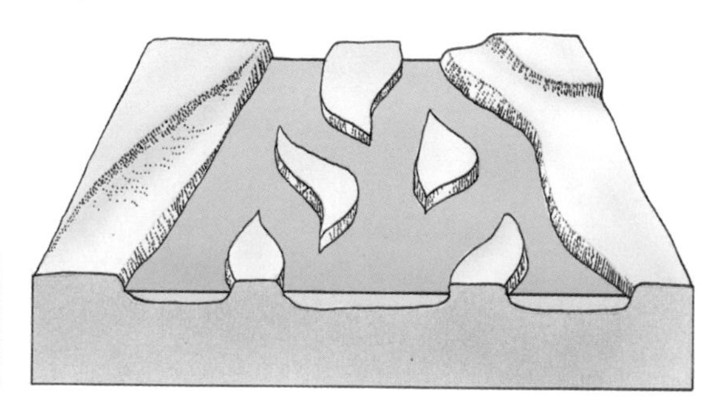

◄ *The continual deposition from a braided stream produces widespread thin beds of sediment, not like the thick point-bar deposits formed by the currents of a meandering river.*

► *A braided stream is quite unlike a meandering river, in that it fills the whole flood plain at the same time. Continual deposition gives shallow temporary sand banks around which the streams flow.*

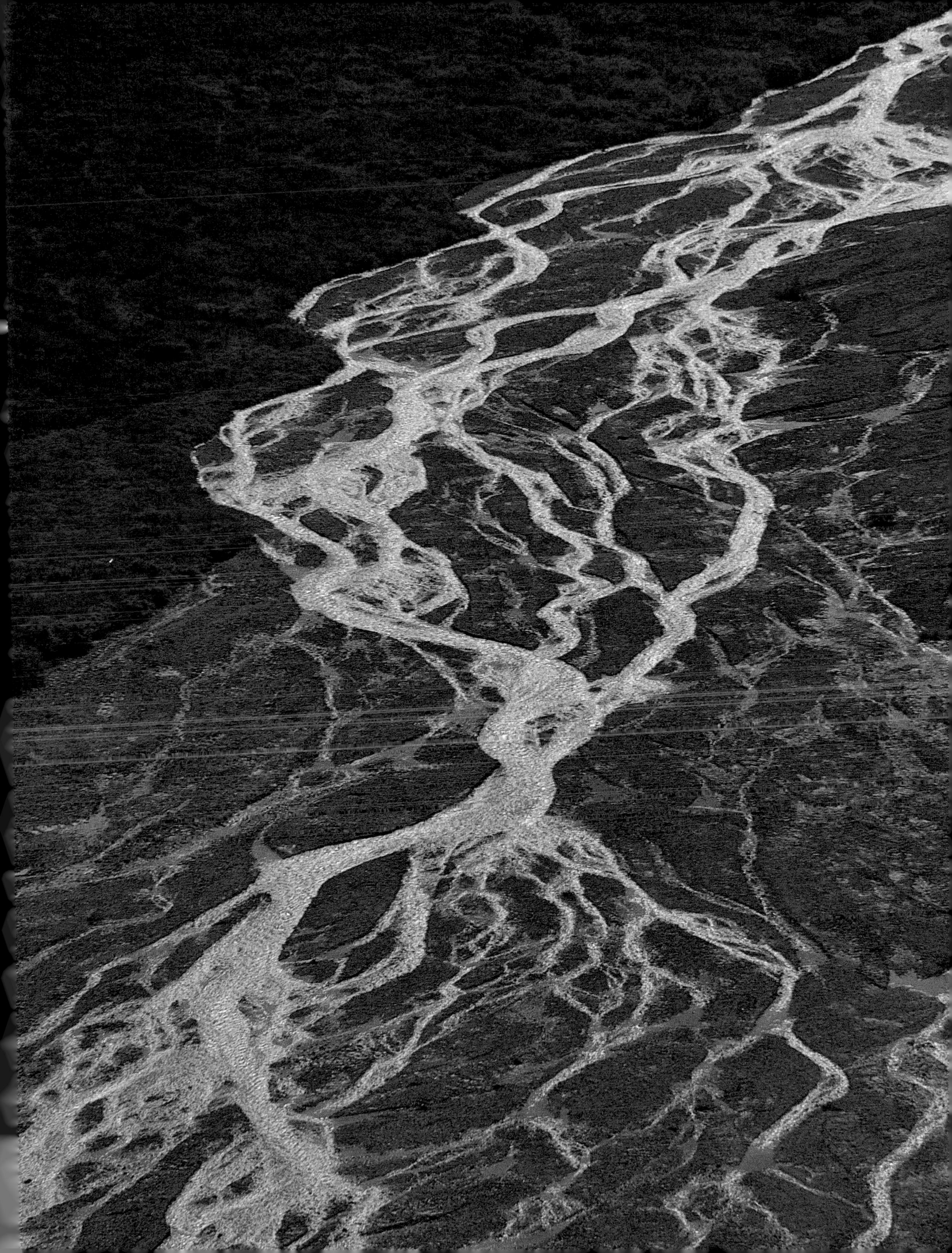

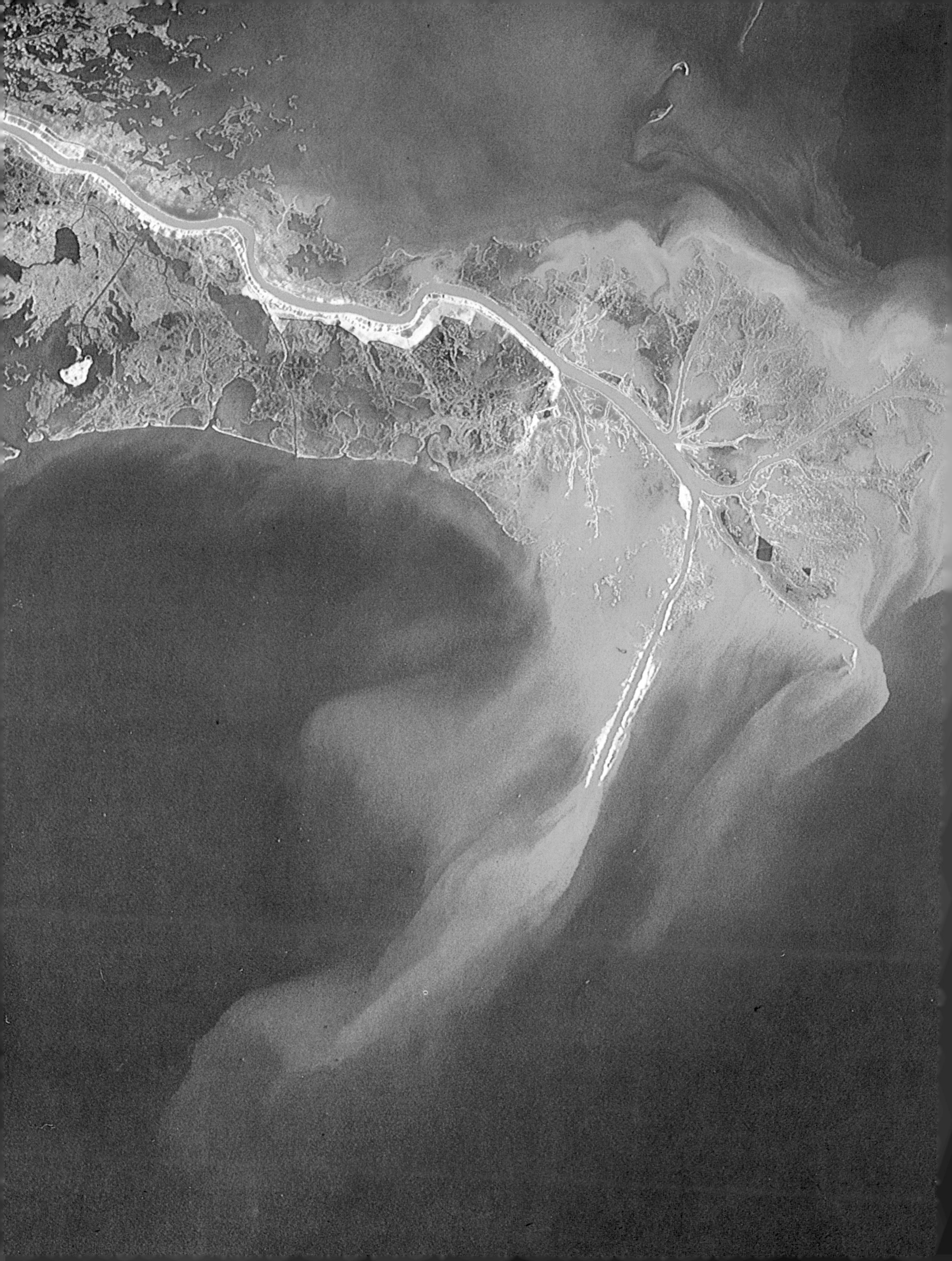

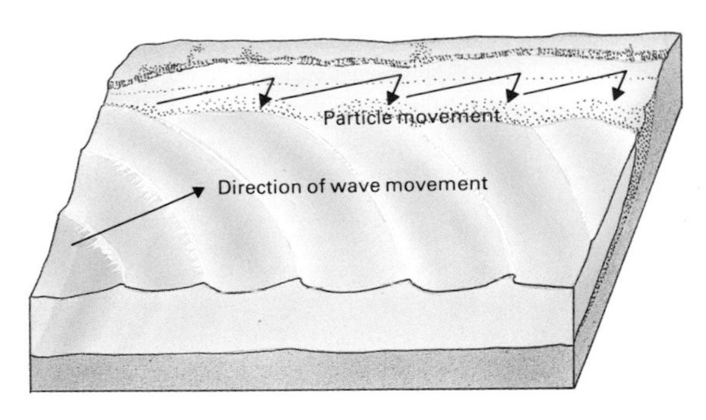

▲ *Beach sand and shingle are transported by longshore drift. Waves tend to strike a beach at an angle, washing the debris diagonally upwards. In the backwash the debris rolls straight back down again, to be washed up diagonally by the next wave.*

▼ *Many places try to halt the process of longshore drift by building groynes – fences running out to sea to hold back the moving sand. The result is a saw-toothed shoreline, with the sand hollowed out from behind each groyne and built up in the front.*

Shorelines and deltas

The world's coasts are marked by features called deltas that occur where river systems debouch into the ocean. Sediment reworked from deltas finds its way into longshore transport systems and is deposited as beaches, spits and bars.

A delta's form is controlled by a number of factors, the overriding one being the tidal and wave energies of the basin receiving the sediments. Where tidal and wave energies are low, the distributary channels are able to build out into the sea unhindered by coastal erosion. This produces a typical "birdfoot" pattern such as the Mississippi delta, in the USA. Where wave energies are strong compared to the river inflows and tides, the sediment delivered to the sea is molded into curved ridges at the delta's front and some is redistributed along the shore as beaches and spits. Deltas of this type, such as the Senegal (West Africa) or the Grijalva (Gulf of Mexico) are roughly arc-shaped and build out from the coast very slowly because of the destructive nature of approaching waves. Deltas strongly affected by tides have tidal channels cutting deep into the coastline. These channels tend to be clogged with sediment because of the to-and-fro motion of the tides, and elongated tidal sand ridges are exposed at low tide. The Ganges-Brahmaputra delta in the Bay of Bengal is of this type.

The nature of non-deltaic shorelines is also controlled by tides. In microtidal areas, where tidal ranges are less than 2m, barrier islands parallel to the shore are very common. In mesotidal areas (2 to 4m tidal range) these barrier islands are increasingly dissected by tidal inlets, and in macrotidal areas (tidal range greater than 4m) there are mud flats and salt marshes in large estuaries. Western Europe shows the transition from the microtidal Dutch coast with its elongated barrier islands to the macrotidal, indented coast of the German Bay.

Where waves approach the shoreline obliquely, sediment is transported alongshore in a "saw-tooth" manner, giving rise to a drift of sand along the shore. It is this process which banks up sand against groynes, built to prevent excessive beach erosion, and which causes the silting-up of harbors, as on the southern Californian coast.

As waves approach the coastline they peak and eventually break, forming a surf zone. On high energy oceanic coasts such as that of California and Oregon, USA, the full force of the Pacific Ocean waves is felt by the shoreline. Offshore the sand forms as ripples. Farther in, it forms larger-scale dunes. Within the surf zone it forms perfectly flat sand beds. In contrast, low to moderate energy coastlines, such as in Kouchibougiac Bay, New Brunswick, Canada, have well-developed offshore bars, sheltering depressions parallel to the coast in which silts and muds accumulate. Waves begin to break on these bars, then reform before dispensing their energy on the beach.

◄ ► *River debris is usually deposited as sediment not far from the river mouth. In sheltered seas the sediment may pile up into a delta through which the river reaches the sea in many channels. Different types of delta depend on the types of current and tide. They range from the birdfoot (1) of the Mississippi (left), the arcuate (2) of the Nile, the tidal (3) of the Mekong and the combined (4) of the river Niger.*

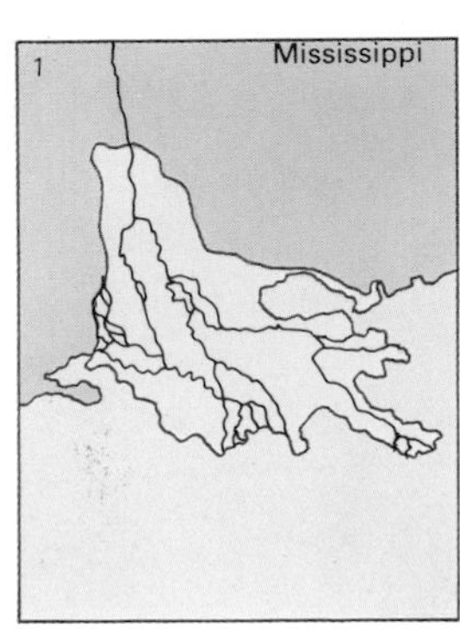

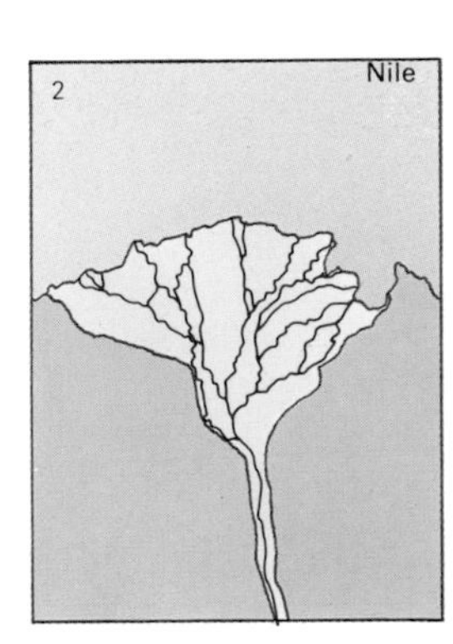

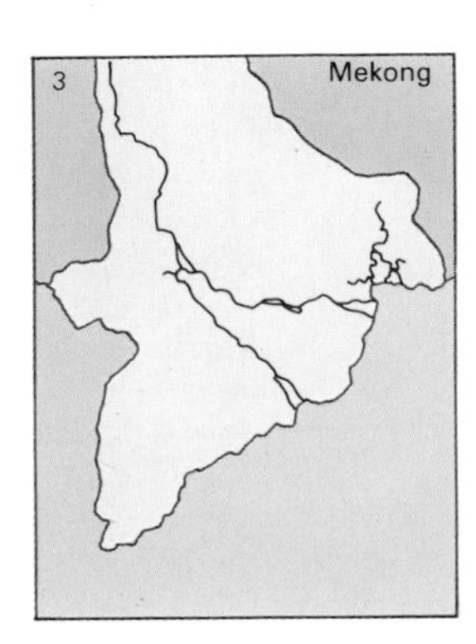

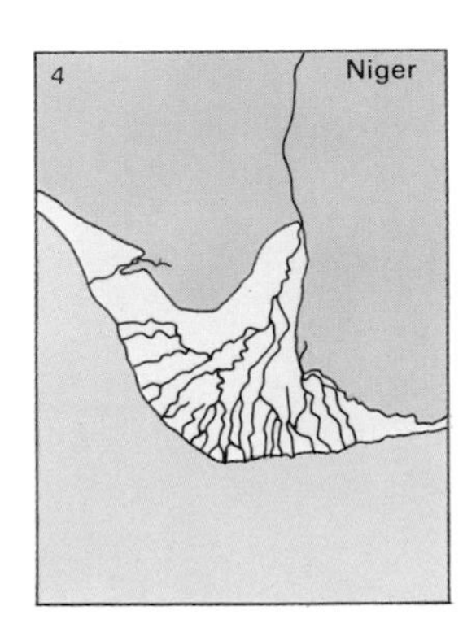

Early Sedimentologists

Major advances in sedimentology
Two Englishmen stand out as contributing significantly to the advancement of sedimentology. The first, Henry Sorby (1826-1908) was the pioneer of modern methods of analysis of sedimentary rocks, while the second, R.A. Bagnold (born 1896) has deepened the understanding of sediment mechanisms.

Throughout his life Sorby sought to apply methods in scientific research which no one had previously tried. He measured flow velocities in a stream running through the grounds of his Yorkshire home and compared these velocities with the bedforms on the stream bed. The earliest statement of the relationship of bedforms to velocity appears in a paper dated 1852, and a more general treatment entitled "On the Structures Produced by Currents During the Deposition of Stratified Rocks" was given in 1859. Sorby showed how structures produced by currents, called cross-bedding, could help in reconstructing ancient geographies. Sorby's innovative studies of microscopy represent the beginnings of disciplines as varied as crystallography, metallurgy and marine biology. During his life, Sorby received many honours from learned societies. His first presidential lecture to the Geological Society in 1879 entitled "On the Structure and Origin of Limestone" was to influence profoundly the way in which carbonate rocks are studied. Sorby studied British limestones and noted the alterations which they undergo when buried by younger sediments.

Ralph Bagnold is perhaps best known for his book "The Physics of Blown Sand and Desert Dunes", published in 1941 immediately before his departure on military service to the Middle East. The full impact of the book had to await its reprinting after the end of World War II. Bagnold's great achievement was to synthesize information from a number of normally disparate fields. His studies allowed sedimentologists and engineers to answer better the fundamental question "why are the beds of rivers or the waterless floors of deserts not flat?" Deserts are marked by an amazing simplicity of form with a geometric order more

▲ Henry Clifton Sorby was the pioneer British sedimentologist who studied the processes of deposition in modern rivers and seas, and then applied the principles to the sedimentary structures he found in ancient rocks.

▼ ► Current bedding in river sediments forms as a tongue of sand builds out in S-shaped beds (1). The next phase erodes the top off the S-shapes and deposits more (2). These curved structures can be seen in ancient rocks (3,right).

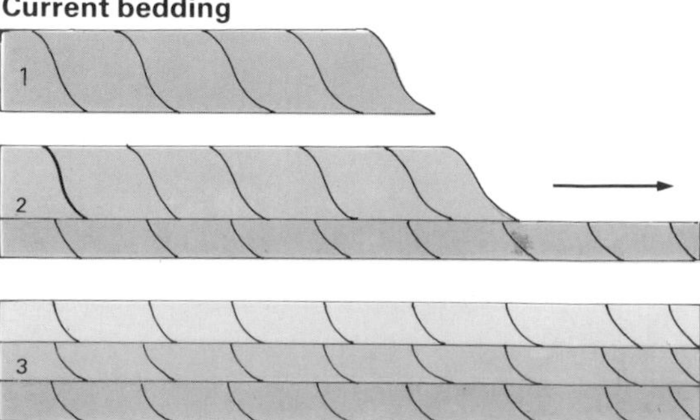

1

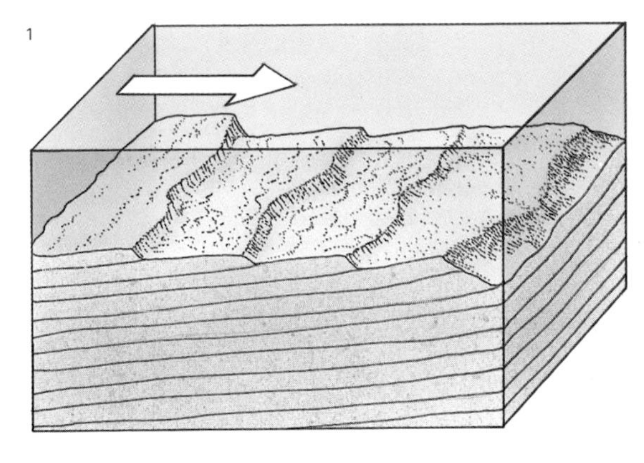

2

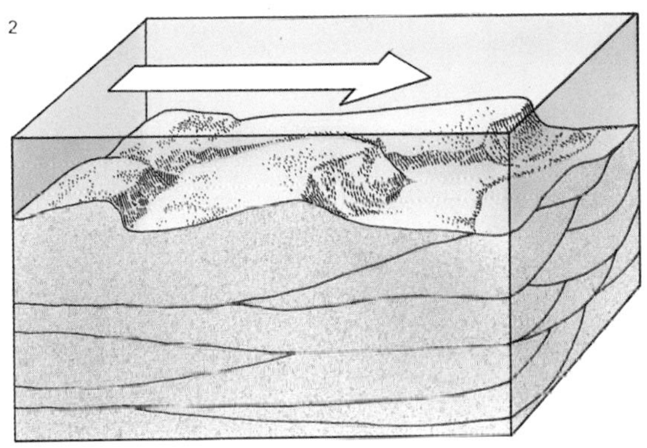

3

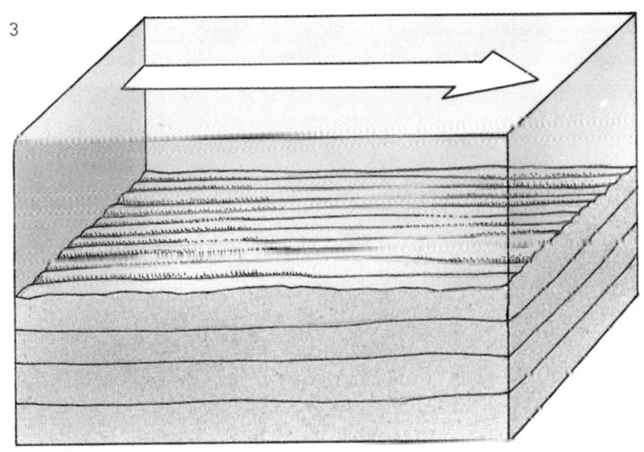

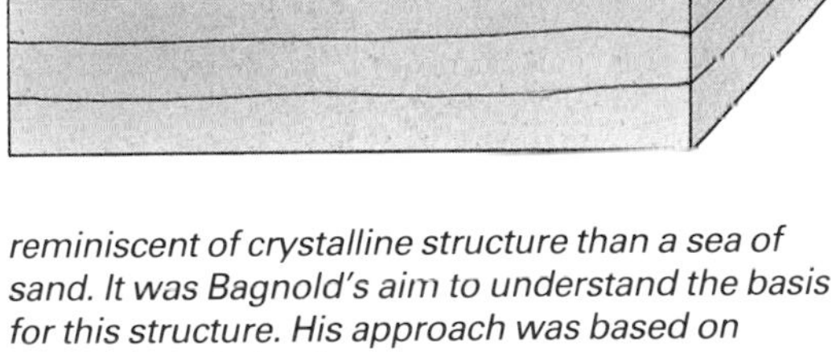

▲ Saltation, the bouncing of fragments along a surface propelled by a current, can only lift sand particles to a height of a meter or so on dry land. Hence, in windy sandy areas, heavy sandblasting close to the ground forms mushroom-shaped rocks.

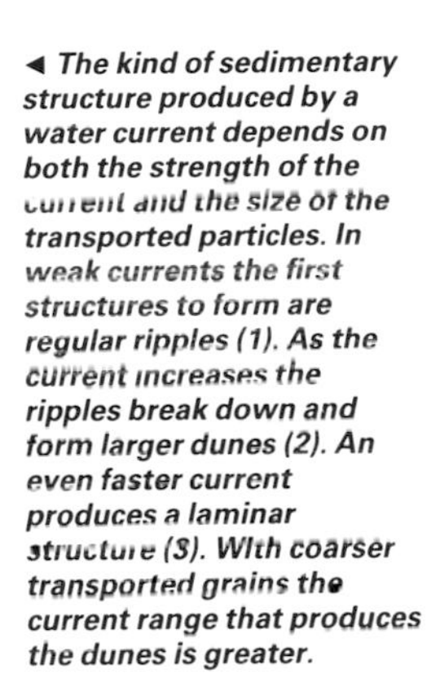

◄ The kind of sedimentary structure produced by a water current depends on both the strength of the current and the size of the transported particles. In weak currents the first structures to form are regular ripples (1). As the current increases the ripples break down and form larger dunes (2). An even faster current produces a laminar structure (3). With coarser transported grains the current range that produces the dunes is greater.

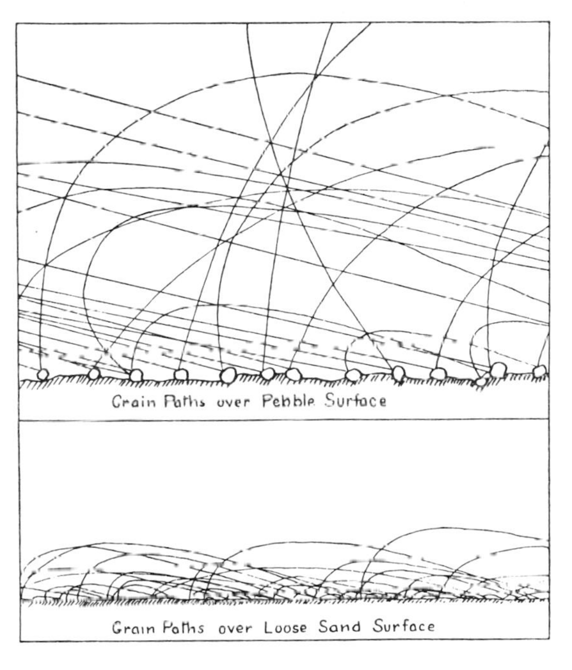

▲ Bagnold's original drawing shows that the height reached by a bouncing fragment during saltation is very much dependent on the type of surface it is bouncing across. A rough stony surface (upper) will send the fragments flying to greater heights than will a smooth level surface (lower).

reminiscent of crystalline structure than a sea of sand. It was Bagnold's aim to understand the basis for this structure. His approach was based on experimental physics.

Bagnold proposed that the movement of sand under a flow of air is an essentially surface effect. Sand grains lifted into the air flow return to the ground along flat trajectories, thereby colliding with stationary sand grains in the bed. The great American geomorphologist G.K. Gilbert in 1914 called this process of sand movement under water "saltation" and Bagnold adopted the same term for sand movement in air, noting that it was far more effective in the latter. The energy of descending saltating grains is spent partly in dislodging other sand grains into the air, but also in setting up a slow surface creep of sand. The rippling of desert sands could then be explained.

Later, Bagnold turned his attention to processes of sand movement under water and provided a widely-used sediment transport equation.

◄ Ralph Bagnold pioneered the study of saltation and other related subjects. His work brought him the award of the Order of the British Empire in 1941.

The world's reefs

Much of the continental shelf between the latitudes of 30°S and 30°N is an area of high organic productivity and is covered not by river derived clastic sediment, but by organic carbonate material.

There are two major categories of subtropical carbonate shelf. "Protected shelf lagoons" are rimmed by coral reefs, as in the Bahamas and the Great Barrier Reef of Australia, and their margins often fall precipitously into the abyssal depths. "Open shelves" on the other hand, such as Yucatan, western Florida and northern Australia, slope gently toward the continental edge and are ravaged by storm waves and tidal currents. The Bahama Platform and South Florida Shelf are the vestiges of a former extensive and continuous rimmed shelf. The Bahama Platform is bounded on all sides by steep slopes which plunge to thousands of meters depth, but the platform itself lies under very shallow waters normally of less than 6m. At the edge of the platform are reefs and a zone of small spherical particles resembling fish-roe. These grains, called ooids, are intermittently rolled around the sea floor by waves and tides and commonly form long submarine ramparts. These barriers are often eroded by storms and redeposited as large lobes ("spillovers"). Inside this belt of mobile sediment is a quieter region where carbonate mud accumulates.

The Yucatan Shelf in the Gulf of Mexico lacks a guarding rim of reef barriers. At the shore are dune ridges and a shell beach, testifying to the high wave energies affecting the coast. These shelter small lagoons and densely vegetated, swampy mudflats. Seaward the inner shelf is dominated by shell debris and farther out a continual "rain" of microfossils from plankton (foraminifera) is dominant.

The three major types of reef existing today are fringing reefs, barrier reefs and atolls. Fringing reefs are directly attached to the shore zone, barrier reefs are separated from the land by a lagoon, and atolls are circular reefs occurring in the open ocean.

▲ *Corals are the builders of tropical reefs.*

The Great Barrier Reef

The 2000km long Great Barrier Reef lies off the eastern coast of Queensland, Australia. The myriad of individual reefs consist mostly of the skeletons of corals, providing a framework bound together by coralline algae and bryozoans. The diversity of organisms is very high, with 350 species of coral and 146 of sea-urchin. At the northern end, the shelf edge is steep and the reef continuous and linear. Towards the south, the shelf edge is gentle and the linear marginal reefs are replaced by scattered ring reefs. There are clear zones within this. The area beyond the growing reef is relatively barren and is marked by wave-agitated sands. The reef slope rises at about 30°, up to a serrated and fast growing reef front. Behind it is a 500m wide zone exposed at low tide that is dominated by red algae; farther inland are isolated corals and coral pools, areas of dead coral and sand flats.

▲ ***Corals will live only under particular conditions of water clarity, depth and temperature. In tropical oceans these conditions are met close inshore around islands where few rivers are bringing sediment to the sea. There develops a fringing reef which is a shelf of coral material growing at sea level outwards from the shoreline of the island. Living corals build up on the dead skeletons of past generations.***

► ***In a barrier reef a lagoon of shallow water separates the reef itself from its island. The living reef forms a ring surrounding the island and within this the lagoon is floored with broken fragments of dead coral skeletons. The reef tends to grow more vigorously and extends outwards farther on the side facing the prevailing current, since the corals' food matter is brought from this direction.***

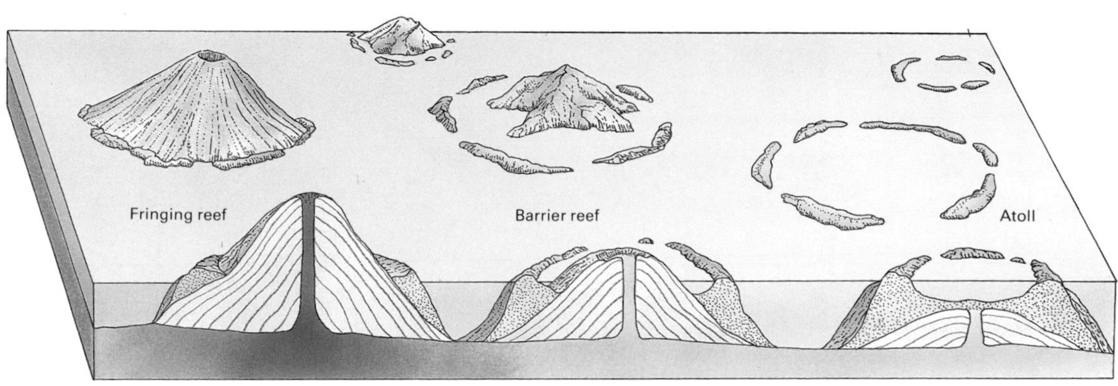

▲ ► In an atoll a coral reef forms a ring with no central island. This sequence of reefs – fringing-barrier-atoll – may result from island subsidence. A reef starts as a fringing reef, then when the island sinks and becomes smaller the barrier reef builds upwards finally producing an atoll when the island disappears. The descent of a volcanic island from an ocean ridge by plate movement would account for this.

Deep-sea sediment

The continental shelf is marked by great incisions which act as sediment transport routes to the deep sea. These submarine canyons feed sediment to submarine cones or "deep-sea fans". They are formed by the scavenging of shelf and coastal sediment by dense, fast-moving sediment and water mixtures called turbidity currents. Near the head of the fan the submarine topography is rugged, being dissected by large channels. Farther down the fan, the main fan valley splits into many distributaries and lobes of sand are deposited where they end. At the toes of the fans the topography is smooth, with only small channels. Turbidity currents progressively drop their sediment load as they travel down toward the deep sea floor, so that the coarsest particles are retained near the inner part of the fan and only silt and clay reach the basin plain. Classic present-day deep-sea fans occur off the west coast of North America. They range in size from a few tens of kilometers radius for the California borderland fans to some 300km radius for the Monterey Fan.

The large sedimentary cones occurring off major river deltas are termed "abyssal cones". They receive sediments from turbidity currents triggered by river discharges. Abyssal cones border vast regions of essentially flat ocean floor, such as the North Atlantic. Instead of possessing a distinct apex, a broad head grades imperceptibly into the continental slope. Channels extend over most of the abyssal cone surface, but the sediments are usually very fine because the sediment load of inflowing rivers is usually held in suspension. The surface of the abyssal cones may bear the scars and slumped masses of huge slices of surface sediment which have slipped downslope. These slumps may involve hundreds of cubic kilometers of sediment, as in the 350m thick Grand Banks slump of 1929, and may travel for tens of kilometers. Abyssal cones widen towards the abyssal plain, and may be 3,000km long, as in the Bengal Fan.

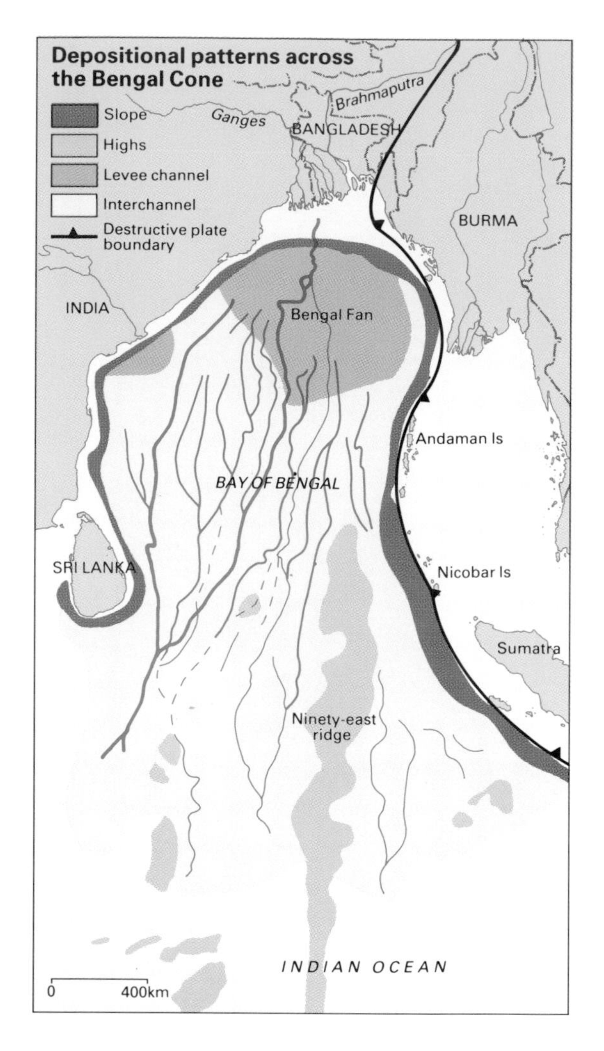

▲ **_The edge of the continental shelf is often ill-defined. This is usually because a vast quantity of river sediment has spilled across the shelf from a particularly big river mouth and has spread down the slope and on to the abyssal plain. The Bengal Cone off the mouth of the Ganges is a particularly good example._**

◄ **_Sediment moving down the continental slope does so in "turbidity currents". A thick suspension of sand and rock fragments moves downwards as a coherent mass and collects on the continental rise. Turbidites – rocks formed of a jumble of different particle sizes and showing current structures – were formed like this._**

▼ **_It is well known that the continental crust extends outwards for some distance beyond the actual coastlines of the continents. This area of submerged continent covered in shallow seas and terrigenous sediments is known as the continental shelf and descends very gradually away from the dry land. At the edge of the shelf the gradient becomes somewhat steeper and produces the continental slope. This slackens off at the bottom to another gentle gradient called the continental rise formed by sediment that has fallen down the continental slope, and this in turn passes to the relative levels of the abyssal plain. In reality all these slopes are quite gentle, only a few degrees._**

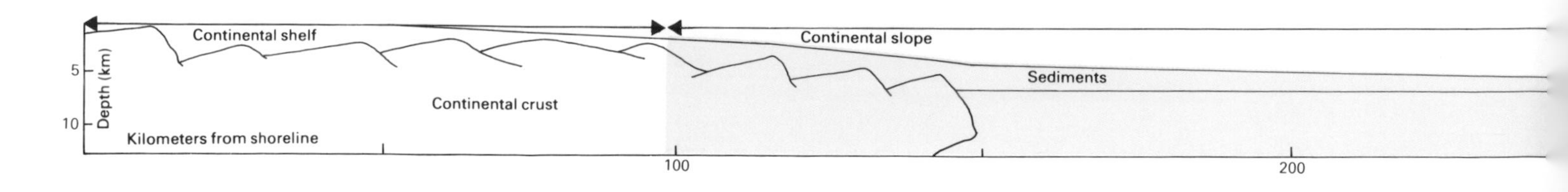

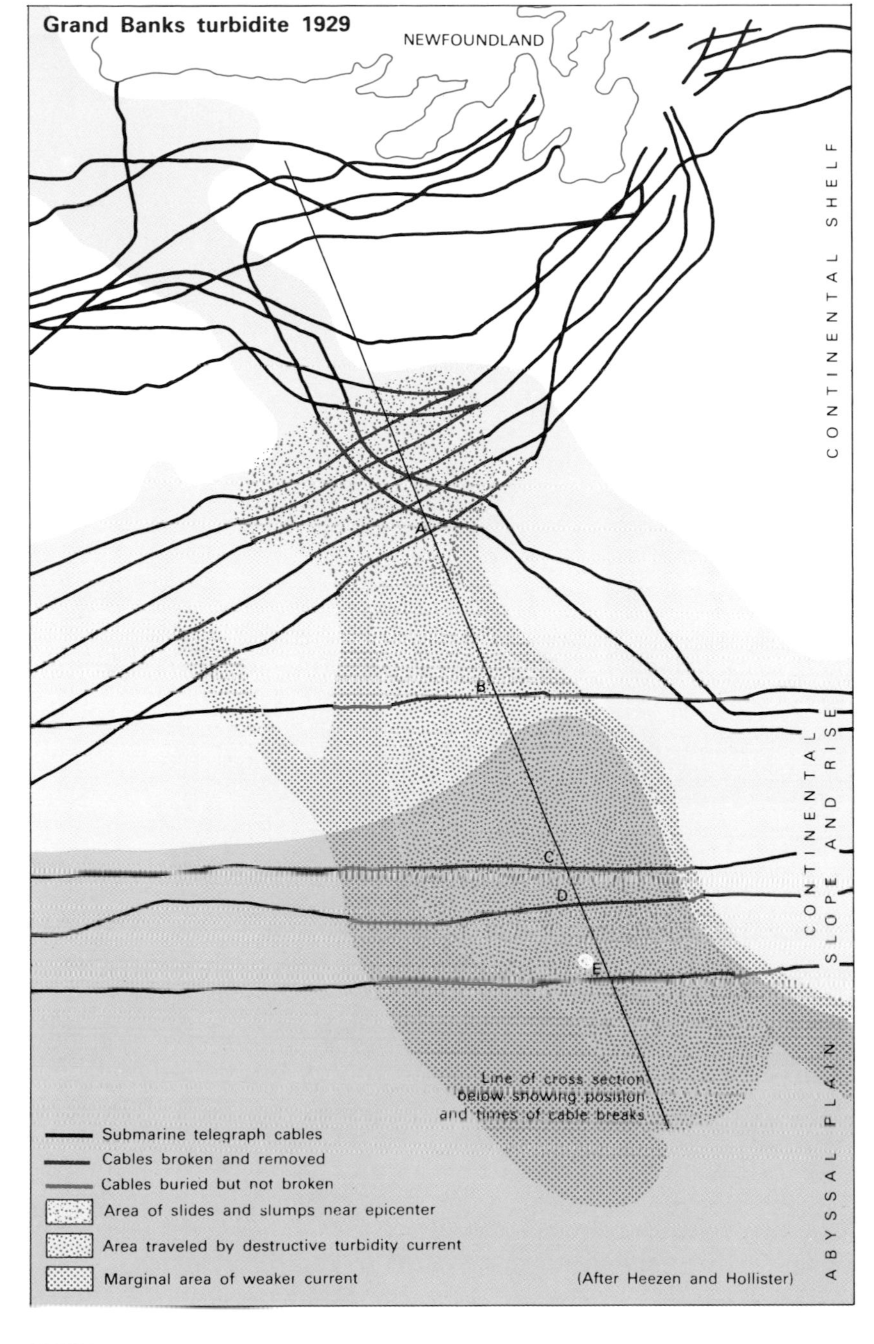

Explaining sedimentation

Prior to the late 1940s and 50s all sedimentation in the deep sea was thought to result from a steady fall-out from the overlying water mass, and the occurrence of sandstones was thought to indicate deposition in shallow waters. Turbidity current theory has revolutionized these ideas.

The Swiss geographer François Forel (1841-1912), in his classic monograph of 1879, recognised that sediment-charged suspensions spread from the delta of the Rhône River as underflows onto the floor of the Lake of Geneva. Forty years later similar results were obtained near the entrance of the Colorado River into Lake Mead, USA. At about the same time, density undercurrents were postulated to be responsible for submarine canyons at continental edges. The first laboratory experiments were conducted by the Dutch geologist Philip Kuenen (born 1902) and the term "turbidity current" was coined for these sediment-charged suspensions. The emphasis in Kuenen's later experiments in the 1950s was to show that turbidity currents were not only erosional agents, but were capable of transporting and depositing large amounts of sand in the deep sea. Identification of the deposits of turbidity currents (rocks called turbidites) in the ancient sedimentary record was a natural progression. The late 1950s and 1960s saw an explosion of scientific inquiry into ancient turbidites and the distribution of rocks formed by them.

The best-documented example of a naturally-occurring turbidity current is from the Grand Banks area off Newfoundland in Canada. The thick sequence of sediment built up by the St Lawrence River failed after an earthquake in 1929. Following the earthquake, submarine telegraph cables were broken in sequence from shallow to deep, recording the downslope passage of an immense slump 100km long and, in front of it, a turbidity current. The broken cables acted as a stopwatch for the progress of the turbidity current, transmissions stopping at the exact time of the breaks, and these showed that the current traveled at a maximum velocity of some 70km per hour! Similar cable breaks have been reported elsewhere, such as off the coast of Algeria, and turbidity currents have been shown to travel for as much as 4,000km before finally coming to rest.

◄ ▲ The speed of turbidity currents was measured in a dramatic manner in 1929 when an undersea earthquake triggered a flow of suspended sand and mud on the continental slope off the Grand Banks near Nova Scotia. The initial slump broke a number of undersea telegraph cables in the area, and as it surged downwards it cut through progressively deeper cables that lay across its path. The time of each cut could be calculated exactly by the time of the loss of transmission through each cable. Cables 500km away were damaged and the speed of the current was estimated as 70 kilometers per hour. Currents such as these cut deep canyons down the flanks of the continental slopes.

▲ **Ancient deep-sea sediments can be found in the geological record. Ordovician black shales are ocean-floor deposits and contain fossil surface-dwelling graptolites.**

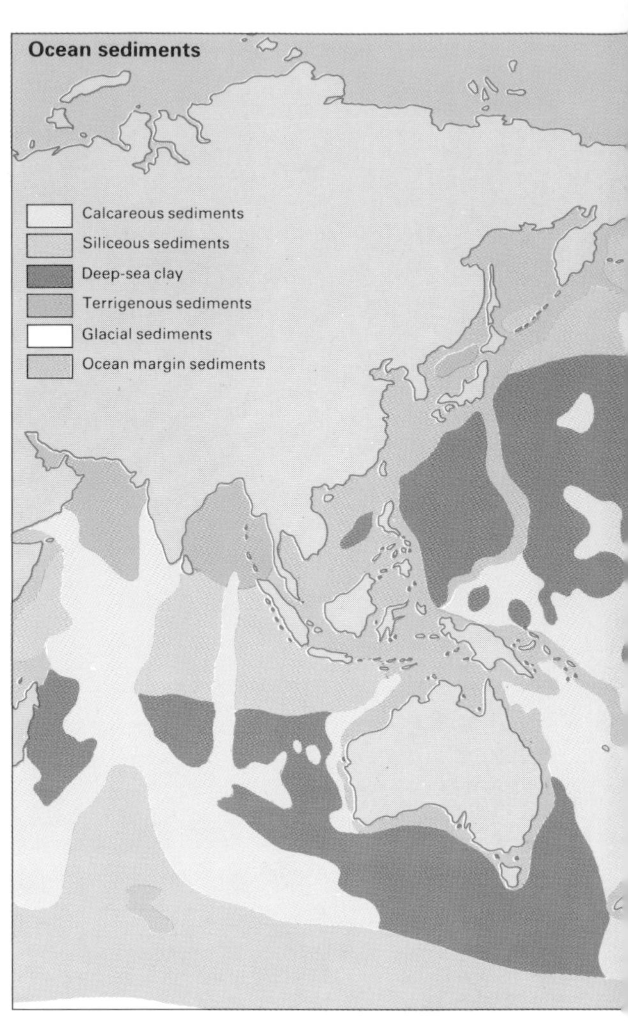

► **Many of the deep-sea sediments consist of oozes made from the skeletons of tiny sea creatures, but where all organic material is dissolved away there may be only a red clay.**

▼ **Radiolaria are microscopic organisms with shells made of silica. At places on the ocean floor the sediment consists of nothing but these shells giving a siliceous ooze.**

The ultimate sediment trap

Flat abyssal plains are the ultimate sediment trap both for material eroded from continents and submarine highs, and for the steady rain of the remains of planktonic organisms which dominate the surface waters. The flatness of the abyssal plain, a result of the smothering by sediment of an irregular ocean floor, is broken only by the submarine mountains marking the lines of rupture of the Earth's crust. Abyssal plains vary greatly in size from the vast floors of open oceans to small fault-bounded basins such as those off California. They also occur at the sites of ocean plate descent, called trenches, where they take on a highly elongated form. Fine-grained sediments are supplied to these deep and remote basins by far-traveled turbidity currents.

Sedimentation in the open sea beyond the influence of the continental land masses is controlled by two major factors, the fertility of the surface water and the presence of a "calcite compensation depth" (CCD) below which carbonate from the skeletons of marine organisms is dissolved. Above the CCD, calcareous oozes derived from micro-organisms such as foraminifera predominate. Below this, the silica skeletons of radiolaria and diatoms produce siliceous oozes, and there are great expansions of the red or brown clays derived from volcanoes, meteorites and dust blown from continents. Deposits of organic origin only accumulate rapidly below areas of high surface water fertility, such as at areas of upwelling of nutrient-rich water and in equatorial zones.

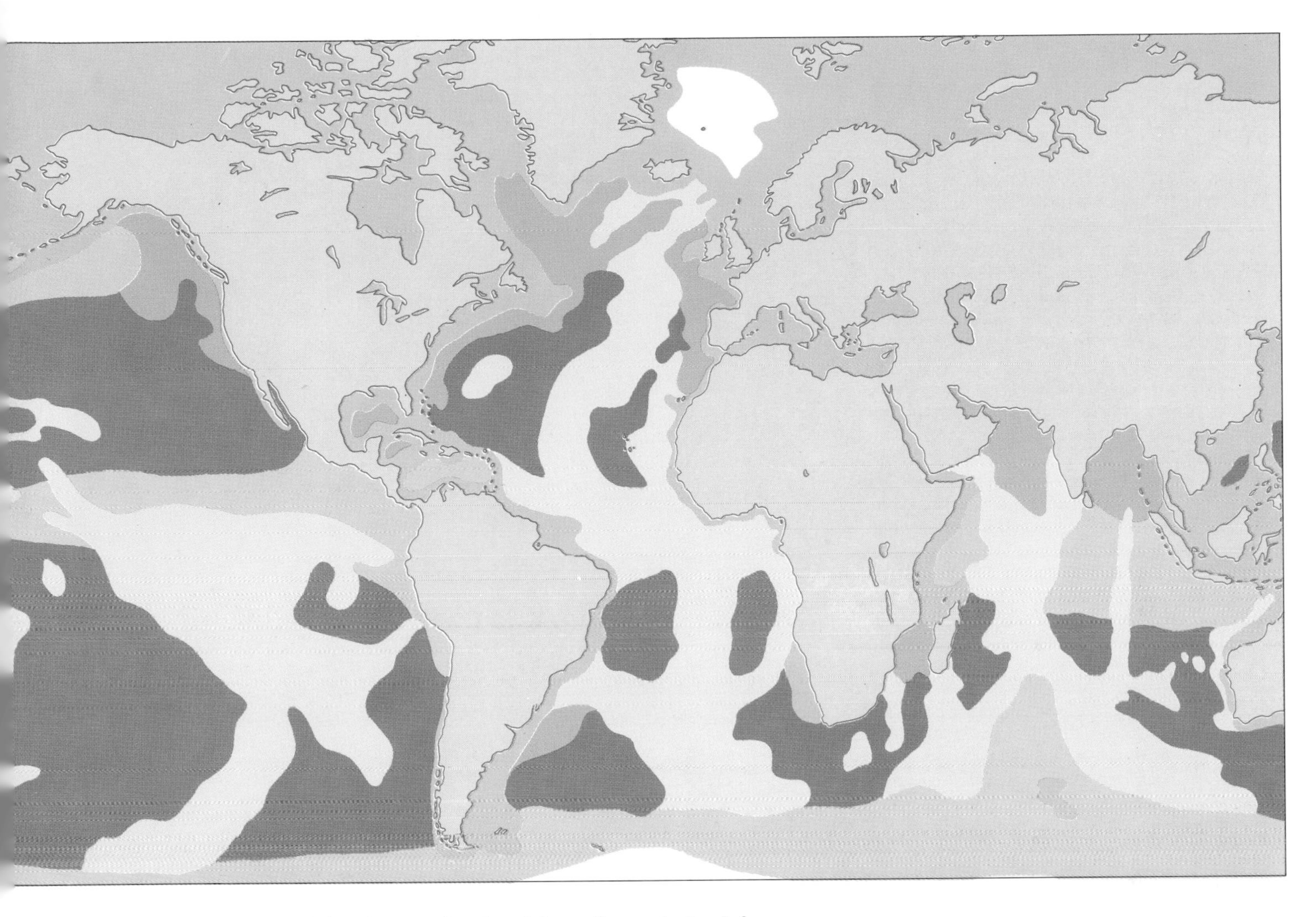

Abyssal currents originating as density-driven flows derived from polar ice-caps are funneled by the oceanic topography and are responsible for mixing of the water mass (◆ page 126). At times of equable climate these currents may have been very weak, allowing the waters to become stratified with a cold bottom layer devoid of oxygen. Combined with greater organic productivity in surface waters at such times, this may have been responsible for the development and preservation of black shales (◆ page 127).

In the vicinity of ocean ridges, sedimentary globules rich in metals, particularly iron and manganese, are found. They are formed either by the upward passage of fluids from deep in the Earth, or by the interaction of sea water and ocean-floor rocks. Away from the mid-ocean ridges, these nodules cover large areas of the deep abyssal plain, for example in the southeast Indian Ocean. Nodules are found in areas of the ocean floor where abyssal currents have swept the sea floor clean of finer sediment.

Because of the movement and collision of the Earth's lithospheric plates, ocean basins are continually being closed and reborn. As a consequence it is possible to view the deposits of the ocean basins together with their basement of ocean crust on land today. Such "ophiolites" are scattered across the globe, but well-studied examples comprise the Troodos massif of Cyprus and the Californian Coast Range Ophiolite. Ocean water sediments are also commonly found dismembered from their oceanic basements.

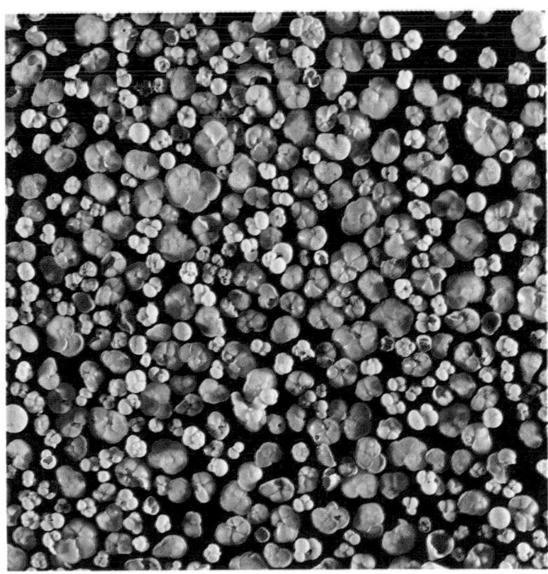

▲ *Much of the ocean floor is covered by a calcareous ooze, made up of the calcite shells of sea organisms. These organisms may be pteropods (floating sea-snails) or, as here, microscopic foraminifera. In some areas the temperature and the pressure of the water are such as to dissolve away all calcareous material. These areas have predominantly clay or silica in their bottom deposits.*

See also
The Dynamic Earth 9-24
Dating the Rocks 73-80
The Earth Through Time 81-100
Development of the Oceans 121-8
Rocks and the Rock Cycle 141-52
Landscape Features 181-96

Drilling the ocean floors

Scientific investigations of the oceans by drilling began only in 1968 with an American venture known as the Deep Sea Drilling Project (DSDP), and since 1975 Britain, France, Germany, Japan and the USSR (until 1982) have collaborated in an international phase of drilling. In the first decade of DSDP 460 boreholes were drilled using the drill ship "Glomar Challenger". The "Glomar Challenger" is not fastened to the sea floor in any way, but simply floats freely at the surface. It maintains its position above the sea floor while drilling by means of a computerized navigational system and powerful thrusters on board the ship.

The scientific results of over a decade of deep sea drilling have been startling and of relevance to almost every branch of geology. For example, between 5 and 12 million years ago the Mediterranean must have completely dried up, an event called the "Messinian Salinity Crisis". Thick-bedded sun-baked salts are found in cores from large parts of the Mediterranean. Other discoveries were that the Atlantic Ocean grew from a series of narrow stagnant basins into the immense basin of today, that Antarctica has been covered with ice for the last 20 million years and that the ice cap was much more extensive some 5 million years ago.

Although DSDP has achieved much, the coverage of the sea floor is still only one drill hole per 800,000 square kilometers and the continental margins and enormously deep ocean trenches are still poorly sampled. The scientific results of the next phase of drilling, the Ocean Drilling Program (ODP) are likely to be just as unexpected and equally exciting.

▲ The "Glomar Challenger", the ship used for the Deep Sea Drilling Project, with its 43m high derrick. Drill string sections lie on the afterdeck.

▼ By 1982 "Glomar Challenger" had drilled nearly 600 boreholes from the deep ocean basins, sampling both deep-sea sediments and crust.

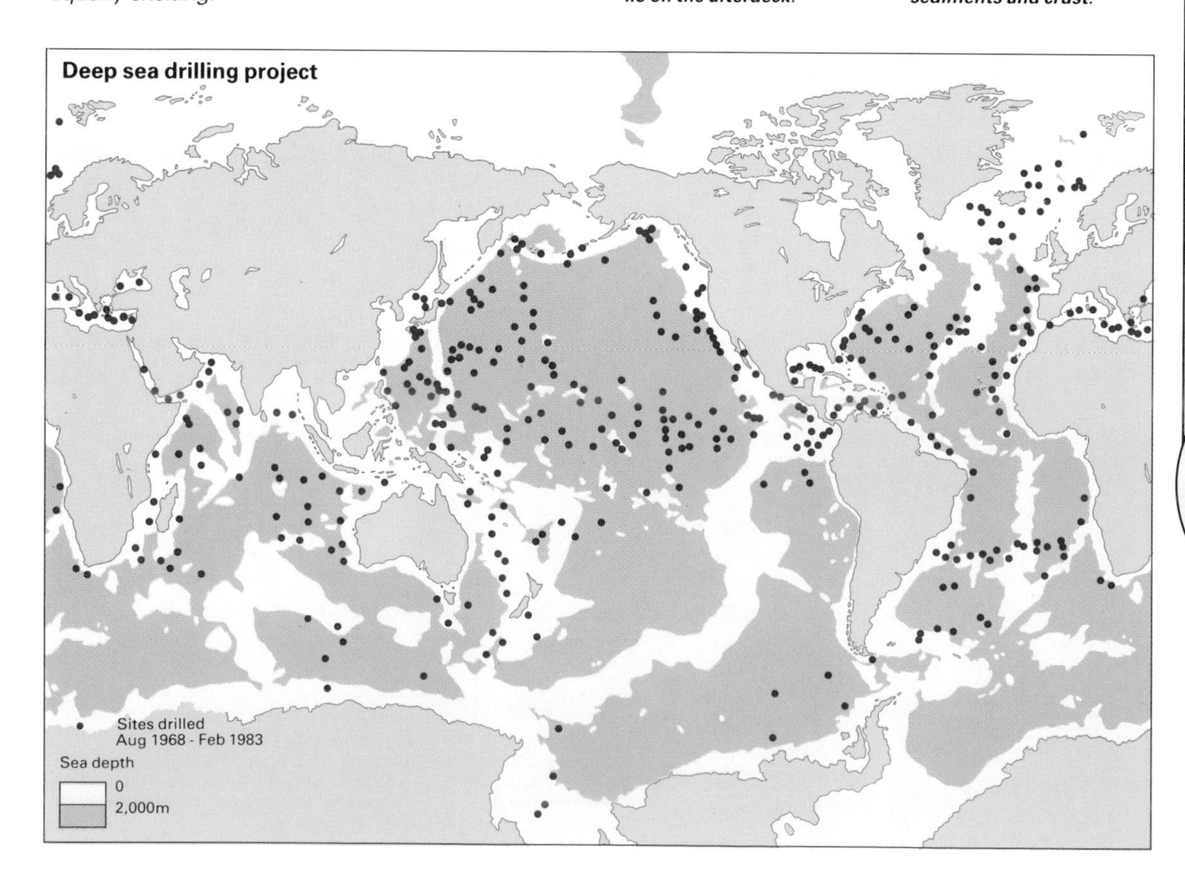

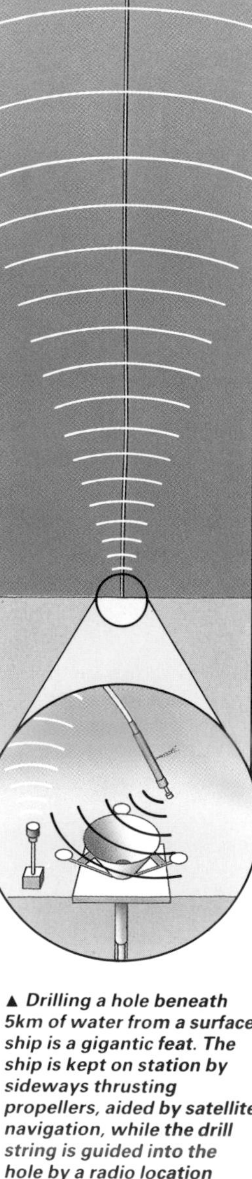

▲ Drilling a hole beneath 5km of water from a surface ship is a gigantic feat. The ship is kept on station by sideways thrusting propellers, aided by satellite navigation, while the drill string is guided into the hole by a radio location device.

Development of Soils

17

Soil – the outer covering...influence of living things – rock destruction, digestion and mixing... Fossil soils – buried and preserved...PERSPECTIVE... Soil profiles...Soils and their development...Types of soil...Growth of plants...Soil profiles around the world...Fossil evidence

Soils form on land surfaces where the hard rock or soft loose sediment of the surface is modified by many physical, chemical and biological processes dependent on the proximity of the atmosphere. The different processes involved in their development are interdependent and may act simultaneously.

The chemical processes include hydrolysis (the breaking up of chemical compounds by water, making new compounds containing hydrogen and oxygen), carbonation (the formation of carbonate minerals by reaction with the carbon dioxide of the air) and oxidation-reduction reactions (in which oxygen is added to, or taken from, a mineral compound). Simple minerals such as calcite or quartz are often completely dissolved, but many of the more complex aluminosilicate minerals are only partly dissolved and leave a solid residue. The carbon dioxide for these reactions comes mainly from the oxidation of humus – the organic substance formed by the partial decay of plant material.

The rates of most chemical weathering processes (◆ page 148) depend on temperature and the removal of the soluble products by downward flow of water (leaching). They are consequently most rapid in warm wet climates and occur slowly or not at all in cold and dry regions. In contrast, physical weathering processes, by which rock and mineral particles are made smaller but are not changed chemically, predominate in arctic regions and deserts. The resulting barren rocky materials are still considered soils, even though they are ineffective media for plant growth.

A soil profile

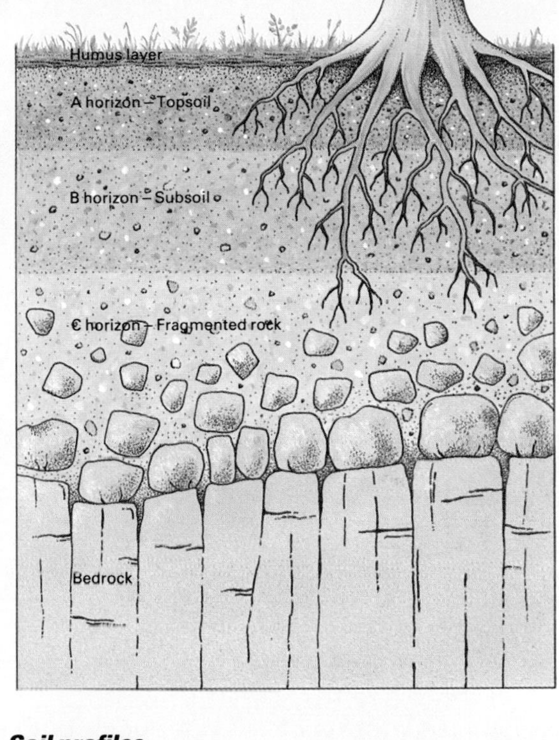

Soil profiles
A soil profile (above) is a vertical section from the ground surface through the altered material into the unaltered rock or sediment below. It can often be divided into a sequence of different layers (horizons), which are parallel to the ground surface but unconformable with structures in the rock, such as inclined bedding.

A humus layer at the top is underlain by the topsoil. Below this is the subsoil, poorer in organic matter but richer in minerals. A horizon of fragmented rock separates this from the parent rock itself.

► The substance known as soil is a complex mixture of fresh and decayed rock, and of fresh and decaying organic matter. It is the product of the weathering of the underlying rock of an area, of the reorganization of this weathered material by percolating water, and of all the biological events that take place in it from the growth of trees to the burrowing of worms. It is the natural product of thousands, perhaps millions, of years of these activities in a particular area and, in its natural state, represents a fine balance. Since soil is the basis of agriculture it is important to civilization. But careless farming can lead to degradation of the soil by erosion or loss of nutrients.

A square meter of soil may contain over a billion individual living things

◄ ***Primitive algae, such as the red "Trentepohlia", can grow on bare rock, and its biological action can begin to break it down.***

▼ ◄ ***Most soil creatures are microscopic, including the tiny mites, such as that seen here in a scanning electron micrograph.***

▼ ***Centipedes are small-scale predators of the soil. They eat mites and the smaller insects that feed on vegetable matter.***

▼ ► ***A larger burrowing predator is the mole, tunnelling through the soil and eating the worms that it finds there.***

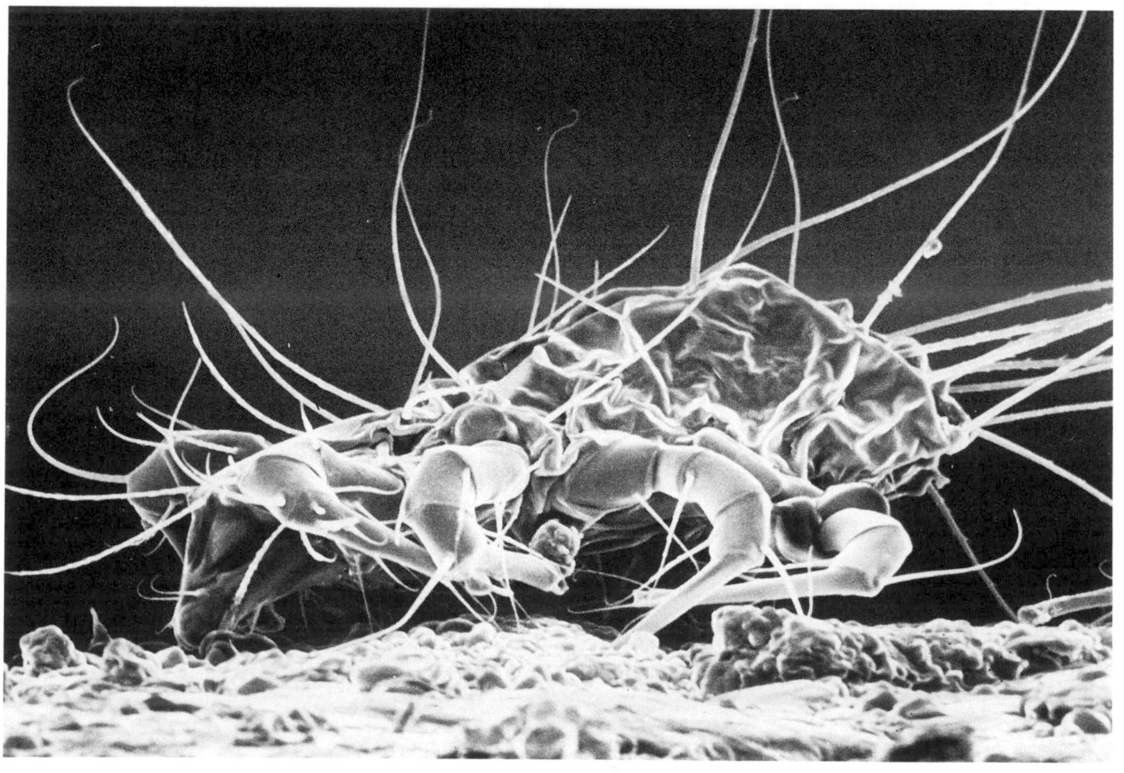

Soils and their development

The way in which soil enables plants to grow has been a matter of speculation and scientific study since at least Roman times. Farmers knew that by adding to the soil farmyard manure and parts of dead animals, such as blood and ground bones, the growth of their crops increased. Yet early scientists thought that the soil merely provided a physical support for plants and acted as a sponge to hold the only substance (water) necessary for their growth.

This was apparently confirmed by the famous experiment of Van Helmont (1577-1644), who grew a willow shoot in a vessel containing a known weight of dry soil to which he added only rainwater. After five years the tree weighed almost 34 times the original shoot, but the dry weight of soil had decreased very little. However, Van Helmont had ignored three important processes. The atmosphere provides carbon as carbon dioxide, the soil provides small but necessary amounts of elements such as potassium derived from the alteration of minerals, and the soil micro-organisms provide nitrogen derived ultimately from the atmosphere but in a form that the willow's roots could assimilate.

Field experiments on the growth of crops, such as those using farmyard manure and artificial inorganic fertilizers begun in the 19th century later showed the importance of nitrogen, phosphorus, potassium, magnesium, calcium and sodium as major plant foods. More recently traces of sulfur, selenium, iron, manganese, zinc, boron, molybdenum and cobalt have also proved to be important for healthy plant growth, though excesses of some of these and of other non-essential elements can be toxic.

The influence of living things

Biological processes are extremely important in soil formation. The first plants to colonize a fresh rock or sediment surface are lichens, algae and mosses, many of which are able to extract nutrients direct from minerals in the rock and to convert atmospheric nitrogen to protein. When these plants die their decomposing remains provide nourishment for a succession of larger plants, often leading eventually to trees. The roots of larger plants grow downwards into soft sediments or through the cracks in hard rock. This action slowly opens up the surface layers allowing water, air and animals to penetrate, and the decomposing plant remains (humus) to mix with rock and mineral fragments.

Soil animals exceed in total numbers those of all other environments put together. The largest of them, such as moles, rabbits and badgers, locally disturb and mix the soil by burrowing. Smaller soil animals are more widely distributed. Earthworms ingest fine soil particles and deposit them on the ground surface as wormcasts. Surface layers of fine soil several centimeters thick may accumulate in this way over several centuries. Termites are social insects common in tropical soils. They eat plant remains, and build nests either below the ground surface or in large mounds above. A range of smaller animals including mites, springtails, centipedes and millipedes constitutes the soil mesofauna. Like earthworms and termites these also help to decompose leaves and other plant remains by eating them. They are consequently less common in cultivated arable soils than in forest and grassland soils, which contain more plant litter. Yet smaller soil micro-organisms include various protozoa, viruses, fungi, algae and bacteria. The most important are the bacteria, which are responsible for further decomposition of organic remains in both aerobic and anaerobic situations. Fungi also help decompose plant remains but usually flourish in acidic soils, where bacteria are less active.

Different soil thicknesses

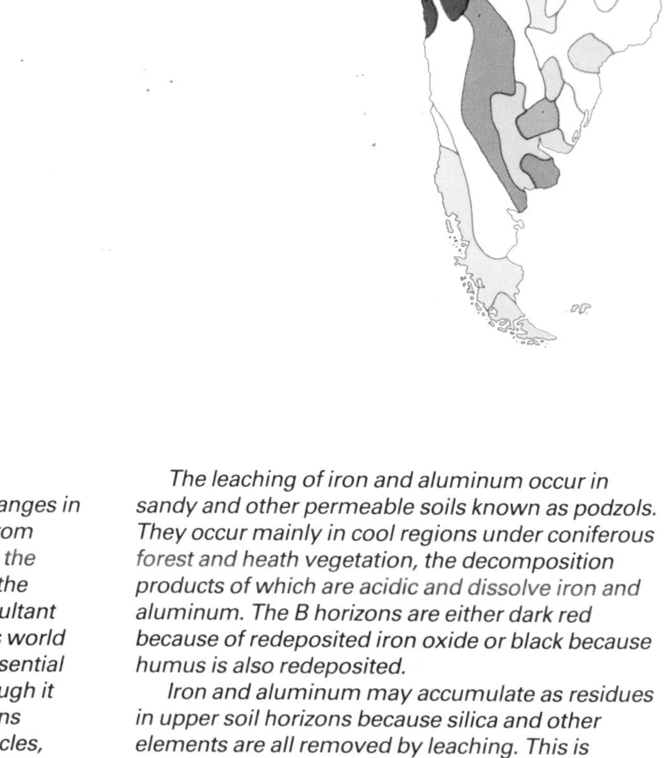

▲ **Different depths of soil gather in different areas, depending on a number of physical factors. The climate (1) is important, since hot humid conditions are more effective in weathering rock and in producing a vigorous vegetation cover than cold arctic conditions. Hence tropical areas have thicker soil than polar regions. Rock type (2) is also important, with hard massive rock being less easily broken down than soft, weak, jointed and bedded rock. Thinner soils are found on slopes (3) than in flat areas simply because of the instability and the downward creep.**

The types of soil

Soil types are distinguished primarily by changes in the bulk chemical composition of the rock from which they originated. The changes involve the removal of soluble minerals (leaching) and the accumulation of insoluble residues. The resultant soil types can be correlated with the various world climatic and vegetational zones. Water is essential to all chemical change. As it percolates through it can leach the surface layers or upper horizons (eluviation) and, in many soils, deposit particles, mostly of clay size, in the subsoil or lower horizons (illuviation).

Calcium carbonate (lime) is leached from upper horizons in water containing dissolved carbon dioxide. Some of the carbonate may be precipitated in lower horizons as nodules of secondary carbonate or thin coatings on fissure walls, but most is carried away in the ground water.

The leaching of iron and aluminum occur in sandy and other permeable soils known as podzols. They occur mainly in cool regions under coniferous forest and heath vegetation, the decomposition products of which are acidic and dissolve iron and aluminum. The B horizons are either dark red because of redeposited iron oxide or black because humus is also redeposited.

Iron and aluminum may accumulate as residues in upper soil horizons because silica and other elements are all removed by leaching. This is typical of tropical and warm temperate regions, and produces soils reddened by iron enrichment, such as the lateritic soils of Africa and India.

Ferric oxides are reduced to ferrous compounds in anaerobic (oxygen-free) conditions. These conditions are caused by waterlogging in soils that are either close to the groundwater table or are only slowly permeable so that rainwater accumulates

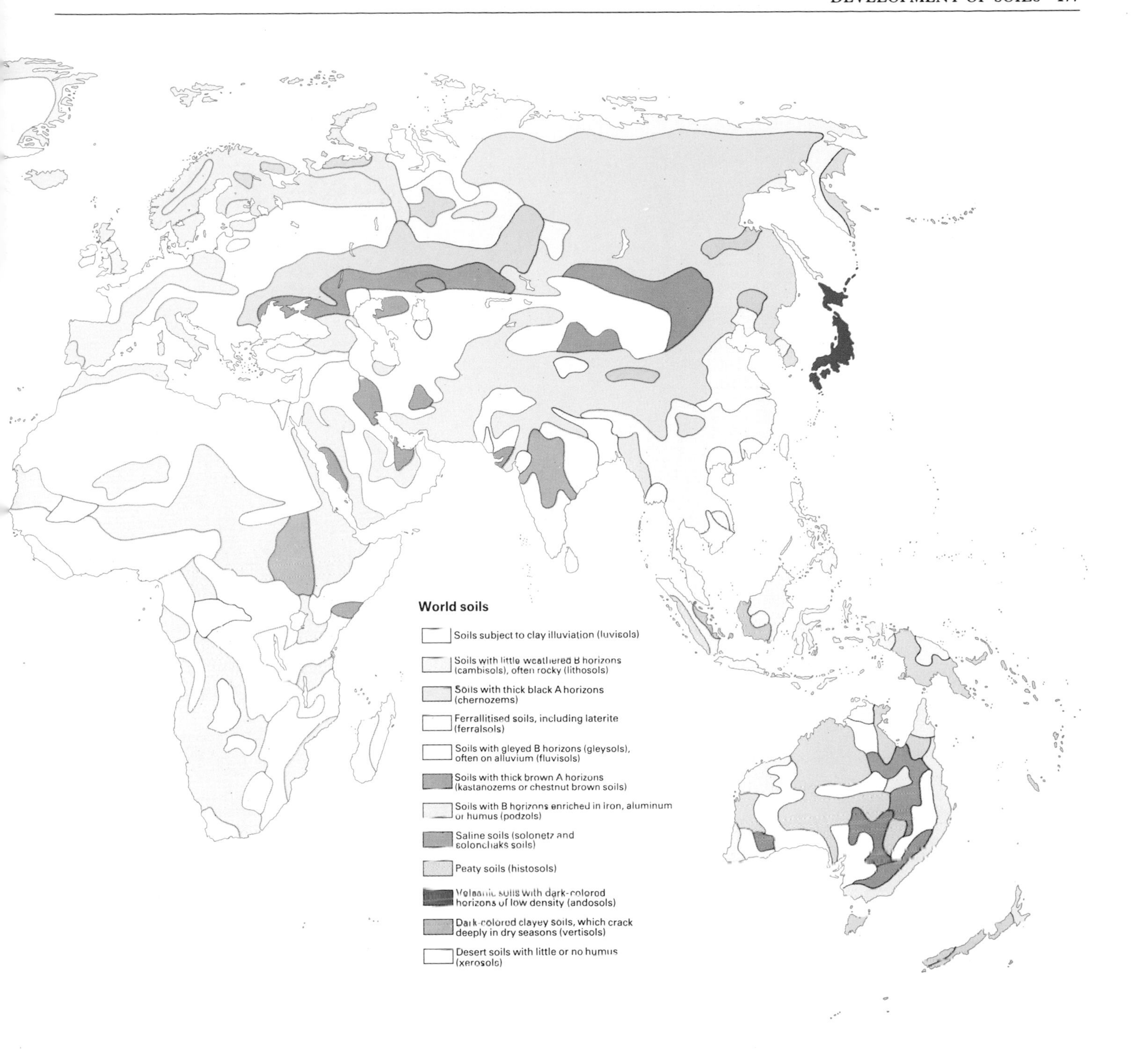

near the surface. Known as gleyed soils, they are gray either throughout or in mottles of different sizes, which are often abundant in lower horizons.

Soils and plant growth

An important difference between soils and the unaltered parent material is their structure. Their interconnecting pores and fissures filled with air or water give soils much smaller bulk densities than either hard rocks or unconsolidated sediments. The presence of the fissures is important for gas exchange with the atmosphere and also affects the strength and stability of the soil.

The mineral composition of parent materials, the nature of current soil-forming processes, and soil structural characteristics, all influence the availability of water and nutrient elements to plants. For example, most plants either prefer alkaline, carbonate-rich soils or acidic, lime-free soils. When the latter are grown in carbonate-rich soils they suffer lime-induced chlorosis – the blanching of the green parts of the plant – because of incorrect iron metabolism.

Anaerobic conditions in waterlogged soils are often bad for plant health, because sulfides and hydrocarbons produced by anaerobic bacteria prevent roots from growing, and some trace elements (iron, manganese, lead and nickel) are more soluble than in aerobic conditions and may be taken up in toxic amounts.

It is important to be able to identify the nature of the various soil types and assess their suitability for a range of agricultural and other purposes. This is achieved through a combination of careful observations of each horizon, supporting laboratory studies and practical experience of the behavior of particular soils in different climates and under different farming conditions.

▲ Soils are such complex substances that, like the world's climates (◀ page 134) there have been many different attempts to classify them. Classifications are based on the chemical content, the color, the texture and other factors such as the content of organic matter. The classification given here is a simplified form of that adopted by the United Nations Food and Agricultural Organization, published in 1974. The original has 106 different categories. Many of the names are based on Russian words because soil profiles and their origins were first studied in Russia.

World Soil Profiles

Luvisol. *Clay washes downwards from the surface horizons and forms coatings on the stones and in pores in the lower levels. This hinders drainage and gives wet lower horizons.*

Cambisol. *A little weathering has taken place, altering the structure of the soil, by removing calcite or by adding clay, to a depth of about 25cm below the surface.*

Gleysol. *A wet and dirty soil formed from unconsolidated sediments, and waterlogged in the top 50cm or so. Usually gray or blue in color and with rusty patches scattered through it.*

Kastanozem. *Chestnut brown soil. A thick brown horizon, at least 15cm thick, lies at the surface. The rest contains redeposited carbonate and sulfate with, possibly, a horizon of clay.*

Histosol. *Peaty soil with a thick wet organic-rich surface horizon, at least 40cm thick. This contains more than 20 percent organic matter. The subsoil may consist largely of clay.*

Ferralsol. *Iron and aluminum minerals have been altered to their oxides to a depth of more than 30cm. This produces a predominantly red and yellow colors. Often found in tropical forest.*

Podsol. *Sandy soils with a consolidated horizon about 2·5cm thick lying deeper than 12·5cm below the surface. This layer is cemented by redeposited organic matter or oxides of iron and aluminum.*

Solonetz. *A salty soil with a clay-rich subsoil horizon containing a large proportion of sodium. Columnar structures often present. These soils are found in arid areas.*

Chernozem. *A thick black organic-rich surface horizon overlays a horizon containing redeposited carbonate or sulfate. The dark upper horizon is at least 15cm thick.*

Andosol. *Volcanic soil. It is light and contains a large proportion of very fine volcanic glass which weathers to clay. It is typical of Japan and the mountains of South America.*

Vertisol. *Arid-climate soil with more than 50 percent clay in the top 50cm or so. Deep cracks formed in the dry seasons enable the surface layers to mix with those at depth.*

Xerosol. *Desert soil with hardly any humus. If any organic matter is present it may be just enough to darken the surface but not to provide a fertile horizon.*

See also
The Earth Through Time 81-100
Climate and Climatic Change 129-40
Rocks and the Rock Cycle 141-52
Landscape Features 181-96
Resources from Rocks 197-212
Engineering Geology 229-40

▲ Reddened fossil soils lie between the individual basalt layers of the Giant's Causeway in Northern Ireland. After each flow had cooled, its surface began to decay into soil. This was then buried by later flows and preserved.

Fossil soils

Soils formed during past geological periods were often very different from those in the same part of the world today. The large climatic fluctuations and associated vegetation changes that affected mid-latitude regions during the Quaternary resulted in successions of very different soils. In the coldest periods, areas that were not glaciated had arctic desert soils, which were strongly frost-disturbed and physically weathered but showed no chemical alteration. In contrast some of the interglacials (◀ page 136) were probably warmer than the present, because in well-drained situations there were acidic soils with evidence for much mineral weathering, clay illuviation and even the slight accumulation or iron and aluminum, giving reddish colors. The nature of pre-Quaternary soils is more speculative. Tertiary soils were probably similar to present-day tropical soils in most parts of the world away from the poles, but Mesozoic and late Paleozoic soils were probably rather different, because the grasses, flowering plants and broad leaved trees, which are responsible for many features of modern soils, did not appear until the late Cretaceous or Paleocene. Before the other main land plants appeared in the late Silurian, soils would have contained some micro-organisms (bacteria, fungi and algae) but few animals and little or no humus. As there was less oxygen in the atmosphere, oxidative weathering processes were less common. Clays were probably formed by carbonation and hydration as the atmosphere contained water vapor and carbon dioxide, but compared with modern soils, they were very weakly structured and prone to erosion.

The fossil evidence

Evidence for the nature of soils formed in past periods is obtained from profiles buried within sedimentary successions, and from unburied soils on older parts of the present land surface that are level and not subject to erosion (such as river terraces). Buried soils are useful in stratigraphy because they indicate episodes of non-deposition and often occur at unconformities. However they are rare, since land areas are normally areas of erosion, not of deposition. Those that are found are usually preserved only in continental deposits, and are studied by the same techniques as modern soils. But many are difficult to identify because during deep burial they lose their typical structure and organic contents. Soils of previous geological periods remaining on the present land surface (relict soils) often contain a mixture of features resulting from different climatic episodes or vegetation types. In temperate regions they often contain wedges, hardened and contorted layers resulting from frost action during the Quaternary glacial periods, and red-mottled clay-rich horizons formed during warm interglacials.

Landscape Features

Slopes – the key to landscape evolution...Mass movement...Running water – the land sculptor... The sea – the strongest force...The wind – the most abrasive...PERSPECTIVE...Physical and chemical weathering...The slope problem...Flow and slip movements...The river's course...Waterfalls and gorges...Valley glaciers...Wave and wind power... Igneous and sedimentary landscapes

The landscape that we see around us is the result of a complex system of natural processes. The bare bones of any landscape are provided by the rocks, which are formed by igneous, sedimentary or metamorphic actions. These are brought to the surface of the Earth by tectonic activity, continental movements and mountain-building. There they are assaulted by all the elements of nature – rain, wind, frost, sunshine – and slowly demolished. They are rotted by the chemical action of the air, and cracked up by the physical onslaught of the weather. The debris of the demolition is then carried away by gravity, wind or flowing water, and is deposited elsewhere. It is as if there were a particular height above which no rock is allowed to rise before nature conspires to bring it down again. At any stage in this process the Earth's surface produces a distinctive appearance that reflects whatever process is taking place at the time. This appearance is what we call the landscape.

Physical weathering

Physical weathering breaks down rocks without changing their composition. Landforms are developed by this type of weathering in two ways.

The first, thermal expansion, most commonly occurs in desert regions where there is a great daily temperature range. The continual heating by day and cooling by night causes mineral expansion and contraction within the rocks often resulting in granular disintegration. Where fracturing occurs parallel to the surface the outer layers of the rock peel off in an onion-like pattern. The Sugar Loaf Mountain of Rio de Janeiro is a well-known example of this "exfoliation" process.

The second process, freeze-thaw action, is concentrated in the cool temperate latitudes where water in rock joints will freeze and expand. When the temperature rises the ice melts and contraction takes place. The alternate stresses which are set up split the rock. Exposed mountain summits may show frost-shattered needle-shaped peaks called "aiguilles" such as Stac Polly in Wester Ross, Scotland.Scree deposits of broken fragments amass on the lower slopes of such mountains.

▼ Vast sweeps of scree slopes on mountain, such as these in Caples Valley, New Zealand, attest to the power of the frost in physical erosion. Water in rock joints expands as it freezes and forces the rocks apart. The broken fragments pile up where they fall at the mountains' feet. The peaks are left characteristically shattered and jagged, and open to further erosion.

A complex series of chemical reactions is taking place on and below the surface

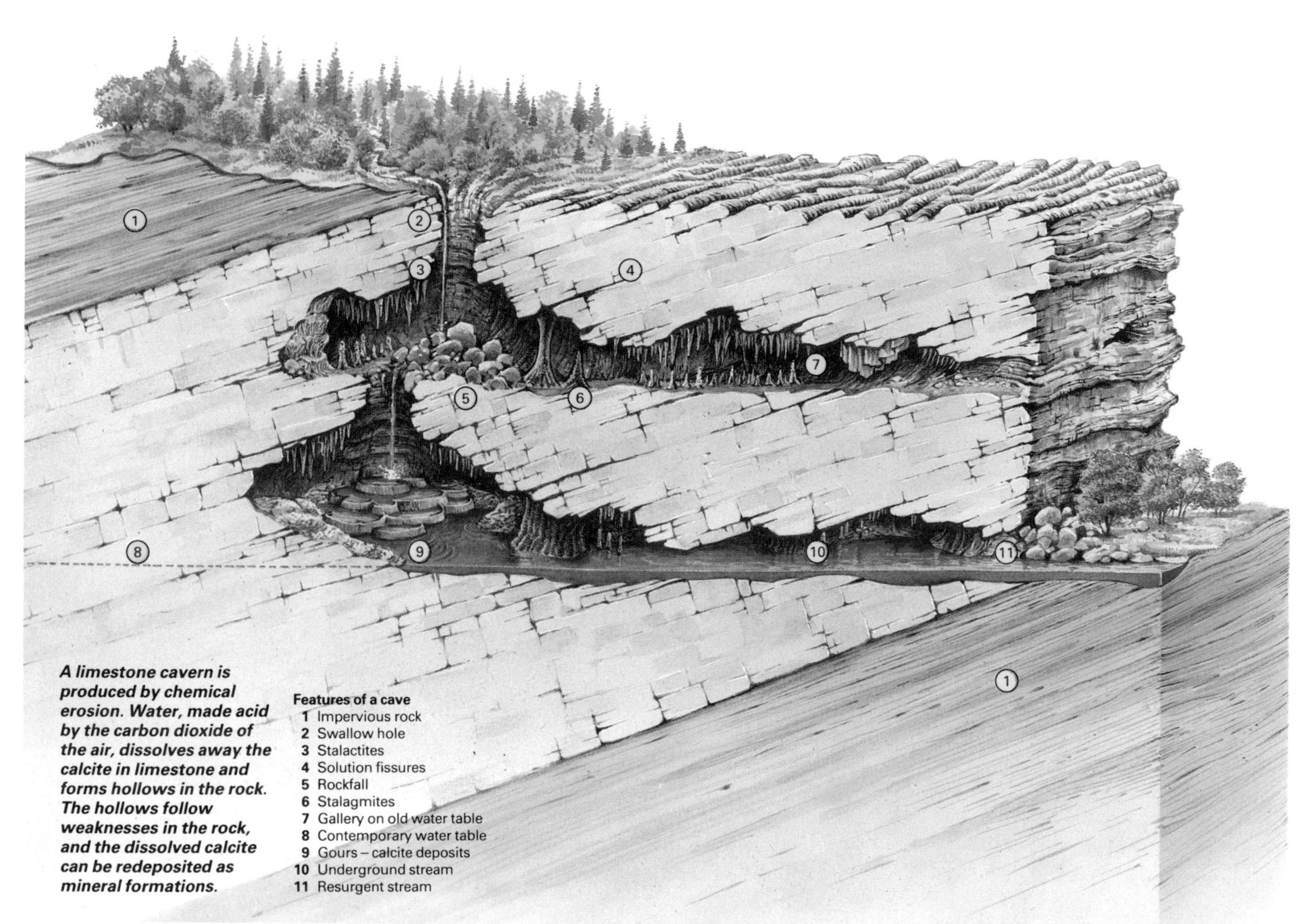

A limestone cavern is produced by chemical erosion. Water, made acid by the carbon dioxide of the air, dissolves away the calcite in limestone and forms hollows in the rock. The hollows follow weaknesses in the rock, and the dissolved calcite can be redeposited as mineral formations.

Chemical weathering

Most chemical weathering takes place when weak acid solutions penetrate and react with rocks. Limestone is most susceptible to chemical change. Rainwater contains carbon dioxide and acts as a weak carbonic acid and the calcium carbonate of the limestone is converted into soluble calcium bicarbonate which can be carried away in the groundwater. When rainwater enters the joints of limestone, solution action tends to widen them and leave a well-defined pavement consisting of clints (blocks) separated by grikes (widened joints). This is well seen above Malham Cove in West Yorkshire where the horizontal beds of Carboniferous Limestone are being subjected to chemical weathering. Deep swallow holes and underground caverns also result from the removal in solution of calcium carbonate in limestone regions. These features are found in what geographers call "Karst topography" after the dry limestone area of Yugoslavia where it typically occurs.

Probably the most spectacular landform produced by chemical weathering is the limestone cavern. Groundwater charged with with carbon dioxide, seeping through limestone beds, may dissolve out hollows beneath the surface. This usually happens along the water table, giving a horizontal tunnel, and along the joints and bedding planes, giving hollows cutting through the rock at an angle. Underground streams, flowing along the water table, produce the characteristic tunnels which are left high and dry if the water table lowers. Groundwater with dissolved calcite, dripping from the ceiling, may evaporate as a stalactite. Stalagmites form on the floor where the calcite is deposited by the shock of the dripping.

Chemical weathering contributes to the decay of igneous rocks when water penetrates the joints and reacts with the more unstable feldspars and ferromagnesian minerals. Well-jointed basalts and dolerites will decompose around the edges until angular blocks are reduced to rounded core stones by a process known as spheroidal weathering. Granite topography is often characterized by the presence of rocks stacks called tors. Their origin is thought to be due to sub-surface chemical weathering along the joint systems. This weathering probably took place under tropical climatical conditions at an earlier stage. Later exhumation and removal of crumbled rock debris has left the core stones standing to form tors. It is significant that where the joints are close together weathering is more complete and few core stones remain, but where the joints are more widely spaced large blocks can survive to form tors. The Bismarck Rock at Mwanze, Tanzania is a good example of a granite tor now partially submerged by the waters of Lake Victoria. Similar features are also seen Otago, New Zealand, although here they are developed in resistant schistose rocks.

Tor formation

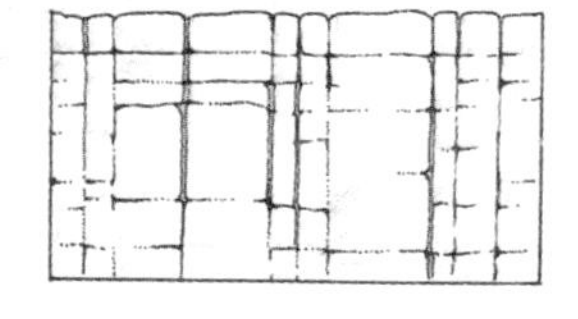

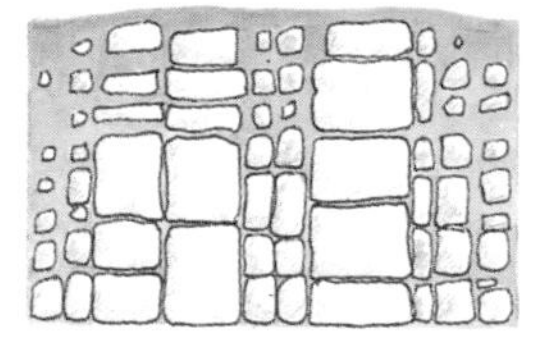

▲ Chemical weathering attacks igneous rocks like granite via cracks and joints in the surface. Susceptible minerals along the joints decompose and the joints widen, breaking the rock into irregular blocks. The smaller blocks decay first and may leave the larger ones as tors.

The slope problem

William Morris Davis the American geomorphologist writing in 1903 enunciated the principle that "landscape is a function of structure, process and stage", meaning that landforms result from the interaction of rock structure, the processes of weathering and erosion and the stage or degree to which the landscape has changed. He also introduced the term "cycle of erosion" – the sequence of stages through which which a landscape and its slopes develop. Davis used the terms "youth", "maturity" and "old age" to describe landscape evolution. He demonstrated that as narrow youthful valleys become wider, their edges tend to decline as they retreat sideways. At the mature stage the initial upland surface is reduced to a series of divides separating adjoining valleys. Ultimately in old age the landscape is worn down to an undulating lowland which Davis called a "peneplain". The opposing interpretation of landscape development was put forward by Walter Penck, the German geomorphologist, in the 1920s. He traveled widely in the tropics and noted the widespread existence of vast undulating tablelands out of which rose isolated steep hills or inselbergs. Ayers Rock, southwest of Alice Springs is a good example of an inselberg. The abrupt slopes of these landforms are thought to have been produced by the parallel retreat of slopes. In other words, the original hill slopes are progressively worn back maintaining a constant angle and leaving at their base extensive gently inclined surfaces known as pediments.

The wearing back process seems to predominate in arid lands, while humid climates may produce the wearing down process.

Slopes form an integral part of the landscape; indeed with the exception of very flat areas such as riverine plains and coastal lowlands, most of the land surface is sloping to some extent. Even so slopes are not merely surface features; their shape and occurrence can provide an indication of the pattern of evolution of landscapes over a long period of time. The analysis of slope profiles has led to the recognition of four main elements.

The "convex slope" which occurs on the top of a hill is usually a soil-covered rounded rock slope. The "free face" refers to the underlying steep rock exposure which is undergoing active weathering. The "debris slope" forms at the foot of the free face being produced by an accumulation of rock waste derived from above and deposited as a scree. In the absence of running water, the debris slope will be straight and will maintain a constant angle of rest at about 45°. The "concave slope" is a much lower-angled slope which becomes progressively less steep as it merges into the valley floor. It consists of finely comminuted downwash material spread by surface water.

The origin of slopes has long been a matter for debate between major schools of thought. On the one hand is the idea that slopes decline or wear down through time; while opposed to this is the concept of parallel retreat or the wearing back of slopes.

Hill slopes

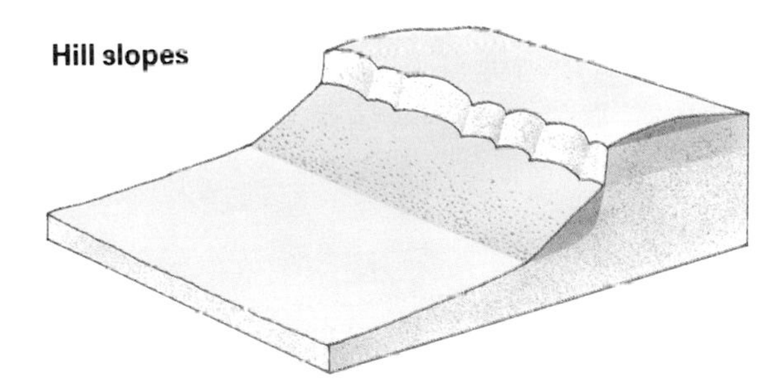

◄ ▲ *A hillslope can be divided into four elements. The top slope is convex because the material has begun to fall off at the lip. The free face consists of exposed and eroding rock. The debris slope is the heap of eroded debris at the base. The concave slope is where the debris slope levels off to the valley floor.*

Valley formation

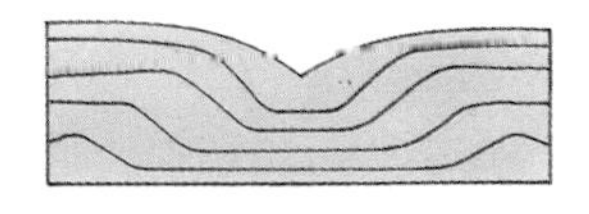

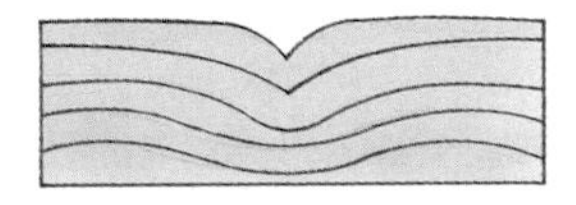

▲ *There are two theories of valley formation. The first suggests that it is worn back – that the valley quickly reaches its valley depth but then the walls erode away from the center. The second suggests that the valley is worn down – that erosion takes place on all parts at an equal rate.*

Downhill movement of material ranges from very sluggish creep to sudden and often disastrous landslips

All forces of landscape formation are dependent, to some extent, on the force of gravity. In some landform processes gravity is the principal mover.

Mass movement is the term given to the removal of weathered material under the influence of gravitational pull. Mass movement varies widely in its form. At one extreme this mass movement is slow and almost imperceptible – requiring sensitive instruments and constant monitoring for its observation. At the other extreme, it occurs with catastrophic suddenness, moving millions of cubic meters of material in a few seconds and causing widespread damage and loss of life.

There is a distinction between the two main types of mass movement – slow flowage and rapid slipping. In the former the movement is most rapid on the surface and dies out with depth. In the latter, all parts of the moving mass travel at the same rate over a clearly defined slip plane.

Flow movements

The slow downhill movement of soil and weathered rock takes place on most slopes and is evident from the presence of tilted posts and trees and leaning walls. The process is known as soil creep which is also responsible for the development of terracettes or "sheep tracks" running horizontally across many hillsides. Surface soil is held together by grass roots, and the terracettes as formed by parallel slabs of solid soil moving gradually downhill one behind the other.

A more rapid form of flowage is solifluction which occurs near the edge of glaciers. When the surface soil layer thaws in summer, the meltwater cannot seep away underground due to the permafrost. Consequently a slow downhill movement of saturated surface debris takes place, lubricated by its own moisture.

Earth flows and mud flows are more rapid movements which occur when masses of water-saturated earth slide downslope. They are commonly activated by sudden rainfall which may turn porous weathered material into a plastic mass of moving earth. The slumping of soil on the upper slopes results in pressure being exerted at lower levels causing the earth to flow and produce bulging lobes of debris along the advancing front. This has happened at Slumgullion Gulch in the San Juan Mountains of Colorado where a vast tongue of weathered volcanic ash has moved down the valley for 10km and dammed the river to create Lake San Cristobal.

When a mudflow spills out of its confining valley on to the lower ground the mud will spread out fanwise often with devastating consequences for the local population. The Roman town of Herculaneum was destroyed by a mudflow in AD 79 when dust from the erupting Mount Vesuvius was converted into liquid mud by torrential rainstorms.

▲ ► Soil creep is the slow inexorable movement of surface material down a slope. It is due to the gradual dislodging of the surface particles. It produces terracettes – step-like soil formations – leaning telegraph poles, collapsed walls, cracked roads and curved tree-trunks. Dipping rock strata are often curved over near the surface because of it.

Soil creep

Mud flow

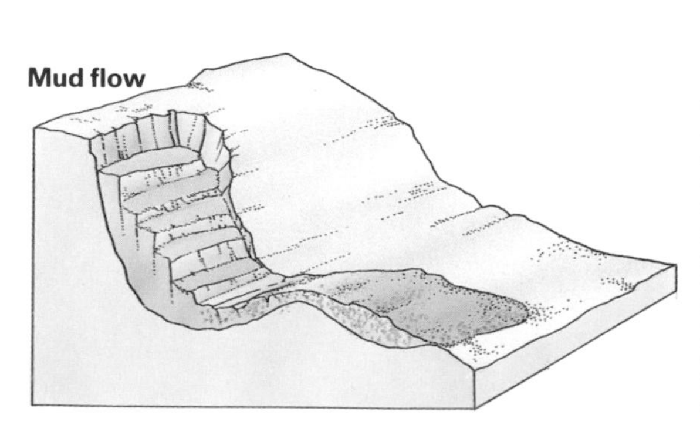

◄ ◄ A mudflow forms when soil or loose material on a slope is soaked, usually by a sudden heavy downpour. The material no longer adheres to the slope and slides downwards as a flowing mass. The collapse may start as a slump structure but then the mass loses all cohesion and becomes a bulbous lobe that spreads out on to the lower flatter areas.

▲ ► *A slump occurs when the falling material holds together as a large mass or several smaller masses. These slip on some internal plane of weakness. Usually the slippage plane is curved so that the slumping blocks rotate somewhat as they fall. A slump produces a series of step-like structures on which the original surface is still visible.*

Slump

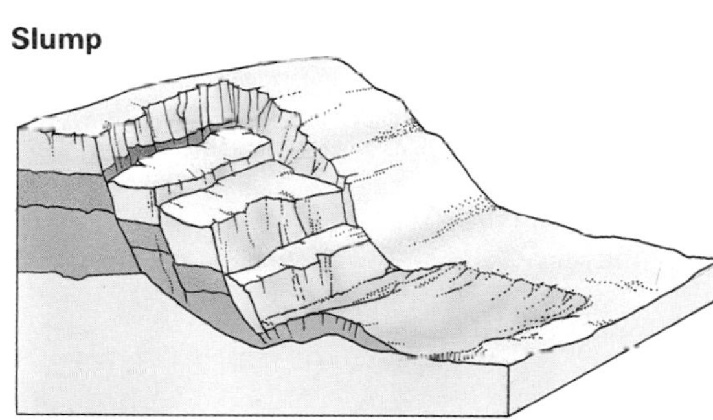

Landslip

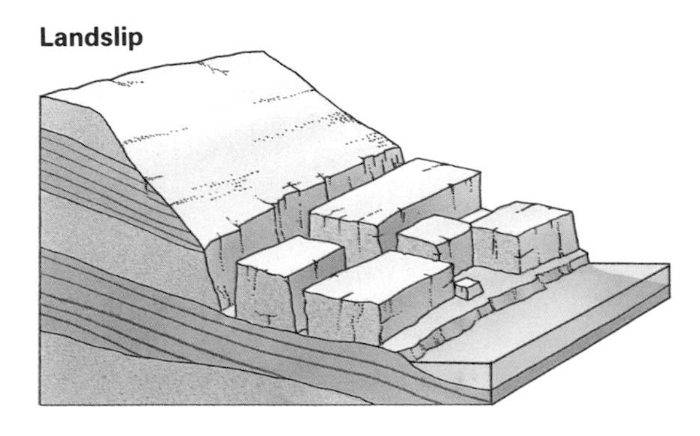

◄ *A landslip is similar to a slump in that large masses move along a plane of slippage as coherent blocks. However, in a landslip the plane is some internal structure in the rock itself, such as a bedding plane. When sedimentary rocks dip towards a cliff, a bed of massive sandstone may slide down a slippery bed of clay or shale.*

Slip movements

The most common type of mass movement is the landslip. This usually occurs on steep slopes in mountain or coastal areas and involves the rapid sliding of large masses of earth and rock. The disrupted mass retains cohesion within itself but slides along a shear surface.

Slumping takes place when slope failure happens along an arcuate or rotational slip plane, a common situation for this being when massive permeable rocks overlie impermeable clays or shales. Classic examples of rotational slumping can be seen on the south coast of the Isle of Wight and at the Folkestone Warren in southern England where the massive chalk has slumped over the underlying Gault Clay. In 1915 the main railway track was displaced by some 150 meters by the seaward movement of the slumped ground.

Where this surface is planar and the strata are dipping at a high angle, slippage will occur as the bedding planes are lubricated by groundwater. The Gros Ventre Canyon landslide in Wyoming took place in 1925 when some 50 million cubic meters of rock plunged down Sheep Mountain to block the river. This dramatic occurrence created a lake 8km in length.

Rivers are the Earth's network for the transport of eroded material over land, and eventually they carry the mountains to the sea

Landscapes shaped by running water dominate the land surface and the most spectacular features are created by the erosive power of rivers. There are several processes involved in fluvial erosion. The force of flowing water, known as hydraulic action, can remove loose material and surge into cracks to force rocks apart. When boulders and pebbles are carried by the current they are used to scour and excavate the river bed by the process of corrasion(➧ page 192). In some streams pot holes are produced by eddies whirling around pebbles which act as grinding tools. This is very marked below waterfalls, where the current is turbulent. Also rock particles themselves are worn down by abrasion as they collide and rub against each other. Solution action is another type of erosion, in this case involving the solvent work of water on the rocks over which it passes.

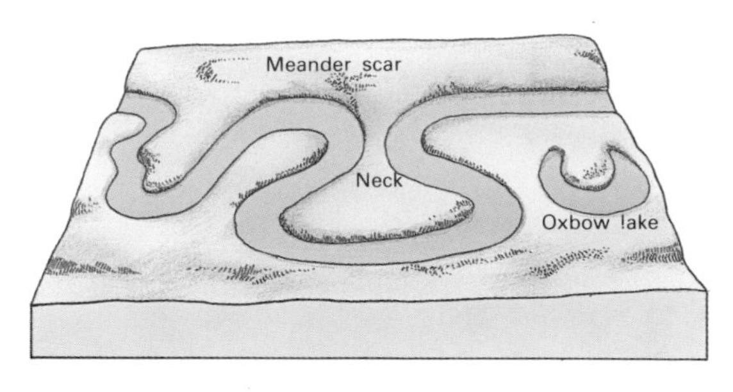

▲ The constantly shifting patterns of meanders, formed by a river in its lower stage, produces oxbow lakes. As a meander accentuates itself the two arms meet up. The river then breaks through the intervening barrier and adopts the new course, leaving the abandoned meander loop as a curved lake. An alluvial plain may have may oxbows, the oldest of which will be filled in by soil and vegetation.

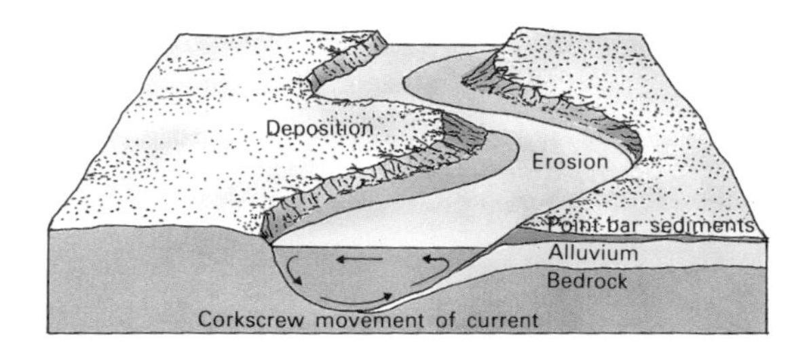

▲ Meanders grow because a current is always greater on the outside of a curve. Here the bank is being eroded away. On the inside the current is slack and a new bank is built up.

► The course of a meandering river such as the Cuckmere in southern England changes all the time. Sometimes it will be at one side of its flood plain, sometimes at the other.

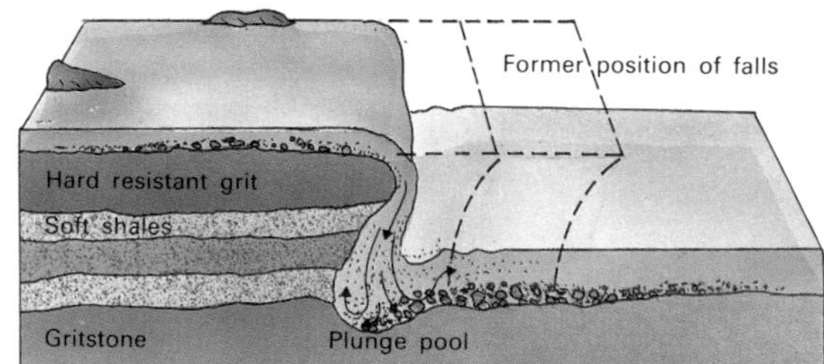

◄ The upper tract shows a river at its most violent and turbulent. The steep slopes give rise to an onward surge of water that carries rocks before it and cuts deeply into its bed.

▲ Waterfalls retreat because of the vigorous erosion that takes place in their beds. The turbulence of the plunge pool erodes the cliff, undermining it and causing it to collapse.

The river's course

It is in the upper tract of a river that erosion is most active. Here the river is chiefly concerned with vertical corrasion, cutting a steep V-shaped valley winding between interlocking spurs of high land. In this stage waterfalls and rapids are common. Downstream the valley opens out as a flood plain develops across which the river starts to meander. In the middle course lateral erosion becomes most effective on the outside of meander bends and where these impinge on the valley sides a river cliff is produced by undercutting. On the inside bends a series of point bar sediments are deposited. Finally in its lower stage the river will meander across a wide alluvial plain bordered on either side by a low line of bluffs, the worn-down remnants of the valley sides.

Waterfalls and gorges

The long profile of a river is most irregular where more resistant bands of rock outcrop transversely across the valley. If hard strata are dipping gently downstream then a series of rapids will develop. A classic example is the Nile cataracts where hard crystalline rock bands cut across the river as it flows through the Nubian Desert north of Khartoum. If the resistant layer is horizontal or dipping upstream and is underlain by softer rock then a waterfall may be produced. In its outlet from Lake Erie the Niagara River plunges 50 meters over a hard limestone ledge. The less resistant shales and sandstones beneath have been eroded by the eddying spray in the plunge pool, so undermining the limestone. The headward erosion has resulted in a gorge of recession 11km long downstream.

River terraces

A river is constantly working to attain the ideal graded long profile by eliminating irregularities in its course. However, the work can be interrupted when there is either isostatic uplift which raises the land relative to the sea level or a fall in sea level itself. In both cases the river is forced to regrade its course to a new base level and in so doing it will cut a new valley into the original flood plain. This process of rejuvenation can result in the formation of river terraces which are remnants of the former flood plain. Successive rejuvenations due to changes in base level can form pairs of river terraces stepped on opposite sides of a valley. The Boyn Hill, Taplow and Flood Plain terraces of the Thames can be recognized in a transect across London from Clapham to King's Cross.

▲ Satellite photography can show rivers in their entirety from their source, through their turbulent youth amongst the peaks and mountains, through their more genteel middle stages, to their old age where they wander slowly across the flood plains to the sea. On this part of the east coast of South Island, New Zealand, the river Hurunui is the second large one from the top, and the broad Rakaia is at the bottom.

It is not just the polar regions that have felt the effects of ice – at some time in the past most regions of the Earth have been shaped by its frozen touch

▲ ***The Malaspina Glacier in Alaska is a vast ice sheet formed by the confluence of a number of mountain glaciers, which then spread their ice and rock load over low-lying land.***

▼ ***Erratic blocks are rocks that have been transported by glaciers and then dropped as the ice melted. Here, Silurian grit lies among Carboniferous rocks in Yorkshire, England.***

Today ice covers some 15 million square kilometers of the Earth's surface mostly in the polar and mountain regions, but during the Pleistocene Ice Age great continental ice sheets extended southwards across North America and Europe (◀ pages 136-8). The climate began to get cooler about 1·7 million years ago when the ice age commenced; there was not simply one continuous period of refrigeration, rather a series of oscillations producing several glacial phases separated by warmer interglacial periods. Glaciers nourished in the mountains of Scandinavia and Scotland supplied the material for the European ice sheets which advanced and receded according to the changing temperature conditions. The position of the ice front during the major advances is shown by a series of terminal moraines. In addition to moraines a variety of depositional landforms were left behind as the ice finally retreated from the lowlands about 10,000 years ago.

In mountainous areas the signs of the Ice Age are the scars of glacial erosion – U-shaped valleys, fjords and rocks worn smooth by the passage of millions of tonnes of ice. In the lowlands the signs are the landforms of deposition, built from the debris torn by the ice from the mountains.

Drumlins, eskers and boulder clay

Behind each terminal moraine there are often groups of low hummocky hills known as drumlins. These were formed as the ice sheet retreated and are usually seen as oval-shaped mounds of sand and clay up to 15 meters high, elongated in the direction of ice movement. They often form a distinctive drumlin topography as in County Down, Northern Ireland, around Strangford Lough where some drumlins actually form islands within the lough itself.

Long sinuous gravel ridges called eskers often wind across glaciated lowlands quite regardless of existing valleys or hills. They are considered as representing the deposits of subglacial streams which flowed in tunnels beneath the ice. Eskers are common in Finland and Sweden where they run across country between the lakes and marshes.. When a delta is formed by melt water issuing from beneath the ice front, it produces a mound of bedded sand and gravel known as a kame. In some areas kames are separated by water-filled depressions called kettle holes. These were formed as patches of stranded ice melted after the recession of the ice sheet.

The chief product of glacial deposition is boulder clay which represents the ground moraine of the ice sheet. It consists of an unstratified mixture of sand and clay containing fragments of various sizes and origins. For example, deposits in East Anglia contain both chalk of local derivation and igneous rock from Scandinavia. Such ice-transported blocks carried far from their parent outcrops are referred to as erratics. The largest blocks are commonly seen resting on the boulder clay surface or even perched on exposed platforms as striking monuments to the earlier passage of ice. The unsorted ground moraine behind the ice front contrasts strongly with the stratified drift of the outwash plain beyond. Here meltwater streams have laid spreads of sand and gravel to form the undulating topography typical of Lüneberg Heath in West Germany or the Geest of the Netherlands.

▲ **Glacial sand and gravel can be deposited as mounds, called drumlins, as the ice retreats. Europe and North America have many drumlins.**

Beyond the ice sheet margin lies the periglacial zone where the subsurface is always frozen (permafrost) but the ground surface may thaw in the warmer season. Repeated freeze/thaw cycles result in the breaking and heaving of the surface and the differential sorting of loose fragments of rock so that a form of patterned ground is produced. On flat surfaces, polygonal arrangements of stones occur, while on sloping surfaces parallel stone stripes are more common. Another periglacial landform is the pingo or ice mound which is formed when a body of water freezes below ground and produces an ice core which raises the surface into a low hillock. Collapsed pingos represented by a central depression surrounded by a rampart can be recognized in areas that consisted of permafrost during the Ice Age. Such an area is Walton Heath in Norfolk, England.

Solifluction occurs in periglacial areas. This is a special type of soil flow on sloping ground where a highly saturated soil layer overlies permafrost. Solifluction debris consisting of frost-shattered fragments in a clay matrix is referred to as head or coombe rock, and is often found lying at the foot of scarp slopes or in chalk valleys in Southern England.

▼ **During the Pleistocene glaciers covered half of North America, northern Europe and northern Asia. Their landforms are found in these areas today.**

Pleistocene glaciation

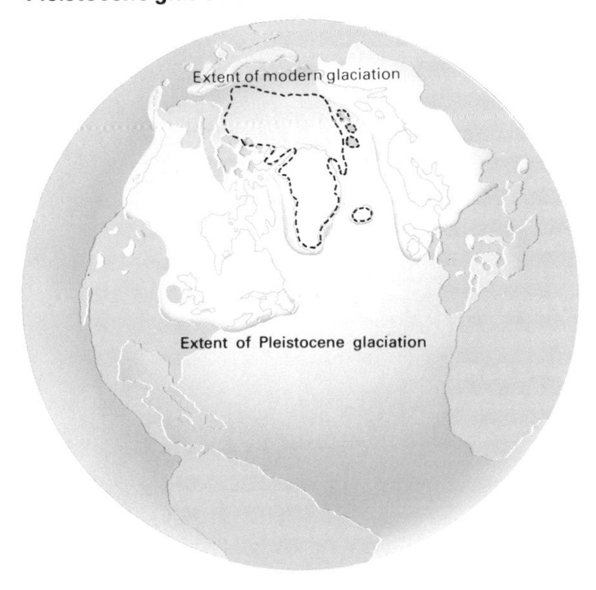

The Glacier

Valley glaciers

In mountain areas, snow may fall faster than it can melt, causing great thicknesses to accumulate. These snows compact gradually into "firn", and then into ice, and begin to flow, eventually forming a slow-moving river of ice called a glacier. Although the speed is slow – a few centimeters a year – the rocks picked up and held by the ice enable the glacier to act as a powerful scraping agent. The sharp ridges dividing glacial valleys are known as "arêtes", and these meet to form typical pyramid-shaped high mountain peaks.

Sediment carried by glaciers is called by the general term "moraine", but there are several types. Subglacial moraines are dragged along under the ice and are responsible for the scouring action of glaciers. Most moraines, however travel on or near the surface of the ice. Lateral moraines are picked up as the valley widens, and avalanches and rockfalls deposit new material on the ice. The merging of two or more glaciers, of course, causes lateral moraines to find themselves in the middle of larger glaciers, and these medial moraines give the characteristic striped look of glaciers. Sediments are dumped as "ablation" moraines at the bottom of the glacier as the ice melts under them. Occasionally, an "ice table" is formed as a large rock slab sits on an isolated pedestal of ice.

The flow of a glacier is seldom uniform, and as the ice flows over irregularities it either cracks to form "crevasses", or buckles to form pressure ridges. At the top of a glacier, where the ice gouges out an armchair-shaped hollow called a "corrie", a large crevasse called a "bergschrund" separates the ice from the back wall, ensuring that fresh rockfalls will supply new subglacial moraine. Further down the glacier, crevasses intersect, forming pinnacles or "seracs", and icefalls.

At lower altitudes, the ice begins to melt and forms streams which cut deep gorges and caves in the ice, before emerging at the front or "snout" of the glacier.

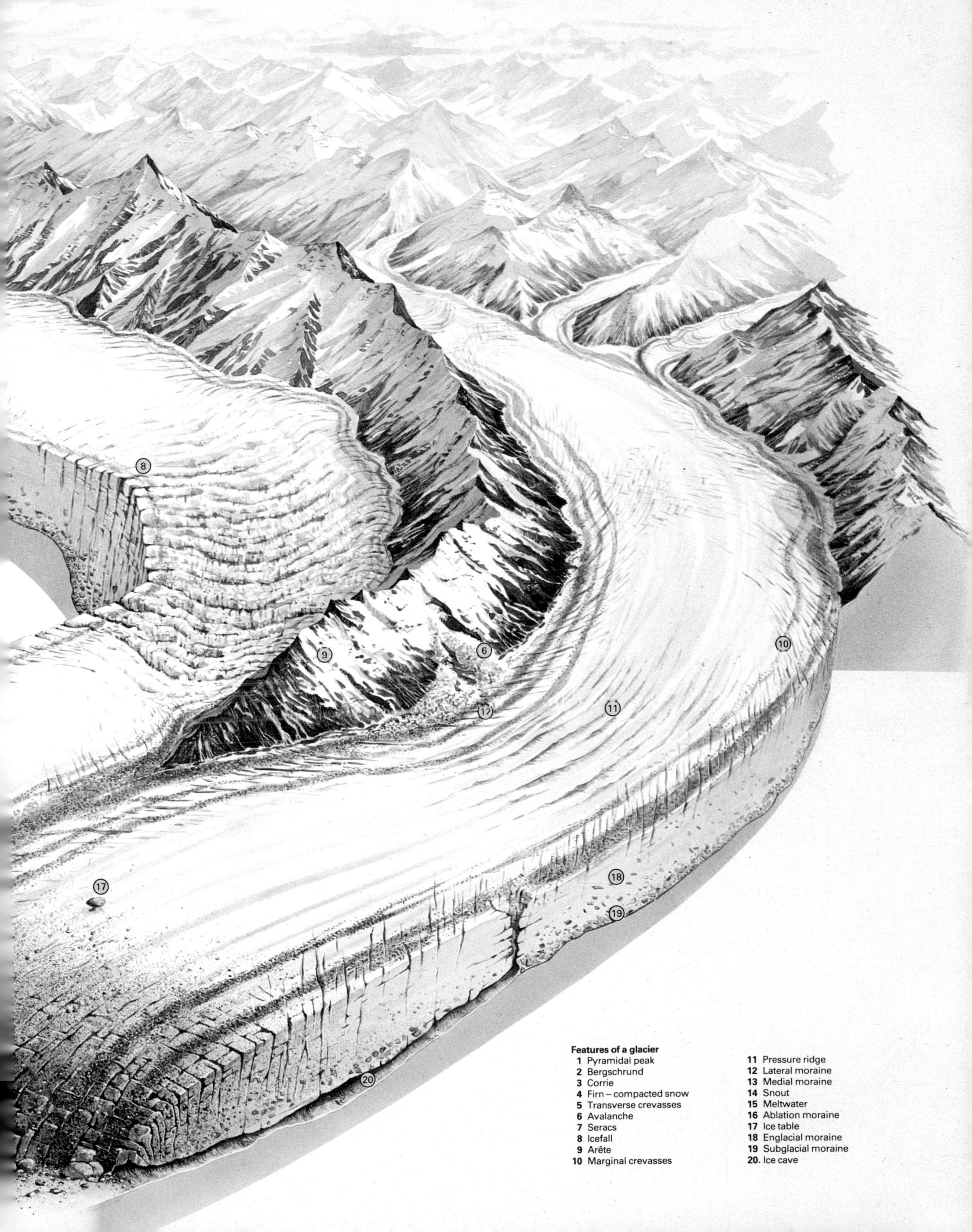

8
9
6
10
12
11
18
17
19
20

Features of a glacier

1 Pyramidai peak
2 Bergschrund
3 Corrie
4 Firn – compacted snow
5 Transverse crevasses
6 Avalanche
7 Seracs
8 Icefall
9 Arête
10 Marginal crevasses
11 Pressure ridge
12 Lateral moraine
13 Medial moraine
14 Snout
15 Meltwater
16 Ablation moraine
17 Ice table
18 Englacial moraine
19 Subglacial moraine
20. Ice cave

The rise and fall of the sea, and the power of the waves, are constantly reshaping the thousands of kilometers of the world's coastlines

Wave power

Most coastal erosion is the result of the power of the waves. Waves are generated by the wind passing over the surface of the open ocean. This puckers up the water surface into a series of ridges and hollows, and the ridges are blown along by the wind. In the process the actual molecules of water do not travel very far. They move up and down and around in a circular motion as the wave passes over them. It is the disturbance in the molecules that moves in the form of the wave over great distances. In this way energy is passed for thousands of kilometers over the surface of the ocean.

The wave energy produces its erosive effect on the shoreline in two main ways. The first is the hydraulic action produced by the sheer weight of water flung against exposed rocks and cliffs. This compresses air in the rock pores and generates forces that can burst the rocks apart. The average Atlantic wave can exert a hydraulic pressure of 10 tonnes per square meter in wintertime. During storms the pressure can reach three times this figure.

The second wave action is called "corrasion". This occurs when rocks and boulders are dragged about on the sea floor and are hurled against the cliffs. This only happens in shallow waters when the circular motion of the waves can actually touch the bottom. Fairly large boulders can be moved in this way if they support a growth of seaweed. Long strings of kelp act as sails and drag the boulders about on the sea bottom. Tillamook Rock lighthouse on the coast of Oregon is over 40 meters above sea level but it still has to be protected from 50 kilogramme rocks hurled up to that height during storms.

The coastline, where the ocean meets the land, is perhaps the most obvious place where the processes of landscape formation can be seen. Anyone who has visited the seaside has seen the movement of the water, and has seen evidence of the changing landforms.

Basically two types of coast can be identified – the coast of submergence, and the coast of emergence. The coast of submergence is produced where the sea level is rising, or the land area is subsiding. The sea encroaches on the land area, flooding the plains and valleys, and leaving hills as islands. The islands, hills and bays around Rio de Janeiro in Brazil are the features of a coast of submergence. A coast of emergence is formed under the opposite conditions. Here the land that was once under the sea emerges as new coastal plain. The Baltic Sea has a coast of emergence and this shows the typical features of raised beaches and inland cliffs.

Coastlines of submergence show the effect of marine erosion. If the original hills and valleys run parallel to the coast, the resulting coastline will be one of strings of islands parallel to the coast. If the grain of the land is at right angles to the coast the result will be long headlands and arms of the sea that reach far inland.

The strongest force of erosion in the sea is produced by the waves. These act at water level and cut away cliffs at the bottom, undermining them and causing them to collapse. They attack headlands from each side, making them narrower and narrower until they are eventually eroded away. Joints, bedding planes and soft areas of the rock are attacked first, and cracks are enlarged into sea caves. The pressures that the waves build up in sea caves may blow a hole in the roof through which jets of water spray as the waves surge about below. Caves eroded at either side of a headland may join to form a tunnel. This will later enlarge into a natural arch. Finally the lintel of the arch will collapse leaving the seaward portion of the headland as a sea stack. The stack will also finally crumble away.

◄ On a coastline of submergence the valleys become inlets and the hills become headlands and islands. A U-shaped valley formed by a glacier will give rise to a fjord. This is a deep inlet with steep sides and often a shallow mouth. There are many fjords in Norway.

► A natural arch, like those in Dorset in southern England, is a spectacular middle stage in the process of coastal erosion. A narrow headland erodes at both sides until a tunnel forms through it. The tunnel enlarges into an arch. Eventually the arch falls leaving a sea stack.

Arch formation

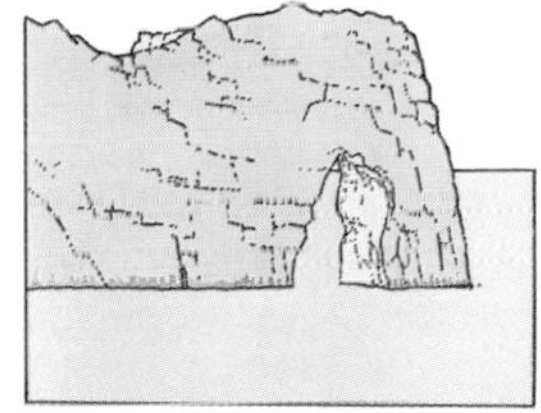

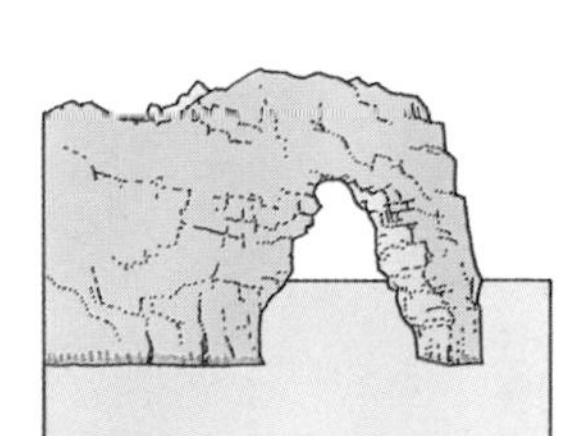

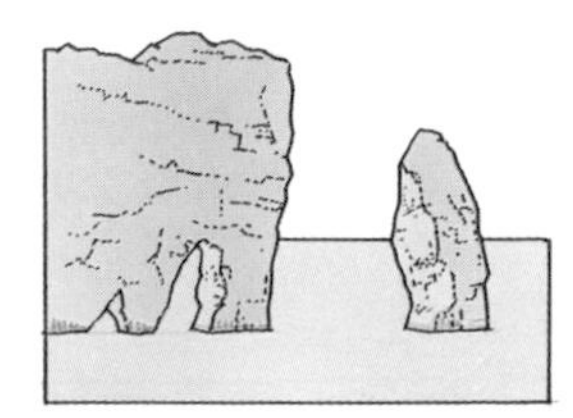

▼ *As waves approach the shore they do so along parallel fronts. On reaching a headland, the waves striking the headland first slow down. This curves the wavefront round it so that it attacks the headland from the sides. Erosion is thus heavier on headlands than in intervening bays.*

▲ *When a coastline of valleys is submerged, the result is known as a ria coastline. In this the river valleys are still present but as a branching network of inlets, the original spurs forming sloping pointed headlands. This spectacular example is Maud Island, New Zealand.*

Erosion of a headland

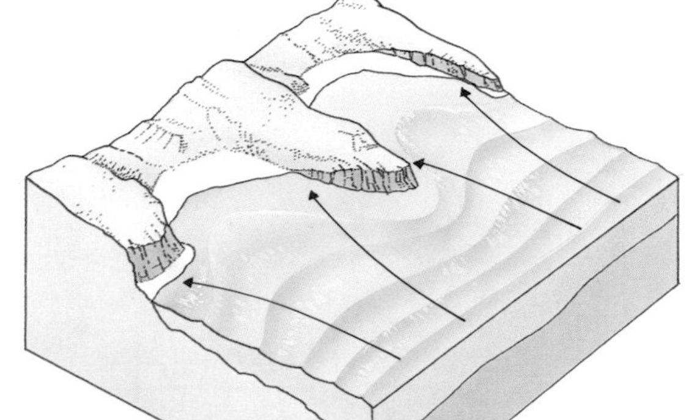

It is the resistance to erosion of the Earth's hardest rocks that gives rise to the most prominent landscape features

The geological structure of a region determines its surface topography; for example, the landforms developed in horizontal strata differ markedly from those formed in an area of intensely folded focks. Equally, the physical features of a volcanic region contrast strongly with those of an undulating boulder clay plain. Thus as weathering and erosion take place landscape features develop in close conformity with the underlying rock structure. The simplest structures are those where sedimentary rocks lie in horizontal formations. In arid climates the vegetation is sparse and so the surface is exposed to the erosional forces of the wind in addition to active downcutting by rivers; thus a wide variety of landforms are produced. The Colorado Plateau in the western USA is deeply cut by canyons and characterized by the occurrence of mesas and buttes. These features are formed by the erosion of the resistant cap rock of the plateau which protects the underlying strata. The mesas are tabular steep-sided hills which represent remnants of the dissected plateau. When the mesas are gradually reduced in size by the parallel retreat of their bordering slopes, they become known as buttes. The soft beds erode more quickly than the hard and a stepped outline develops. In well-jointed horizontal rocks erosion may carve out intricate castellated shapes such as are seen in Monument Valley, Utah.

Igneous landscapes

When igneous rocks are exposed at the surface, they form distinctive scenery because they are generally harder and more resistant to erosion than surrounding country rocks. For example, batholiths are large intrusions of coarse-grained rocks such as granite which are often associated with orogenic zones; the Rocky Mountains of British Columbia, Canada, are underlain by a single batholith about 1,000km wide. However, smaller batholiths like those of southwest England are more common. Dartmoor, with its moorland plateau capped by rocky granite tors, is a monument to the permanence of granite.

By contrast, basic gabbros sometimes occur in very large saucer-shaped intrusions called "lopoliths"; these features are seen in the great Bushveld Complex of South Africa and at Sudbury in Ontario, Canada. Magmatic blisters which cause doming of the overlying strata are called "laccoliths", and some excellent examples of these landforms occur in the Henry Mountains of Utah.

On a smaller scale, minor intrusions in the form of "sills" are horizontal injections of fluid between the bedding planes of sedimentary rocks. The Palisade Cliffs overlooking the River Hudson in New Jersey are formed of a dolerite sill some 300 meters in thickness. "Dikes" are vertical wall-like intrusions which cut discordantly through the pre-existing rocks. They often radiate outwards from ancient centers of volcanism, or form cone-shaped sheets around them, such as in the islands of Skye, Mull and Arran of the west coast of Scotland.

Landforms in present-day volcanic regions range from the low shield volcanoes of Hawaii or Iceland, to the high peaked cones of Fujiyama in Japan or Mt. Etna in Sicily. The Puy country of the Auvergne in central France, however, shows many steep-sided domes which are the exposed centers of ancient extinct volcanoes, and these "plugs" have formed from highly viscous silica-rich lavas. Finally, in Northern Ireland the spectacular Giant's Causeway on the edge of the Antrim plateau, and also at Fingal's Cave on the island of Staffa, Scotland, a large widespread basalt lava flow has cooled into a series of hexagonal columns exposed along the coast.

Erosion of igneous rocks

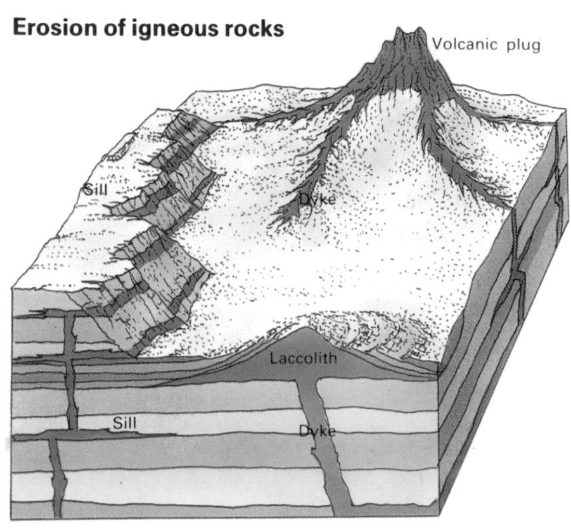

▲ An old volcanic vent full of lava will normally be harder than the rock round about, and will be left upstanding as a volcanic plug when the surrounding rock has eroded away.

◄ When a terrain containing igneous rocks is eroded, the harder igneous rocks will form prominent outcrops on the landscape. These include dikes, sills, plugs and laccoliths.

► Sills are sheets of igneous rock intruded between the bedding planes of sedimentary strata. Where they outcrop they may resemble very hard sedimentary beds.

▲ *In any sedimentary sequence it is the softest beds that will erode most quickly. The buttes and mesas of Monument Valley, Utah, are formed of a cap of resistant massive sandstone. Softer beds beneath and round about have worn away leaving the distinctive horizontally-bedded landscape. The mesas are the flat topped hills and the buttes are those that have a narrow summit. In places they have been eroded from the side until they are mere pillars.*

Sedimentary landscapes

Where sedimentary strata are inclined in one direction they are said to exhibit "uniclinal" structure. Where such rocks have differing resistance to erosion, they will form a series of "cuestas" – ridges, with a steep scarp slope where the hard rock strata are exposed, and a gentle dip slope following the bedding planes. Southern England has many good examples of cuestas between the Cotswolds and the Chiltern Hills. The drainage pattern in this region follows the topography, with major rivers cutting through the ridges in gaps, and their tributaries following the softer beds of the vales. The scarp faces erode and retreat, sometimes leaving remnants isolated in the intervening vales. As a scarp slope erodes back, the water table in the shrinking ridge will become lower. This may account for the numerous dry valleys that occur in chalk and limestone scarplands. In steeply-dipping rocks the ridge will be quite narrow and will form a "hogback".

In regions of symmetrically folded strata, the topography may be one of parallel ridges and valleys, as in the Jura Mountains on the border between France and Switzerland. Here the anticlinal folds form the ridges and the synclines form the valleys.

A far more common fold pattern is a structural dome where the rocks dip away on all sides. However, since an anticline is structurally weak because of the tension caused by uplift, its crest will be rapidly eroded and underlying rocks exposed. There will be inward-facing scarps surrounding the older rocks of the core. This distinctive relief pattern can be seen in the Black Hills of Dakota where a central igneous dome is exposed, the overlying sedimentary rocks having stripped back by river erosion.

See also
Internal Composition 57-68
Climate and Climatic Change 129-40
Rocks and the Rock Cycle 141-52
Faulting and Folding 153-6
Development of Soils 173-180

In very dry areas, such as deserts, the main agent of erosion is the wind. The wind can act on the dry, loose rocky material and can produce distinctive landforms in a number of different ways. Deflation is the process by which the wind carries away unconsolidated material like sand and small pebbles. The latter may be rolled along the ground by wind eddies – a movement called saltation (♦ page 165). One of the main landforms resulting from sand removal is the deflation hollow. Thus wind erosion is responsible for the formation of the Qattara depression in the Libyan desert which is 300km across and its floor is 134m below sea level.

Abrasion is the sand blast action of the wind on rock surfaces. A strong wind armed with quartz grains is capable of powerful erosion particularly within 50cm of the surface. Ventifacts are polished faceted pebbles. These are found on the desert floor and have been shaped by abrasion. The several faces of such pebbles are produced either by the changing direction of the wind or by movement of the pebble itself. When layers of rock of differing hardness are horizontally bedded, sand blast will erode the softer layers below the hard cap rock and a series of fluted ridges called zeugen may be formed. Alternatively, where the dip of the strata is steep, upstanding ridges of resistant rock will be separated by wind-cut troughs of softer rock; this landform is referred to as yardang. Wind abrasion will also seek out structural weaknesses like joints and bedding planes and widen them. Thus selective differential erosion can produce a variety of shapes in the rocks. Cliff faces may be pitted and honeycombed to form rock lattices, and upstanding rocks can be undercut producing pedestal or mushroom rocks. Even rock arches may be excavated by the wind although Rainbow Bridge, Utah, one of the most famous and spectacular examples, was probably initially cut by a meander of Bridge Creek, which eroded both sides of the ridge.

Wadis and canyons

Rainfall is low in deserts but occasionally torrential rainstorms turn dry river beds into raging torrents for a few hours. This flash flooding will transport large amounts of debris down the valleys and enable rivers to undertake rapid vertical erosion. The result is deep steep-sided valleys called wadis, often where the mountains meet the arid plains. The Hadhramaut region of Saudi Arabia is dissected with spectacular wadis formed in this way. Where rivers maintain a constant flow through a desert area deep gorges or canyons may be excavated. The Grand Canyon, Arizona, was first cut in Miocene times (26 million years ago) as the Colorado Plateau was slowly uplifted by Earth movements. The canyon has a maximum depth of some 2,080 meters from the plateau top to the Colorado River. Differential erosion of the horizontal strata has formed a spectacular terraced cross-section up to 24km wide. Erosion continues today with the river maintaining its flow across the arid plateau fed by the Rocky Mountain snows.

Desert basins are often surrounded by steep-scarped mountain ranges where parallel retreat of slopes is taking place. This mass wasting results in the formation of a rock pediment which inclines gently down to the sandy bajada and central basin of inland drainage with its salt-encrusted flats and playa lakes. The pediment is often covered with gravel, and alluvial fans usually mark the outfall of wadis as they open out on to the pediment.

▼ Ayers Rock, central Australia, is an inselberg – the result of onion-skin weathering. Chemical actions, and daily heating and cooling, split off the outer layers of a rock resulting in a rounded outcrop.

Resources from Rocks

Igneous ore bodies – metals concentrated by heat...Sedimentary ores – winnowed and collected by currents...Fossil fuels – oil, coal and gas... Water – the principal resource...Bulk materials – stone and cement...Prospecting and exploitation... PERSPECTIVE...Water management...Prospecting techniques

▲ *Probably civilization's most practical interest in Earth sciences relates to the finding and extraction of natural resources. A motor car's construction materials and fuel attest to this.*

Metals are found in every rock of the Earth's crust. However, only rarely are these metals concentrated enough to be worth the effort and expense of extraction. During the formation of such ores, natural processes have combined together and concentrated metals to hundreds or thousands of times their usual abundance. The huge range of processes in the geological cycle redistribute the chemical elements which comprise the crust. Ore deposits are the results of these processes acting on chemical differences between the elements, and differences in density, hardness, solubility and melting point between the minerals containing them.

In igneous activity several factors determine whether ores form or not. The depth, temperature and vapor pressure associated with partial melting of the rocks, determine the composition of the magma. The crustal conditions through which the magma then passes control the chemistry of igneous materials and the order in which they crystallize. Each new mineral that forms extracts its component elements from the magma, thereby changing the chemistry of what remains.

Most igneous minerals are silicates of only a few metals. Some, like olivine, may concentrate rarer metals, such as nickel. But silicates are difficult to smelt and they rarely constitute ore minerals.

▼ *Economic minerals are irregularly distributed. Fold mountain areas are generally the best for metallic minerals because of the Earth processes that concentrate them.*

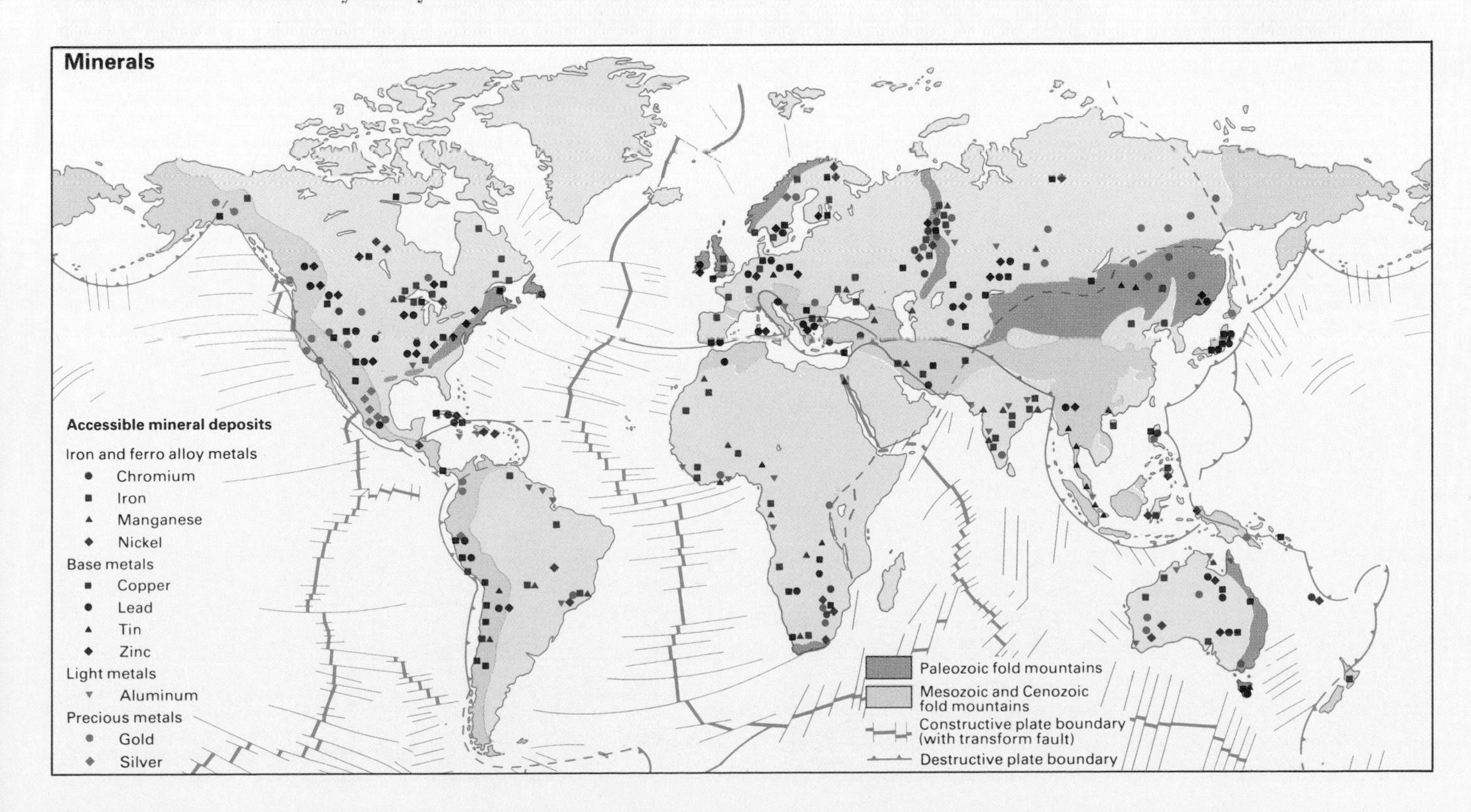

The valuable minerals in a rock may be concentrated by natural processes

◄ *Some concentrations of metal ore, like the nickel in the Thompson Mine in Manitoba, derive from magma – different minerals separating at different times.*

Magmatic separation

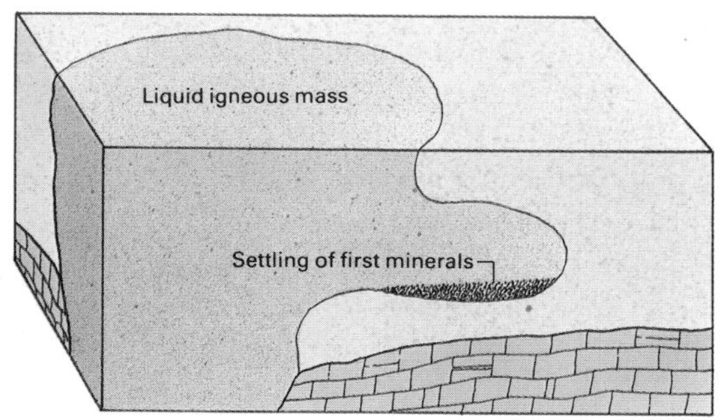

▲ *As a molten mass hardens the first minerals to form may settle as a separate layer from the rest. This may give a deposit of ore minerals at the base of an igneous body.*

Sources of igneous ore bodies

The most important economic minerals are those of metals that do not enter silicates. Some, like chromium, appear as oxides at high temperatures. Being dense they sink through the magma to accumulate in metal-enriched layers. The major sources of chromium in South Africa and Zimbabwe formed as gravity segregations from huge basic intrusions. Sulfur-rich magmas can separate into two liquids which do not mix, forming silicates and sulfides. Those metals which combine readily with sulfur, like iron, copper and nickel, enter the sulfide liquid. This dense sulfide accumulates deep in the magma chambers, forming nickel and copper deposits such as those at Sudbury, Canada.

Most igneous silicates are anhydrous. As they crystallize, the water content increases in the residual magma. This then becomes enriched in all those elements which could not be absorbed by the silicates. Towards the end of crystallization, what remains is a hot watery silicate fluid charged with those elements which were incompatible with the main igneous minerals. The nature and composition of these late stage fluids depend on the starting composition of the magma. A granitic magma gives fluids are rich in potassium, beryllium, lithium, uranium and tin. They crystallize slowly in cracks in the granite and country rocks to form coarse pegmatites containing many complex ore minerals. With more basic starting magmas, the residual fluids can contain copper, molybdenum and gold. High vapor pressures force open cracks in the host rock, alter it and deposit ore minerals in an intricate network of tiny veins. Such porphyry deposits form the main source of copper.

All rocks become saturated with water. Natural permeability allows the water to circulate, particularly if it is heated by igneous activity or deep burial. If it contains chlorine, fluorine and sulfate ions, and is acid and hot enough, the water can dissolve and transport metals from the rocks through which it passes. Normally this produces merely a noxious fluid which emerges at the surface as hot springs. But if, during percolation, it contacts natural reducing conditions, such as in rocks containing hydrocarbons or on an ocean floor seething with life, some remarkable chemistry occurs. The sulfate ions are reduced to hydrogen sulfide gas, which in turn precipitates the dissolved metals as sulfides to form a hydrothermal mineral deposit.

Pegmatite emplacement

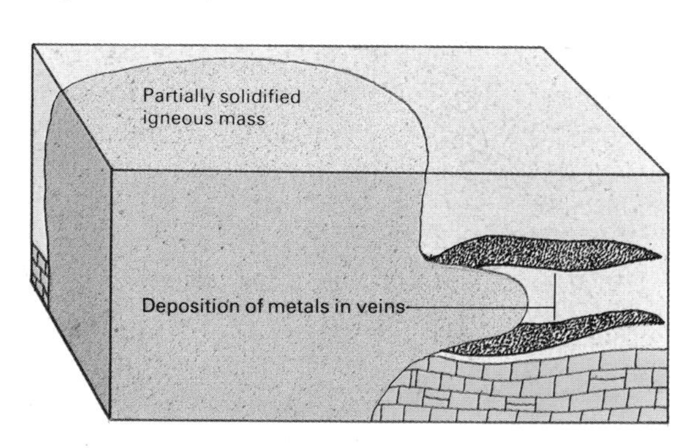

▲ *When an igneous body has almost completely solidified, the liquid portion that is last to harden contains the metals that were not taken up by the early crystallization of the silicate minerals. These late stage fluids work their way into cracks in the igneous body and the country rocks round about. There they crystallize slowly giving metal-rich ore veins in the vicinity of the igneous intrusion.*

Metasomatism

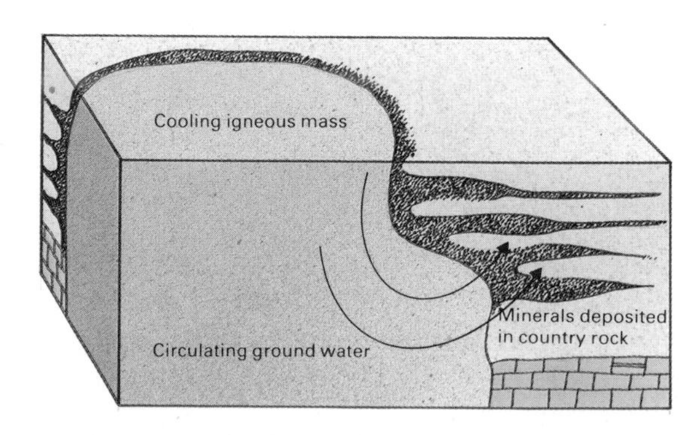

▲ ► *Ground water is heated up as it circulates through a cooling igneous mass. It can then dissolve metals from the newly-formed rock and transport them in solution into the rocks round about. The metals may then be deposited in the country rocks or, if conditions are right, around hot springs on the ocean floor. The molybdenite ores in Colorado (right) are deposited by this metasomatism process.*

Sedimentary ores

Sooner or later most crustal rocks reach the Earth's surface and are eroded (◀ page 144). In tropical climates minerals are rotted by organic acids carried by percolating surface water. Silicates break down leaving insoluble clays from which soluble compounds including silica have been leached away. Usually the residue is a red, iron-rich soil called laterite. Aluminous rocks, however, produce a residue of aluminum hydroxides which is known as bauxite, the main source of aluminum. Where they have developed on ultrabasic rocks in the New Hebrides, laterites themselves now contain concentrations of nickel which originally resided in olivines.

Dissolved metals ultimately reach the ocean where they merely add to the low metal content of sea water, and usually stay in solution forever. However, shallow seas have bloomed with life for billions of years. In stagnant conditions, bacteria can rot organic matter and generate hydrogen sulfide, and dissolved metals entering such seas can be precipitated as sulfides. These fetid basins were sources of the great stratiform deposits of copper in Zambia and the lead and zinc deposits of Broken Hill in Australia.

Physical weathering merely frees minerals in rocks from one another. The minerals are winnowed into relatively pure fractions according to their size, hardness and density, and are transported by water, wind and ice. These can be deposited where currents slacken, for example in the bends of slow-moving rivers, or on beaches. Economic deposits that have accumulated like this are called placer ores. A dense mineral, like gold, which can resist comminution and being transported in suspension, becomes concentrated in pockets. This is the source of placer deposits of Klondike gold rush fame. More important are the tin-bearing placers of Malaysian rivers and the titanium-rich black beach sands of Australia.

▼ Panning for metals is the same process that concentrated them in the river sands in the first place.

▲ Sedimentary deposits of tin minerals are harvested from an artificial lagoon in Malaysia using dredgers.

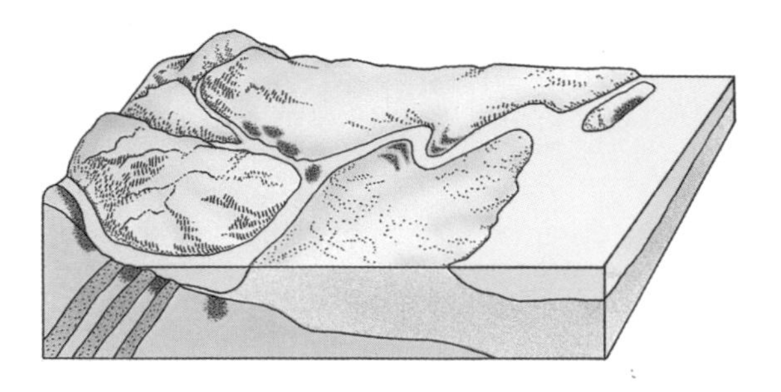

▲ When a rock containing some metallic minerals is eroded and the fragments are washed away by streams, the heavy metal grains tend to sink to the bottom first and collect in particular areas where they can be panned or dredged easily.

▲ *Hydrocarbons – oil and gas – represent one of the most valuable natural resources of our time. It is only in this century that they have been exploited and prodigious feats of engineering are often needed to extract them.*

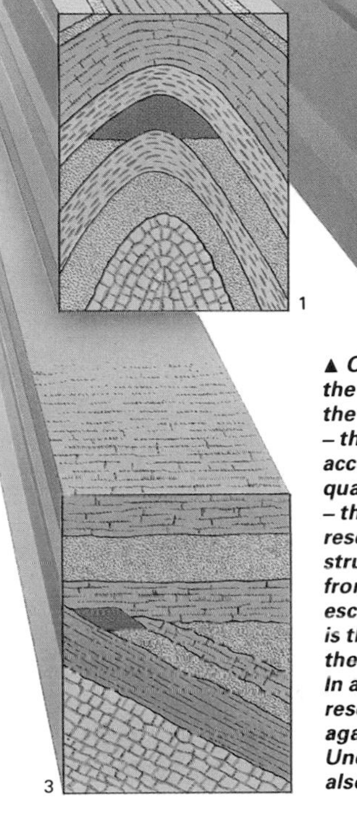

▲ *Oil deposits occur where the oil has migrated from the rock in which it formed – the source rock – and has accumulated in large quantities in a porous rock – the reservoir rock. The reservoir rock must form a structure known as a trap from which the oil cannot escape. The simplest trap is the dome (1) in which the oil gathers at the apex. In a fault trap (2) the reservoir rock is faulted against another rock. Unconformities (3) can also form traps.*

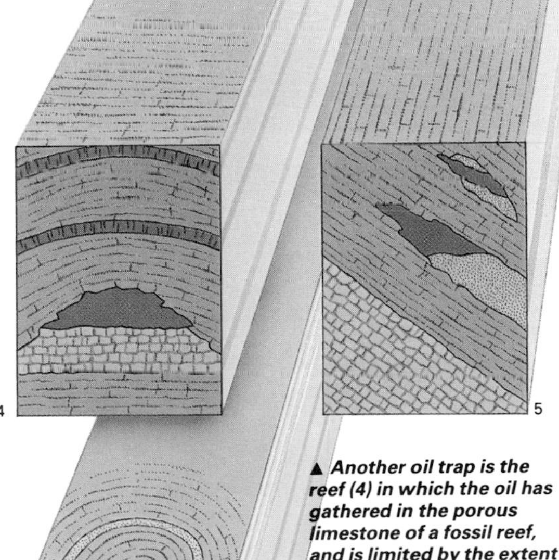

▲ *Another oil trap is the reef (4) in which the oil has gathered in the porous limestone of a fossil reef, and is limited by the extent of the reef itself. Stratigraphic traps (5) occur when the reservoir beds pinch out upwards. Above a salt plug (6) the beds are bulged upwards forming a dome. In all cases the oil floats upwards through the groundwater, and when the reservoir rock is overlain by something impermeable, that is where the oil gathers.*

Fossil fuels – oil

Oil is formed by processes on the ocean floor where minute floating plants or phytoplankton form the intermediary between solar power and fossil fuel. Dead plankton accumulate in stagnant oxygen-free water and are quickly buried by clays. The peak periods for oil formation coincided with blooms of phytoplankton in the Ordovician to Devonian and Jurassic to Cretaceous Periods. Flooding of the continental lowlands formed vast shallow seas rich in nutrients derived from the land and in which marine life exploded. Accumulation of the remains in sea-floor muds resulted in hydrocarbon-rich source beds.

The transformation of hydrocarbons in impermeable muddy source beds to resources of oil and gas in permeable reservoir rocks is a multistage process. At low temperatures bacteria metabolize the organic remains to form methane and a viscous complex hydrocarbon called kerogen. The temperature rises at depth until, at about 100°C, the kerogen breaks down into simpler, lighter hydrocarbons. This is the stage where low viscosity oil is formed. Above 150°C the oil too decomposes producing gases and residual carbon. These conditions depend on the heat flow from the crust and mantle, which, together with rate of subsidence, governs a "window" in space and time when oil can exist. If source beds descend below this window, or do not release their oil in time, their potential for oil formation is destroyed.

Under the right circumstances the oil can escape into permeable sediments, such as sandstone or limestone. If the rock is saturated with water the oil can then migrate upwards and away from hostile conditions. To form an oilfield this migration must be halted by a trap structure, where the escape route to the surface is blocked by a barrier of impermeable rock. If no traps intervene, oil reaches the surface and loses its lighter components to produce a tarry cement to the reservoir rock and form oil sands or bitumen lakes.

Fossil fuels – coal and gas

Coal and natural gas, like oil, represent a fraction of the Sun's energy which biological and geological processes have combined to trap. Each is derived from incompletely decayed plant tissues engulfed by sediments and then slowly decomposed by heat and compression. All other commodities depend on energy for their extraction and processing, and these aptly named fossil fuels are the most fundamental resources of modern civilization.

Photosynthesis is the process in which a plant converts water and carbon dioxide to carbohydrate food and oxygen. The power for the process comes from sunlight, and the green chemical chlorophyll converts it into a form that the plant can use. About 30 times the total energy demand is extracted from solar power and locked into plants each year. There is a balance though – decay, combustion and the metabolism of plant-eating animals releases roughly the same amount, together with water and carbon dioxide. Since life appeared, there have been many breaks in this equilibrium, when organic matter built up in oxygen-free environments and was buried by sediments before it could decay. Not only did this produce fossil fuels, but these disequilibria allowed oxygen to accumulate in the atmosphere and sponsored the evolution of animal life.

Most buried organic matter is too thinly distributed in sedimentary rocks ever to be extracted. The fossil fuel bank amounts to less than a thousandth of this total amount, and coal, oil and gas fields represent only those occurrences in space and time where the rate and volume of preservation were accelerated. Coal had a striking peak during the Carboniferous to Triassic Periods.

The Carboniferous coal deposits form a band along the ancient equator running through the continent of Pangea (♦ pages 90-2), and

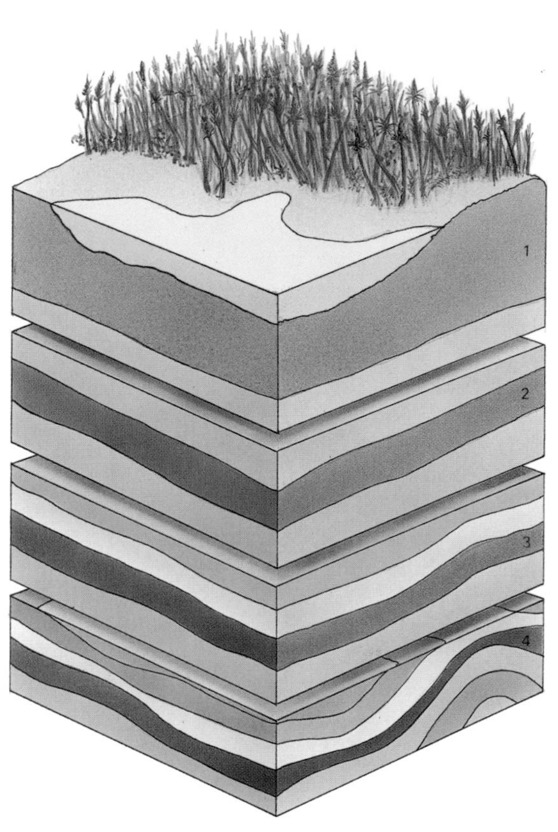

▲ ▼ *Coal forms where accumulated vegetable matter cannot decay completely. It firstly forms thick beds of peat (1). Once buried and compressed it becomes lignite (2). Further compression produces bituminous coal (3), and light metamorphism turns it into anthracite (4). Most of the world's coalfields lie in basins of late Carboniferous (Pennsylvanian) sedimentation (below).*

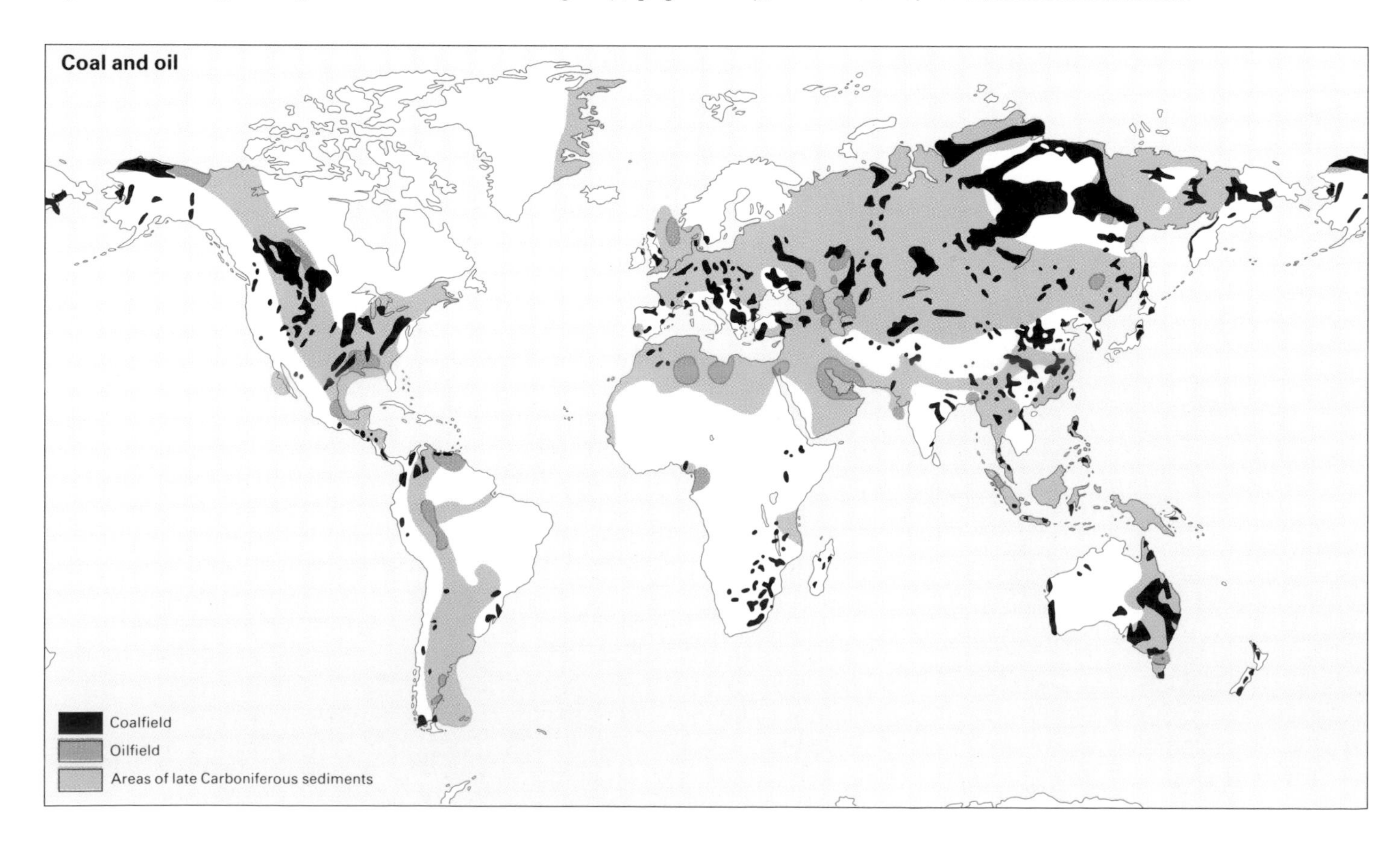

▲ *Modern deep-mining of coal involves the use of huge semi-automatic cutting machines, such as this shearer, that travel along coal seams ripping out the coal as they go.*

their formation in the tropics is confirmed by the lack of seasonal growth rings in fossil logs preserved by the coal. On the other hand Permian and Triassic coals in the southern continents and in Asia lie in the contemporary temperate latitudes, and clearly formed from vegetation communities different from those of the Carboniferous Period.

In most cases coal formed as thick peaty mats derived from forests growing in swamps. It therefore occurs in seams and contains large tree-like fossils. Each seam is overlain by a muddy, sometimes marine, sediment which grades upward into deltaic sandstone, showing the changing delta conditions. The top of the sandstone is infested with roots and was the foundation on which another, younger coal seam developed. This sequence represents a cycle of subsidence, sediment accumulation, re-emergence and swamp formation, and is repeated many times in every coalfield. There are two main theories to explain this cyclicity – one invokes periodic fluctuation in sea level, possibly related to glacial melting and regrowth, the other suggests sporadic subsidence of the sedimentary basins in which the coal sequences accumulated.

The formation of coal is controlled by temperature and pressure, and is related to the rate of burial. The first stage is the breakdown of cellulose in the soft tissues to form methane (CH_4), water and carbon dioxide, leaving a peat enriched in woody material. Increased pressure and temperature drive off more water and oxygen to boost the carbon content, forming lignite and eventually bituminous coal. In areas of high heat flow or at great depth, high temperatures drive off methane, leaving anthracite with the highest carbon content. Much of the world's natural gas resource formed in this way, methane having risen from anthracite to become trapped in suitable structures.

▼ *In open-cast coal, or strip-mining, the overburden is removed and the coal worked from the surface. The overburden is usually replaced once the seam is worked out.*

Water – the principal resource

The first need of any society is a dependable supply of drinking water – about two liters per person each day. Today this volume can reach over 2,000 liters, the bulk being used in agriculture, services and industry. With growing populations and standards, water supply poses greater problems than all other resources. The problems involve both quantity and quality.

For an exceedingly wet planet, the Earth has little easily used water. Less than one millionth of surface water is fresh and in streams or rivers. A hundred times more is held in lakes, but 80 percent is held in only about 40 of these, like the Great Lakes of North America. More than 20 million cubic kilometers is locked in ice sheets. River water threatens huge populations with disease, flooding and droughts, and is responsible for most major disasters.

Not all rain or snow becomes surface water. Some is intercepted by vegetation, some creeps into the soil. Depending on temperature, humidity and wind speed, a proportion is evaporated back into the atmosphere or transpired by plants. The loss by evapotranspiration exceeds precipitation in arid areas. In humid areas rivers are very efficient at draining the remaining precipitation to the sea. Except in areas of continuous high rainfall, they are unreliable. They must be managed by dams and canals to arrest their flow and divert it on demand. Construction of dams itself poses geological problems (▶ page 229) and may even induce earthquakes (▶ page 246). Despite the hazards, however, dams form one important solution to water shortages. Another lies in extracting water stored deep beneath the surface.

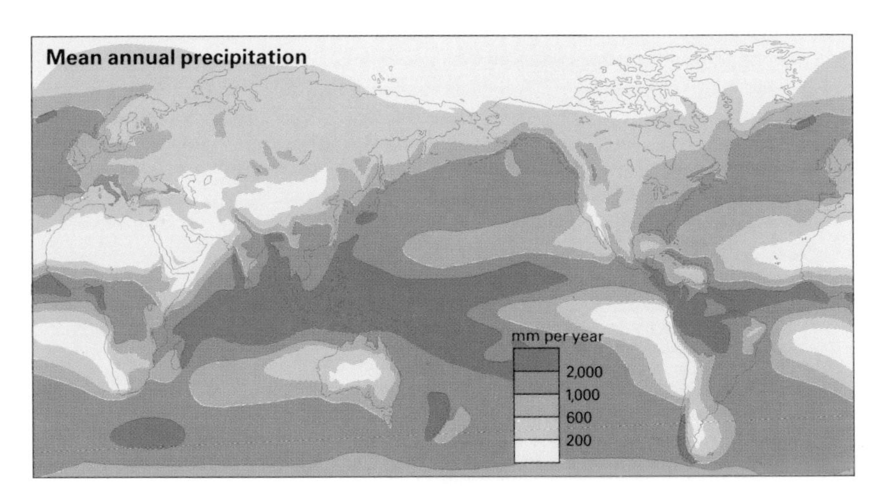

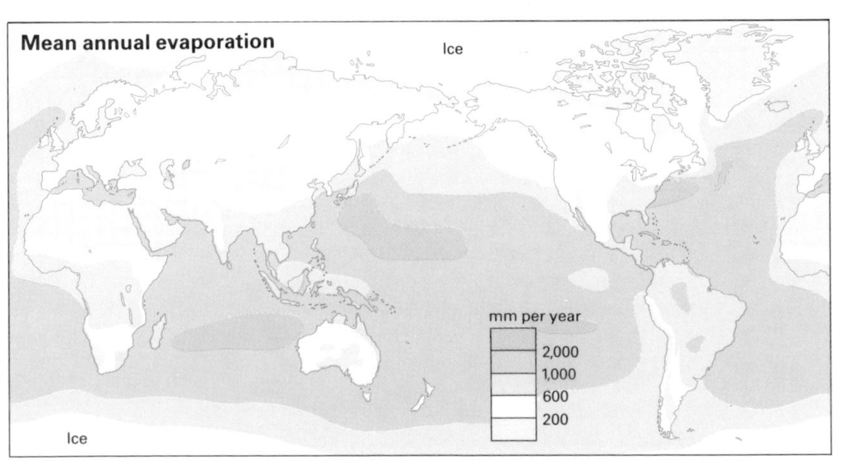

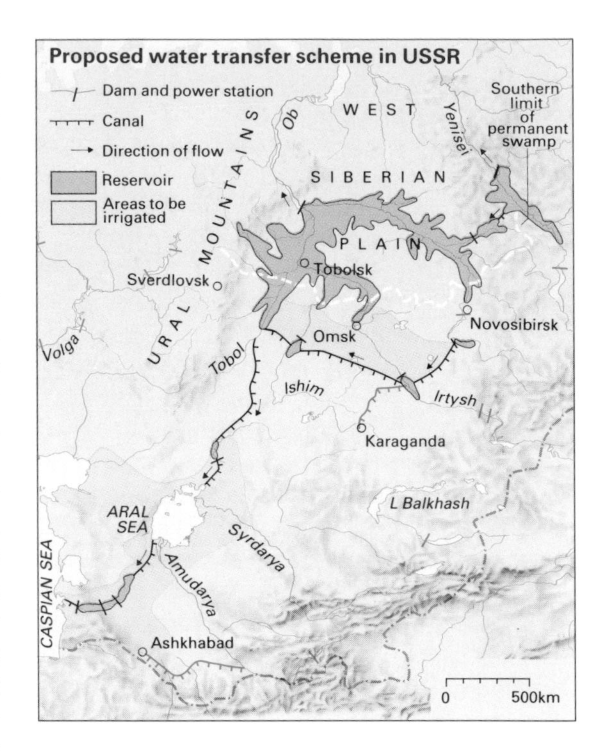

▲ ***The unproductive northward flow of the Ob river may one day irrigate the Caspian deserts.***

► ***The Romans understood the need for a reliable water supply, shown by the aqueducts that they built.***

Water management

Today, usable water is rarely where it is needed most, whether it be in an industrial city, a drought-stricken peasant community or an agricultural development zone. The traditional solution is to drill wells. Where groundwater abstraction is less than natural recharge this is ideal, as quality is often good and supply is constant. The problems arise when demand begins to grow.

As groundwater moves relatively slowly, extraction from wells depresses the water table (▶ page 207) in the form of an inverted cone. The more a well is pumped, the deeper and broader the cone becomes. Other users may suddenly find their wells run dry. The main aquifer west of the Jordan serves both Israel and the West Bank. Increasing Israeli abstraction during the 1970s and 80s so depressed the water table that the once fertile West Bank became drought-prone. The change in gradient of the water table may cause a reversal in flow. A source of pollution in an aquifer will be harmless if it is downstream from a well. Once the well forms a cone of depression, however, the polluted water may run down the side of the cone and appear in the water supply. Coastal groundwater normally discharges to the sea. High abstraction near the coast allows salt water to move inland destroying the resource temporarily.

Loss of water supply is not the only problem. Water in rock helps bear the load of overlying material. If it seeps from a clay during abstraction, this support is lost and the clay shrinks. The sinking of Venice is due to groundwater use by industry on the adjoining Italian mainland.

As supply grows more acute, so the solutions

considered become more ingenious. Rain making by seeding of clouds with chemicals seems to work only when rain is going to fall anyway. Because of abundant cheap energy, desalination of sea water is operated in the Middle East. The Antarctic ice has proved irresistible to some theorists. It is feasible, if a little bizarre, to tow icebergs rapidly enough to the Middle East so that enough would remain to create substantial reservoirs in gigantic dry docks.

The Romans and Persians transferred water from source to demand in aqueducts, and intricate tunnel systems. In Britain, which has shortages in the southeast but an abundance in the west, transfer is by pipeline. However, plans exist to augment river supply by aqueducts, and by using aquifers both to transport and purify river water pumped into them. The most grandiose plans involve transfer across continents. The waters of the Brahmaputra and the Ganges have been very destructive, and largely flow underused to the Bay of Bengal. India has long planned to construct aqueducts to tame their flow and divert a proportion to drought-prone areas. North America and the Soviet Union have vast, potentially fertile, but semi-arid central plains. Yet the great rivers, the MacKenzie and the Ob, flow uselessly across the tundra to the Arctic Ocean. Both have schemes to reverse this flow for irrigation. The low relief makes this feasible, but some consequences may be dramatic. At present the northern areas are warmed by the rivers as indicated by their control of the tree line. Diversion would remove this effect and transfer cold water south. No one has estimated fully the effects, but it could upset the continental climatic balance.

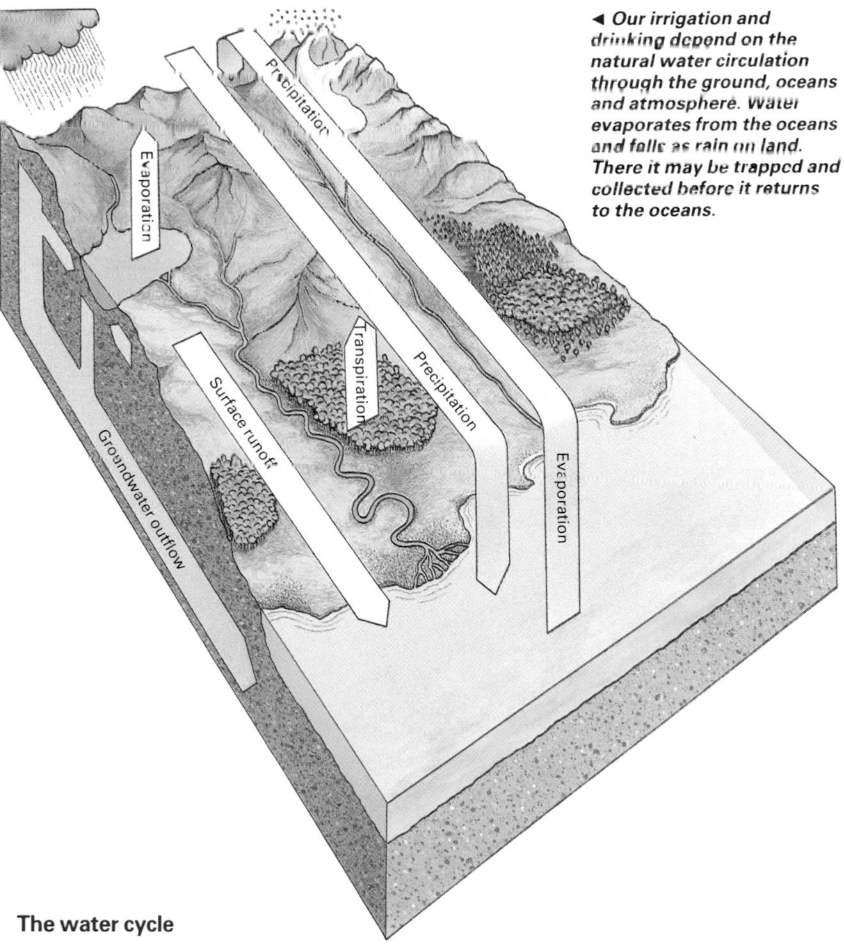

The water cycle

◄ ***Our irrigation and drinking depend on the natural water circulation through the ground, oceans and atmosphere. Water evaporates from the oceans and falls as rain on land. There it may be trapped and collected before it returns to the oceans.***

The solution to water shortages lies deep beneath the surface

▲ ***An oasis is a fertile spot in a desert occurring where water tables reach the surface, such as here, in Taghit, Algeria. Towns develop using the water for drinking and irrigation.***

The location of springs, wells and oases

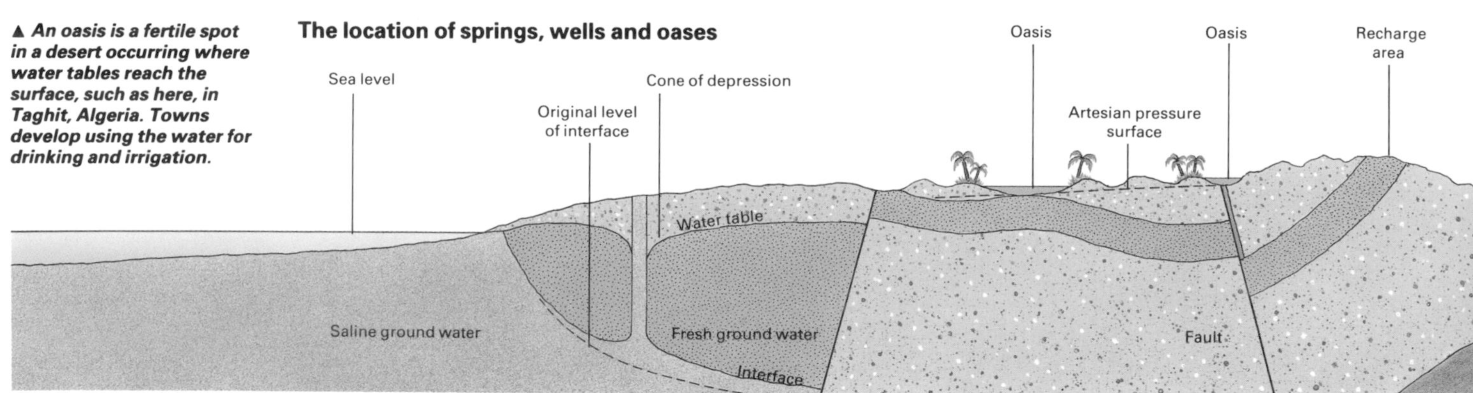

▲ *Water emerges under pressure from an artesian well. The head of water in the aquifer in the surroundings mountains is higher than the well, ensuring a constant flow.*

▼ *Where the water table reaches the surface, the water rises from the ground to form a spring. The world's greatest rivers develop from springs such as these.*

Subterranean water storage begins with infiltration. Once beneath the surface, water is protected from evaporation and moves slowly, over years rather than days. Much of it becomes trapped. Groundwater is the second greatest store of water after the oceans, but it becomes hot enough at depth to dissolve salts and so only about 8 percent is useful.

The amount of rainfall that infiltrates rocks depends on topography, vegetation, rock and soil permeability, and the time available. Heavy rain waterlogs the soil causing rapid runoff, so intermittent showers are better for recharging groundwater. Steep, vegetated slopes are inefficient at collecting water, so ironically flat deserts with little rainfall have the highest potential for water absorption.

Groundwater movement is controlled by permeability, and its amount by porosity – the proportion of voids to solid material in a rock. An aquifer is a rock that can both store water and allow it to flow. An aquiclude may be very wet, but will not allow percolation, and can act as a flow barrier. The best aquifers are poorly cemented sandstones, limestones whose weaknesses have been widened by solutions, and sometimes jointed and fractured crystalline rocks. Aquicludes are usually mudstones or rocks produced by crystallization.

The water table is the level at which groundwater stands in boreholes and marks a balance between supply and slow percolation. It tends to follow the ground surface but has a gentler gradient. The easier the flow, the more nearly horizontal the water table.

An aquifer which outcrops at the surface can only supply water from pumped wells penetrating the water table. Where the water table intersects the side of a valley, simple springs occur, which often fail in dry weather as the water table falls. If the aquifer is underlain by an impermeable stratum, the water table is perched higher than it would be normally and springs develop where the contact meets the surface. Faults can throw aquicludes against aquifers thereby sealing off underground flow and forming a line of springs along the fault.

Aquifers that dip beneath the surface and are confined between aquicludes contain water under pressure. The pressure depends on the depth and the level of the water table, and is reflected in the height that water rises in wells. Where the water table is higher than the ground above a confined aquifer, deep wells bring the water directly to the surface. Such artesian conditions occur in synclinal or faulted basins. They are responsible for oases in deserts where the head of the water in the mountains produces springs where aquifers are folded to the surface. Similar distant water transport can be controlled by major faults. They fracture the bedrock and provide a subterranean channel from areas of high rainfall. Faults generally outcrop as straight lines and show up clearly on satellite pictures.

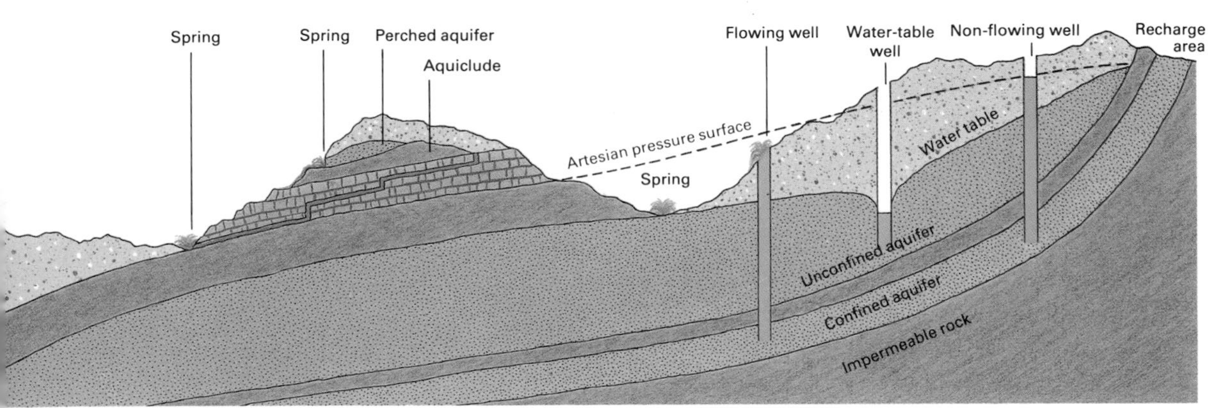

◄ *Springs are formed by different geological and topographic features. Rain falling in the recharge area fills the aquifer and where the aquifer breaks the surface, springs and oases will form. Artificial wells are dug or drilled down to reach the water table – the upper limit of the saturated rock. Water extraction lowers the water table immediately around a well, while near the sea, the removal of freshwater from the aquifer may bring up deeper salt water.*

Building materials, fertilizers and chemical feedstocks are common rocks produced by simple processes

◄ *The vast quantity of concrete and building stone that goes into the construction of a city the size of New York is a potent reminder of the importance of building materials to our way of life. Nearly all of these building materials are the products of geological action and have to be mined, quarried or extracted from the ground in some way.*

► *China clay is the product of the decay of granite. The feldspar minerals in the granite decompose slowly on exposure to the atmosphere, loosening the other minerals and causing them to be washed away as sand. The feldspars decompose into clay minerals and give rise to deep deposits of white clay on granite moors. These famous clay pits are on Dartmoor in southwest England.*

Industrial bulk materials – the most familiar resource

The most familiar of the Earth's physical resources are those used in construction. Together with the feedstock for the chemical and fertilizer industries, they outweigh all other resources taken together. All are common rocks, produced by simple processes. Crystalline silicate rocks are tough and take a high polish. Their internal structure makes them attractive for ornamental facings, and broken crystalline rocks are a near-ideal surface for roads. Sedimentary rocks make poor surfacings, but have been popular building materials. However, the costs for masonry are now so high that the trend is increasingly towards preformed or cast synthetic materials, such as bricks, concrete, metals and glasses.

Surface processes act to separate the different components of rocks. Some minerals, such as quartz, resist weathering, others break down to clay minerals (◄ page 148) and soluble compounds. The most obvious products are sands and gravels, dominated by resistant grains of quartz. Such deposits are used in the building and glass industries.

To produce useful clays requires more stringent processes. The common denominator of all industrial clays is their plasticity – the ability to be shaped when wet. Of those clays, three are noteworthy – kaolinite (china clay) from broken down feldspars, illite from decomposed mica and montmorillonite as a by-product of weathered ferromagnesian minerals. Clays rich in kaolinite and illite form from granites and thus are dominant on or near continents. Clay concentrations are winnowed from sand and deposited in quiet basins, cut off from contamination by coarser material. Montmorillonites derive from basaltic rocks and so dominate the ocean basins.

For brick and pottery manufacture plasticity must be conditioned by retention of shape before and after firing. Illites and some kaolinites satisfy these conditions. The more plastic montmorillonite absorbs oils and coloring matter, and is used for cleaning wool. The sodium variety swells when wetted, remains solid when standing but turns liquid when agitated. It is thus an important lubricant in drill-

▲ *Limestone is one of the most traditional building stones. Massive beds of it exist, particularly those dating from the lower Carboniferous (Mississippian) and Jurassic Periods. These have been quarried for centuries.*

▲ *Guano is the name given to accumulated masses of bird droppings. Certain islands off coast of South America are covered in it, and it represents an extremely valuable source of phosphates and nitrogenous chemicals.*

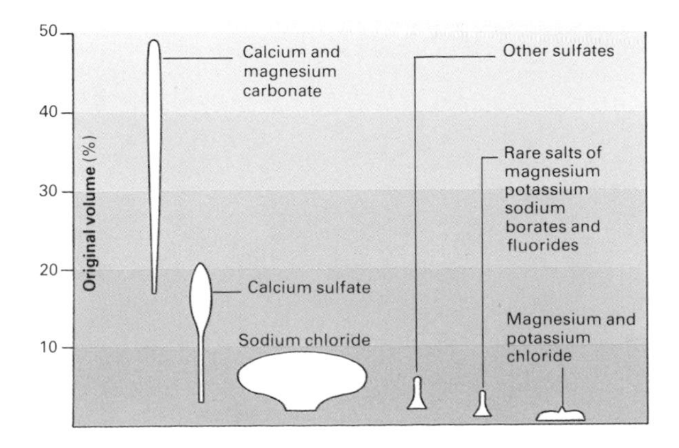

▲ *Dissolved salts precipitate from evaporating seawater in a particular order. Carbonates are deposited first, giving limestone. The great masses of salt do not form until the water has evaporated to a tenth of its original volume.*

ing, and is pumped into rocks to seal or grout fractures.

Cement, the binding material used in concrete, begins as a powdered mixture of anhydrous calcium silicates and calcium-iron aluminates. Calcium is held in solution in seawater as the sulfate or bicarbonate. Many organisms fix dissolved calcium in their hard parts as the carbonate, and most limestones consist of accumulations of these skeletons. The most productive areas for such biogenic limestones have always been clear tropical seas. When mixed with water, these minerals react to form interlocking hydrated alumino-silicates which bind together the gravel aggregate and impart strength to concrete. These cement minerals are produced artificially by roasting a ground-up mixture of limestone and clay, or clay-rich limestone.

Where land-locked seas have dried up, the evaporation of water has provided those compounds that contain the fertilizer element potassium. They are also the source for the sodium used in the chemical industry and the calcium sulfate from which plaster of paris is derived. Precipitation by evaporation is simple. Compounds crystallize in the order of their solubility, theoretically to form a predictable sequence. However, the water in shallow desiccating basins has to be replenished continually to build up thick accumulations of evaporites. Consequently, most natural evaporites contain only the most abundant or least soluble compounds – sodium chloride, calcium carbonate and calcium sulfate. The highly soluble materials were not deposited, simply because the seas never dried completely. Only the most arid environments, like the Atacama Desert and Death Valley supply exotic salts such as borates and nitrates of sodium.

The other bulk materials for the chemical industry and fertilizers are derived from different sources. Sulfur, although once mined on the slopes of volcanoes, is now produced as a waste product of oil refining. Nitrogen is extracted from the atmosphere by combination with hydrogen to form ammonia. Phosphorus is derived from the fossilized skeletons of marine organisms which filled shallow pockets on the sea bed, and from accumulations of the droppings of sea birds.

Prospecting Techniques

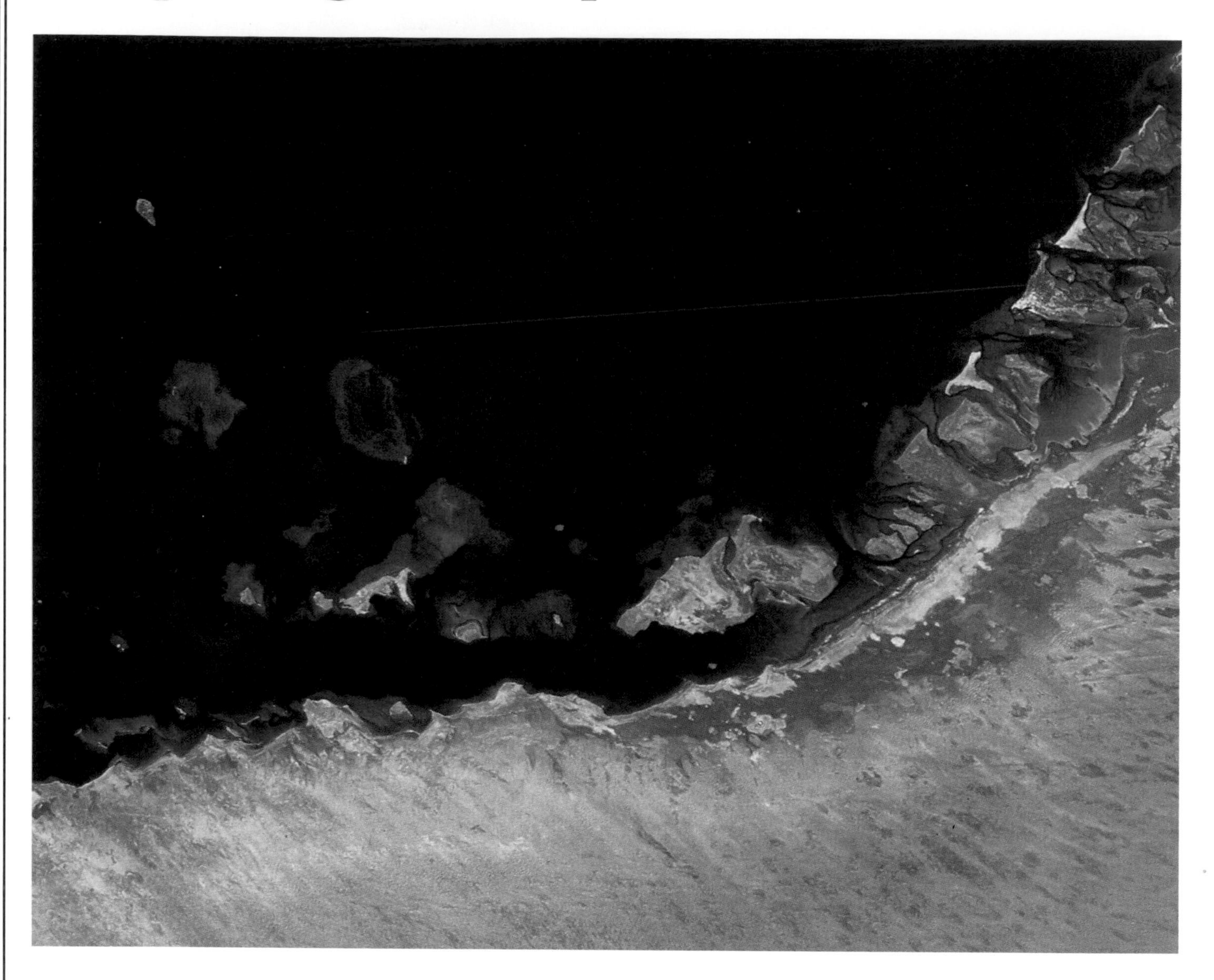

▲ *The first stage in prospecting for a valuable resource is to take a long distant look at the area to be studied. Nowadays this is done by satellite surveys, from which likely-looking large scale rock structures may be found and noted for further investigation.*

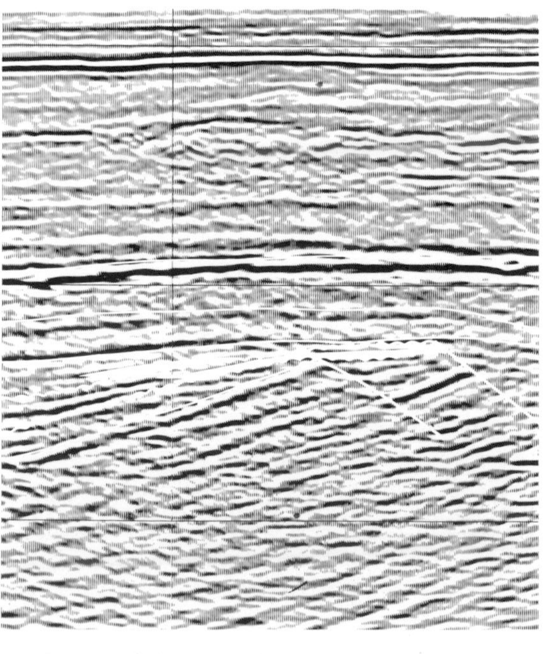

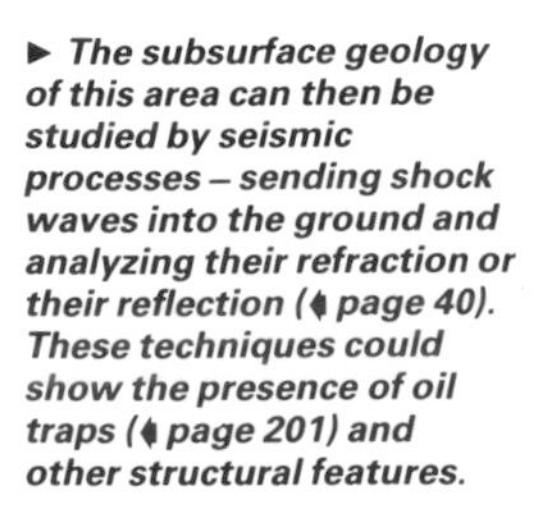

► *The subsurface geology of this area can then be studied by seismic processes – sending shock waves into the ground and analyzing their refraction or their reflection (➧ page 40). These techniques could show the presence of oil traps (➧ page 201) and other structural features.*

Targeting a resource

Physical resources are products of universal geological processes that have gone to extremes. They comprise concentrations of metals, common minerals that have been purified, and pore fluids that can be pumped to the surface. They are unusual, and exploration seeks the anomalies in rocks and materials derived from them.

To prospect comprehensively every square meter of the Earth's surface would be both tedious and hugely expensive. Each resource forms through various unique permutations of geological processes, some of which are mutually exclusive. Brick clays are never found on volcanoes, nickel and tin deposits do not occur in the same rocks, metamorphic rocks only rarely contain oil and little drinking water can be tapped from shale. So the first stage in exploration is deciding on a goal, say an oil field, and then on a favorable target area. Because finding an elephant in elephant country is easiest, exploration focuses on producing areas.

Oil provides a good illustration of targeting. The prospect area must comprise a thick, rapidly deposited sequence of tropical marine sediments.

▼ ***After some promising subsurface features are found by seismic methods, a hole is drilled into them and the physical properties of the rocks – moisture, electrical resistivity, radioactivity etc – can be studied by lowering instruments down. This is called well-logging.***

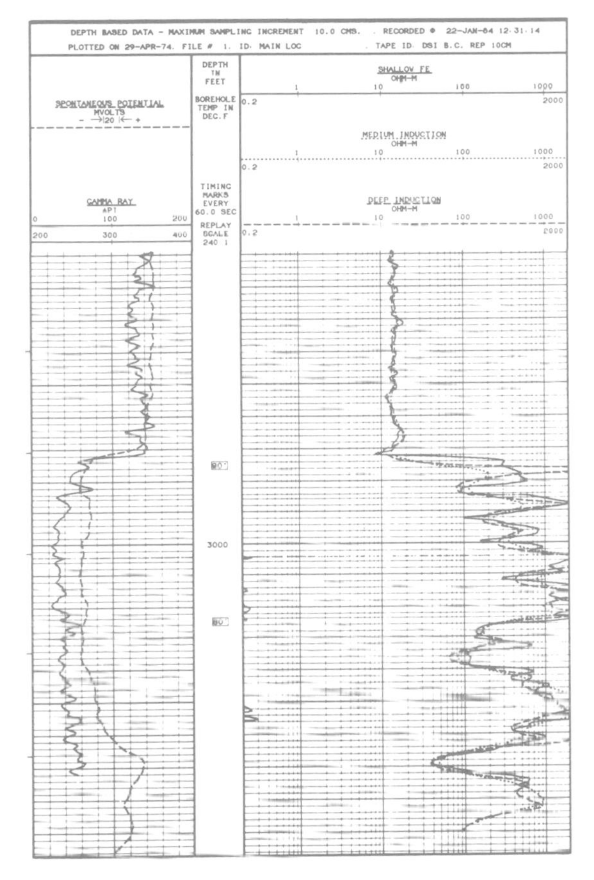

► ***Nowadays computers are used in many stages of the exploration of the Earth's crust for reserves of economic minerals. The information to be analyzed comes from so many different sources that electronic aids are needed to enable all the data to be processed.***

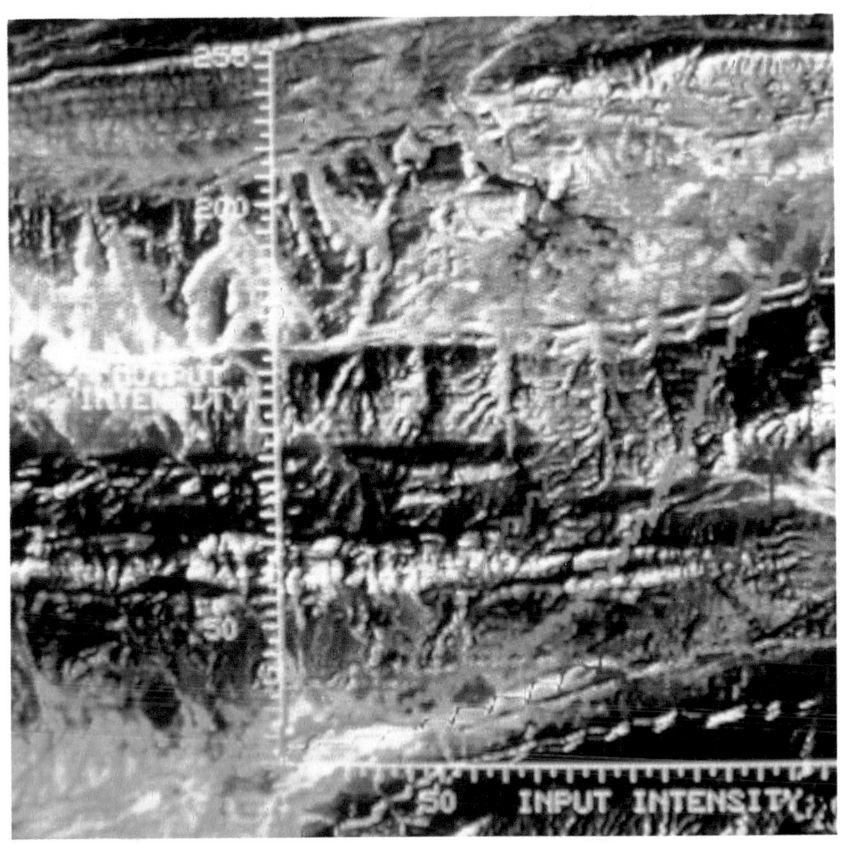

▲ ***Remote sensing can give an indication of what the underlying rock is like. But to find out for sure it is necessary to drill out core samples. The fragments brought up from a drill are laid out in sequence in a rack to represent the succession of rocks beneath the surface.***

Among them must have been organic-rich mudstones. The rocks had to be heated, but no higher than 150°C. Oil can only migrate in suitably tilted and permeable reservoir rocks. It is preserved only if such sandstones or limestones are configured with impermeable rocks into some form of trap structure. Such fundamental information may already exist in geological records. For instance, deep sedimentary basins often have negative gravity anomalies and smooth magnetic fields. New information is always needed, to define potential trap structures by geological mapping and seismic profiling.

The next and most expensive stage involves exploratory drilling and collecting every scrap of information about the rocks that are penetrated. Drilling stikes oil relatively rarely, but it is essential in building a three-dimensional picture of all the geological variables upon which final success or an aborted program depends.

Unlike oil, every metal can be concentrated in many broad geological environments and several metals may be present in one deposit. Consequently, target areas are chosen on the basis of their likely ore deposit rather than a specific metal (◀ page 198).

Most modern mines owe their discovery to individual geologists. They were obvious and easily found. The future demands that discoveries be made of remote or less anomalous surface deposits, and deeply buried ores. This will rely on two general techniques, that hitherto have been adjuncts to general prospecting. One depends on detecting subtle concentrations of metal in rocks, soils, water and plants. The other stems from the abnormal physical properties of ore minerals themselves.

An ore deposit is a tiny but extreme geochemical anomaly. They can be overlooked. However, surface processes conspire to weaken and broaden these geochemical high points, and disperse them over a larger area than the source itself. Simple, accurate means of analysis now enable these broad, weak signs to be mapped out systematically for most metals. For those which are intractible, other metals with which they are commonly associated can be used as signatures for possible ores.

See also
Rocks and the Rock Cycle 141-52
Sedimentary Environments 157-72
Development of Soils 173-80
Changing the Landscape 213-28
Engineering Geology 229-40

From prospecting to exploitation

There is little profit in merely identifying metal-infested streams or vegetation that is groping for survival in naturally polluted soil. Ore deposits extend to depth and some means of penetration is clearly needed. The molecular structure of some ore minerals provides a clue. Many of them are oxides and sulfides. Oxides incorporating iron in their structure are frequently magnetic, whereas many sulfides have structures akin to metals and conduct electricity.

Prospecting which exploits the magnetic properties of minerals measures deviations from the geomagnetic field (◀ page 45). Though positive magnetic anomalies can indicate iron ores, they may relate to iron minerals that accompany ores of other metals, such as chromium or nickel. Magnetic anomaly maps also give a crude outline of regional geology and thus a clue to mineralized sites, because iron is one of the most variable common elements in rocks.

Geochemical dispersion

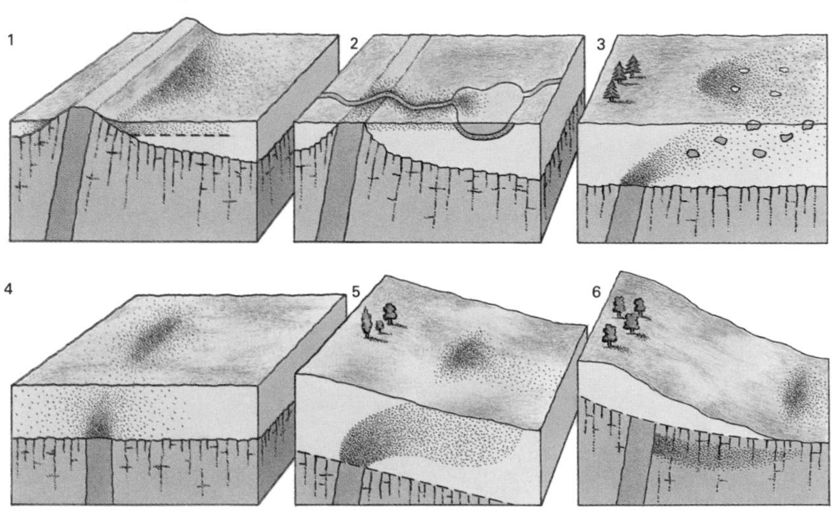

Geophysical prospecting

Zones of buried conducting sulfides can be detected and outlined by passing currents through rock, and finding fields of low electrical potential at the surface. If any conductor is exposed to radiowaves, it acts as an aerial and alternating currents flow through it. These artificial currents generate radiation too, so electromagnetic prospecting relies on detecting secondary radiowaves. They are usually induced by pulses from an aircraft, but can be produced by high energy long wave broadcasts from nuclear submarines. Some sulfides react with oxygen dissolved in groundwater, and in doing so generate electrical currents. They are natural batteries. The reactions may have to be triggered by high voltage pulses through the rock, but in either case detecting the presence of sulfides merely requires monitoring electrical fields at the surface. Electrical survey methods have become so refined that they can now give a measure of depth and size, as well as location of rich sulfide ores.

Magnetometers can be used in prospecting to detect variations in the Earth's magnetic field due to the presence of metal ores. Analysis of the results from an area can eliminate the regional magnetism of the country rocks and the background magnetism due to the topsoils. The remaining anomaly may be due to buried ores.

Gravity surveys, too, may reveal potential economic structures below the surface. Salt-domes are masses of salt that have bulged up the overlying strata. Since many salt deposits are contemporaneous with oil, and the bulged up strata make a perfect trap, the negative gravity anomaly produced by the lightweight salt is often sought by oil companies. Alternatively, a positive gravity anomaly may reveal a heavy ore-body at depth.

▲ *Geochemical study of soil can lead to the discovery of ores. A buried ore body will shed particles into the overburden, and these can be detected by soil analysis. Then, by taking into account the factors that are moving the soil it is possible to trace the particles back to their source. These factors include the prevailing wind (1) which would then move the particles in one direction. River deposits (2) and glacial deposits (3) could be followed upstream to their source. Soil water (4) spreads the substances around. Soil creep on a slope (5) moves them downhill as does prevailing groundwater flow beneath the water table (6).*

► *Remote sensing has become so important in the last decade that it occupies whole departments of survey organizations. This is a radar image of water resources near the Red Sea.*

Changing the Landscape 20

Human population – the increasing pressure...Human development – its impact...Human influences – landscape building, erosion and subsidence...Flood control and irrigation – water foe and friend... PERSPECTIVE...A pioneer conservationist...Coastal erosion...Manmade landforms...the Dust Bowl... Desertification...Weathering...Sand dunes... Permafrost...Dams and reservoirs...Lava flows

As in recent centuries the number of human inhabitants of the Earth has grown exponentially, and as their technology has dramatically expanded, so their impact on the environment has become increasingly significant. Such changes are pervasive and complex, for in nature it is impossible to change just one thing. Every action has a series of consequences. One of the first people to appreciate this was a New England polymath, George Perkins Marsh (1801-1882), who in 1864 wrote a pioneer book on conservation, *Man and Nature*. The following brief extract illustrates the geomorphological and hydrological ramifications which he identified as following from human interference by deforestation:

"Vast forests have disappeared from mountain spurs and ridges, the vegetable Earth accumulated beneath the trees by the decay of leaves and fallen trunks...washed away... Rivers famous in history and song have shrunk to humble brooklets...the bed of the brooks have widened into broad expanses of pebbles and gravel, over which, though in the hot season passed dryshod, in winter sealike torrents thunder...harbors, once marts of an extensive commerce, are shoaled by the deposits of the rivers at whose mouths they lie; the elevation of the beds of estuaries, and the consequently diminished velocity of the streams which flow into them, have converted thousands of leagues of shallow sea and fertile lowland into unproductive and miasmatic morasses."

In the few million years that Man has been an inhabitant of the Earth his potential for causing change has changed.

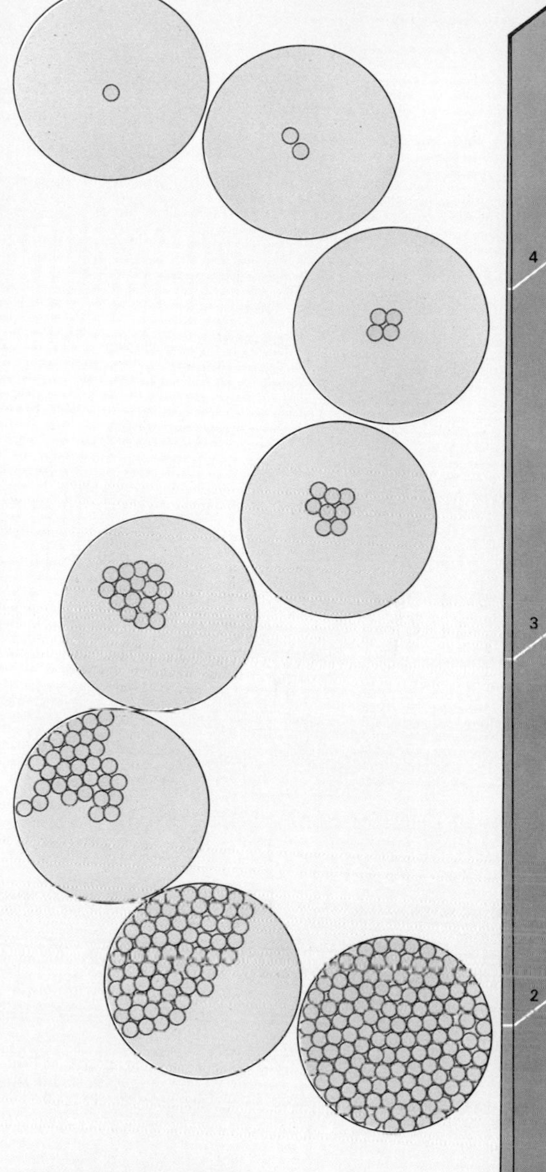

▲ George Perkins Marsh, an American pioneer of the conservation movement, was one of the first to recognize the influence of Man on the environment.

A pioneer of conservation

With his appointment as a US Diplomat in the Near East and Italy, George Marsh's perception of the perils of Man-induced erosion in the USA was corroborated by his travels abroad. The same destructive processes – removal of forests, overgrazing, extinction of wildlife, and so forth – recurred wherever civilizations had flourished. The denuded hillslopes of the Mediterranean were forlorn monuments to human greed or improvidence. It was in his classic monograph "Man and Nature" that he proved that Man is a uniquely potent creature. It was to the felling of trees, above any other causes, that he attributed the derangements Mankind had unwittingly wrought: violent fluctuations in river discharge, desiccation of the soil, dried up springs, depleted fauna and flora, and severely accelerated erosion.

▲ ► Population has been increasing since civilization evolved. With early advances, like farming and animal husbandry, the finding of food became easier. As Man came to understand disease, medical science began to influence survival. Fewer individuals were being killed off by natural processes than would have been the case in the wild. More people could then raise more people, who would themselves survive. Hence the world population has risen exponentially. From AD1 it doubled in size over 1,500 years – it is now likely to double within 40.

"Population . . . increases in a geometrical ratio. Subsistence only increases in an arithmetic ratio."
Thomas Robert Malthus (1766-1834)

Ten thousand years ago, at the close of the Ice Age, the total world population may have been about five million people, and large areas would only recently have witnessed human migration. The Americas and Australia, for example, were not inhabited by Man to any significant degree until after about 30,000 years ago. By the time of Christ, the total population may have reached about 200 million. It was not until the Victorian era that the total passed 1000 million, but the fifth billion will be achieved before the end of the 1980s.

Milestones in human development

Certain technological changes have occurred during the history of Man's tenure of the Earth which have modified his ability to cause geomorphological changes. Fire may have been used deliberately as long as 1·4 million years ago in East Africa, providing the potential for substantial removal of vegetation. The development of stone tools permitted both the slaughter and, in some cases, extinction of wild animals and the cutting down of wood. However, until about 10,000 years ago Man was still essentially a hunter-gatherer living mainly in small, nomadic groups. After that time the domestication of plants and animals in many parts of the world caused revolutionary developments, including the possibility of large, settled populations existing in considerable densities. Draft animals gave Man extra power to shape the land surface. About 5,000 years ago irrigation was first practised, involving the partial control of river systems, and the plow was developed, thereby providing a major tool for soil disturbance and erosion. Mining of ores and the smelting of metals developed rapidly at much the same time, creating a need for excavation and for charcoal. Finally, since the late 17th century there has been the development of major industries and of sprawling urban areas. Man has acquired enormous physical power to change the surface of the Earth with the invention of steam power, the internal combustion engine and nuclear energy. This technology has also permitted Man to change an even wider range of environments. For example, it is only in the last few decades that desert, like those of the Middle Eastern oil states, or fragile tundra, like those of Siberia and Alaska, have become subjected to major environmental stress.

▼ Almost everything that civilization does has an effect on the surface of the Earth. Human activities can be categorized by their effect on the environment.

► Landscape modification, its alteration to suit human needs, was practised in ancient Egypt, as shown in this tomb painting of an irrigation system.

Direct anthropogenic processes	Constructional	Tipping and grading
	Excavational	Digging Cutting Mining Blasting and trampling
	Hydrological interference	Flooding Damming Draining Channel modification Canal construction Coastal protection
Indirect anthropogenic processes	Acceleration of erosion and sedimentation	Clearance of vegetation Plowing Grazing Engineering (especially road construction and urbanization) Modification of hydrological regime
	Acceleration of weathering	Pollution of rainfall Introduction of salt through groundwater level change
	Subsidence	Mining Fluid removal Permafrost degradation Organic soil decay
	Slope failure	Loading Undercutting Shaking Lubrication
	Earthquake generation	Loading by reservoirs Lubrication along fault planes
Deliberately created landforms	Moats	Defense
	Reservoirs	Water management
	Canals	Transport, irrigation, defense
	City mounds	Defense, memorials
	Craters	War, qanat construction
	Dikes	River and coast management
	Embankments	Transport, river and coast management
	Cuttings	Transport
	Ridge and furrow	Plowing
	Terracing, lynchets	Agriculture and soil conservation
	Spoil heaps	Mining
	River cutoffs	Mineral and peat extraction
	Drainage networks	Flood control and navigation
	Lakes	Irrigation and drainage

◄ ▲ ► Population pressures vary. The Scottish highlands, left, are overpopulated since the land cannot support all the people. However, a city like London, above, is not since the area can sustain a large population. In the ancient Middle East similar population factors led to the first cities (right).

More soil was moved by bombing during the Vietnam War than was moved to construct the farmlands of Holland

Landforms and human activity

Almost all spheres of human activity create landforms. The most important are construction, excavation, hydrological work and farming. Some such landforms are very ancient. In Norfolk, eastern England, for example, there are bell-shaped pits that represent neolithic flint mines, there are innumerable small pits and ponds where generations of farmers have dug out lime to improve their sandy, acidic soils, and there are 25 large lakes (the Norfolk broads) which are now known to be flooded medieval peat diggings. All these forms were produced without the benefit of modern mechanical aids but are still a significant part of the English rural landscape.

The most important reason for excavation is mineral extraction, producing open-pit mines, strip mines, quarries for structural materials and similar features.

In the 20th century there has been dramatic growth in the production of aggregates for concrete. In Britain, demand for these aggregates (such as sand and gravel) grew from 20 million tonnes per year in 1900, to 276 million tonnes in 1973, an increase in *per capita* consumption from 0·6 tonnes per year to about 5 tonnes per year. One of the consequences of this extraction is the creation of many manmade lakes in low-lying areas of river terrace gravels.

▲ Mining has always had an impact on the landscape. Grime's Graves in the chalk of Norfolk, England, are a series of neolithic flint mines. They were the source of material for blades and the basis of local trade. The pits are a feature even today.

▼ Modern mining has a proportionally greater effect. The Bingham Canyon Copper Mine is the world's largest excavation. It reaches a depth of 774m over an area of 7·21km² and has involved the removal of 3,355 million tonnes of material.

Accelerated coastal erosion

Because so much economic activity is concentrated by the sea, the consequences of accelerated coastal recession can be very serious.

One of the best forms of coastal protection is a good beach, and so if a beach is depleted, the land behind may be attacked by the sea. Such removal may take place for construction purposes as when over 600,000 tonnes of shingle were taken from the beach at Hallsands in Devon in 1887 to provide material for the construction of dockyards at Plymouth. The shingle proved to be undergoing very little natural replenishment and in consequence the shore level was reduced by about 4m, 6m of cliff erosion occurred, and the village was demolished by the waves.

Another common cause of beach and cliff erosion at any one point is coast protection at another. Although a groyne may cause a beach to be built up at one spot, downcoast, because the source of beach material is interrupted, erosion will occur. Piers or breakwaters can have similar effects.

A further cause of beach depletion may be the damming of rivers, for in some areas sediment-laden rivers bring material into the coastal zone which becomes incorporated into beaches through the mechanism of longshore drift. If a large dam traps such sediment before it reaches the coast, coastal erosion may result. This is believed to be one of the less desirable consequences of the construction of the Aswan High Dam on the Nile. The Nile sediments, on reaching the sea, used to move eastward with the general anticlockwise direction of water movement in the Mediterranean, generating sand bars and dunes which contributed to delta accretion. Approximately a century ago an inverse process was initiated, and the delta began to retreat. For instance, the Rosetta Mouth of the Nile Delta has lost about 1·6km of its length from 1898 to 1954. The imbalance between sedimentation and erosion appears to have started with the Delta Barrages of 1861 and to continue with later dams. The Egyptian coast is now "undernourished" with sediment, and, as a result of this overall erosion of the shoreline, the coastal sand bars bordering some of the Delta lakes are eroded and liable to collapse. If this were to happen, the lakes would be converted into marine bays, so that saline water would come into direct contact with low-lying cultivated land.

Many different types of stabilization measures have been made, including groynes, sea walls and offshore breakwaters. However, the performance of these expensive and often unsightly structures has been poor in protecting coastal communities from beach retreat and destruction. Even more damaging, however, is the fact that they frequently enhance erosion by reducing beach width, steepening offshore gradients and increasing wave heights.

▲ Much of the human influence on the landscape is quite unintentional. The removal of shoreline shingle at Hallsands in Devon, England, for constructional work led to the increased coastal erosion and the destruction of the town in 1887.

▼ Modern warfare has a serious environmental impact. During the Vietnam War 26 million craters, covering an area of 171,000ha, were produced by bombing between 1965 and 1971. This represented the displacement of 2·6 billion m² of soil.

A quarter of the present area of the Netherlands has been reclaimed from the sea

The process of constructing mounds and embankments, and the creation of dry land where none previously existed, is also of great antiquity, but is rapidly increasing at the present time. Many schemes of hydrological modification involve dams over 300m high and huge embankments. Many rivers are "channelized" by great embankment systems such as those that run for more than 1,000km alongside the Nile and over 4,500km in the Mississippi valley.

Probably the most important features are those resulting from the dumping of waste materials derived from mines and from cities. In the Middle East and other areas of long-continued human urban settlement, the accumulated debris of life has gradually raised the level of the land surface and "tells" (occupation mounds) are a fertile source of information for the archeologist. The discharge of refuse from the New York Metropolitan Region is comparable to the estimated suspended sediment yield of all rivers along the Atlantic coast between Maine and North Carolina.

Manmade landforms

Major manmade landforms of accumulation include schemes to fill coastal swamps and marshes and to reclaim land from the sea. The most spectacular example of this is provided by the polderlands of the Netherlands. The coast from Calais to the Frisian islands consists of sand bars and dunes, and these provide the natural protection of the lowlands behind. However, heavy gales and spring tides posed a continuous threat, as did river flooding. Thus a large series of dikes have been constructed and the original coast length of the Netherlands has been reduced by some 300km. The Zuidersee Project, which started in 1919, and covers just under a quarter of a million hectares, has reclaimed some areas that are as much as 5m below sea level. It was based on the construction of a 32km long dike separating this Zee from the North Sea.

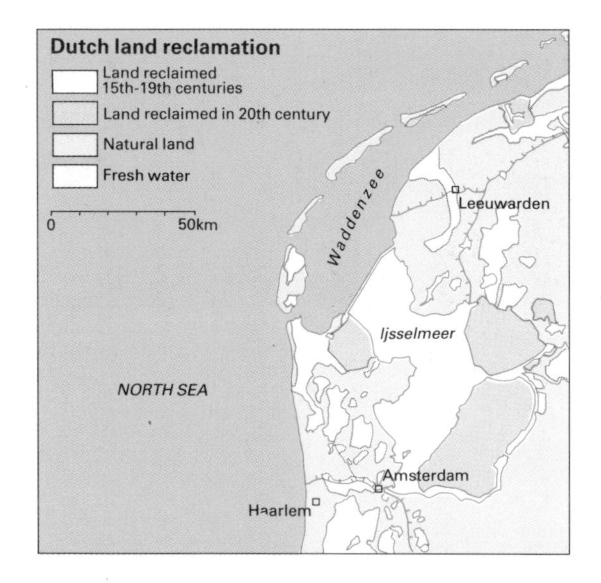

▲ ► The most spectacular example of landscape creation is the formation of the polders of the Netherlands. The local people have been draining the shallow coastal waters since the 7th century. The dike across the Zuider Zee turned it into a freshwater lagoon, and this is now being drained.

◄ The low-lying newly created polder farmland is criss-crossed by drainage channels through which the water is constantly being pumped away.

► Ancient examples of landscape construction are the tells of the Middle East. These are mounds of debris left from centuries of human occupation.

In the agricultural areas of the USA, soil is being eroded eight times faster than it is being formed.

▲ ***A farmland is a totally unnatural landscape. Agriculturalists have taken an area of land and modified it so that it becomes something more useful to human society. Often, however, this activity upsets the natural balance of physical, chemical and biological properties of that land and disaster results. Soil erosion is the most likely unwanted outcome of inefficient farming practices. Perhaps the most famous case of soil erosion in history was the Dust Bowl of the 1930s in the United States, in which widespread poverty followed farmland erosion.***

The American Dust Bowl

During the 1930s the drier parts of Texas, Oklahoma, Kansas and neighboring states were subjected to severe wind erosion which produced the famous dust storms – "black blizzards". They reduced visibility to a few meters and removed valuable top soil from the fields. In part, these dust storms were caused by a series of especially hot, dry years, which depleted the vegetation cover and dried the soil. However, the situation was gravely exacerbated by years of overgrazing and unsatisfactory farming techniques. After World War I, there was a great expansion of wheat cultivation into the Great Plains, helped by a combination of government assistance and the development of the internal combustion engine. Tractors, combine harvesters and trucks enabled land to be cultivated with greater speed than ever before.

The consequences of human activity – erosion

Of all the unintended consequences of human activity on the landscape it is probably soil erosion that is the most serious. In the USA $15 billion has been spent on soil conservation since the mid-1930s – the famous Dust Bowl era immortalized by the novel *The Grapes of Wrath*, by John Steinbeck – but in spite of that, the erosion of croplands by wind and water remains one of the biggest, most pervasive environmental problems the nation faces. It has been estimated that, on agricultural land in the USA, soil is being removed eight times more quickly than it is being formed, delivering 4 billion tonnes of soil each year to the rivers. This in turn leads to sedimentation in reservoirs, thereby shortening their lives and reducing their capacity (◆ page 160).

The prime causes of accelerated soil erosion are deforestation and agriculture. Forests protect underlying soil from rainsplash, bind the soil with their roots, produce a good soil structure, and generally reduce runoff. When they are removed this protection is lost and runoff and soil loss are greatly increased. In the USA, for example, the rate of sediment yield appears to double for every 20 percent loss in forest cover.

Urbanization can also create significant changes in erosion rates. During the construction phase, when there is much bare ground and disturbance produced by vehicle movements and excavations, the equivalent of many decades of natural or even agricultural erosion

Desertification

The spread of desert-like conditions into climatic zones where they would not naturally occur is a process called "desertification" or "desertization". It is regarded as one of the most intractable and serious environmental problems facing Man. Desert margins are fragile areas, but in recent decades they have been subjected to ever-increasing human pressures. Grazing and cultivation are extending into new areas and are being intensified in old. Improved veterinary services allow larger herd sizes to be maintained, as does the provision of water from boreholes. Vast tracts of woodland and scrubland are being cut down to provide firewood. Thus in drought years, such as those experienced in large parts of west Africa between 1968 and 1983, the land surface is subjected to such pressures that the soil is compacted and eroded, dust storms occur with greater frequency, increased runoff of water occurs rather than penetration into the soil, and general deterioration sets in.

◄ There is eye-witness evidence that the present extent of the Sahara desert is much greater than in the past. The desert plateau of Tassili-n-Ajjer in Algeria has rock paintings of cattle, sheep and other grassland animals. The area must have been much moister and more fertile three or four thousand years ago. There may have been changes in climate since the Ice Age, but some of the Sahara is due to the relentless grazing of domesticated animals.

▼ Deforestation is the inevitable result of an expanding population trying to cut a living out of a tropical forest region. Once an area of trees is cut down the soil soon loses its fertility to the crops grown in it. When the settlement moves on and cuts down a fresh area, the original forest can never grow back. The result of this so-called "slash-and-burn" method of agriculture is a scrubby landscape with bare hillsides as here, in northern Sarawak.

may take place during a single year from areas cleared for construction. Soil loss from an urban area may reach 55,000 tonnes per square kilometer per year, while in the same area rates under forest may be around 80-200 and under farming around 400. However, construction does not go on for ever, and once the disturbance ceases, roads are surfaced, and gardens and lawns are created. The rates of erosion go down and may be of the same order as those under natural or pre-agricultural conditions.

One consequence of the acceleration that occurs in rates of erosion under human influence is the development of gullies. In the south-western USA, many broad valley bottoms became incised with gullies, locally called "arroyos", in the late 19th century. This cutting had a rapid and detrimental effect on the flat, fertile and easily irrigated valley floors, the most desirable sites for settlement and economic activity in a harsh environment. Many students of this phenomenon believe that miscellaneous types of human activity could have caused this and the apparent coincidence in time of European settlement and the arroyo trenching support this view. Among the human actions that could have been responsible are timber-felling, over-grazing, hay-cutting and other agricultural activities, soil compaction along well-traveled routes, channeling from trails and railways, trampling by animals' feet, and the invasion of natural grasslands by miscellaneous types of scrub. Alternatively, natural changes in rainfall over the same period may have contributed to the phenomenon.

The uncontrolled erosion of soil on one area is balanced by the equally uncontrolled deposition of material in another

Accelerated erosion produces accelerated sedimentation. This has been exacerbated by the deliberate dumping of mining waste and other debris to stream channels. For example, gold mining in the Sierra Nevada mountains of California in the 19th century was undertaken by sluicing. This led to the addition of great quantities of sediments into the river valleys draining the range. This in itself raised their bed levels, changed their channels and caused flooding. Furthermore the rivers transported great quantities of debris into the San Francisco bays, causing serious shoaling. Similar shoaling occurred on the other side of America in Chesapeake Bay – the result of large quantities of soil arriving in the rivers because of the spread of European farming into previously forested areas.

Some mass movements can be triggered by Man, often because waste soil and rock are piled up into unstable accumulations that fail spontaneously. At Aberfan in Wales, 150 people died when a 180m high coal-waste tip began to move as an earthflow. The tip had been constructed not only as a steep slope, but also upon a spring line.

In Hong Kong, where a large proportion of the population lives on steep slopes developed on deeply weathered granite debris, mass movements are a severe problem, especially along road cuttings and where slopes have been artificially modified through construction and cultivation.

In 1881 in Switzerland a disaster occurred below the Tschingelberg mountain which had been deeply quarried. After several weeks of rain, it became so unstable that its northwestern face collapsed, engulfing the village of Elm in which over a hundred people lost their lives. In 1950 at Surte in Sweden a pile-driver working in clay set up vibrations that weakened the surrounding slopes and caused a landslide that destroyed the village.

However, one of the most serious of all the mass movements triggered by Man was that which created the Vaiont Dam disaster in Italy in 1963, in which 2,600 people were killed. The waters of the impounded reservoir made the rock masses on its margins unstable, so that 20 million cubic meters of ground slipped precipitately into the reservoir causing a great rise in water level which overtopped the dam. This caused flooding and horrific loss of life downstream.

Human effect on weathering

Little is known about Man's effect on weathering. However, it is clear that as a result of increased emissions of sulfur dioxide because of the burning of fossil fuels, there are increased levels of sulfuric acid in rain over many industrial areas and in regions downwind of them. Similarly, atmospheric carbon dioxide levels have been rising steadily because of the burning of fossil fuels, and carbon dioxide may combine with water, especially at lower temperatures, to produce weak carbonic acid which can dissolve limestones. Weathering can also be accelerated by changes in groundwater levels caused by irrigation in arid areas. In parts of the Indus Plain of Pakistan the water table has been raised since 1922 by 6m, bringing salty groundwater high enough to evaporate under the influence of high air temperatures. Salt crystals composed of sodium sulfate precipitate out and as they do so cause stone and brickwork, such as that at the great 4,000 year old archeological site of Mohenjo-Daro, to decay at a catastrophic rate.

Sand dune reactivation and fixation

The world's coastlines and the world's deserts are often characterized by large areas of sand dunes. Many of these dunes are naturally active, moving slowly across the landscape propeled by the wind, especially where there is a lack of binding vegetation or a very ready source of mobile sand *(◀ page 158)**. Elsewhere they are relict features that have been reactivated by the loss of the vegetation that had trapped the sand particles and held them immobile. Recognition of the menace of shifting sand is indicated by a decree in 1539 in Denmark which imposed a fine for destroying certain species of plants on the coast of Jutland. The fixation of coastal dunes by planting appropriate plants was initiated in Japan as early as the 17th century, while attempts at the reafforestation of the spectacular Landes dunes in southwest France began as early as 1717. Within deserts, attempts to reduce the movement of desert dunes have involved the erection of barriers made of palm fronds and chicken wire, or the application of oil or salty water.*

▲ *Drifting sand dunes can be controlled to a limited extent by building fences or, as here in the Majd region of Saudi Arabia, by planting grasses and trees.*

► *A disastrous mass-movement took place at Aberfan in Wales in 1966 when a wet spoil heap collapsed in a slurry engulfing a school.*

Consequences of human activity – subsidence

The inhabitants of London and Venice, of interior Alaska, of the English coalfields, of the Cheshire salt-mining areas, and of the English Fens and Florida Everglades all have something in common – the areas where they live are subsiding because of their own activities. Such subsidence arises from a variety of very different causes – the removal of underground resources of oil, gas and water; the removal of solids by mining them (such as coal) or dissolving them (such as salt or sulfur); the disruption of ice-cemented ground (permafrost); and the compaction or reduction of sediments because of drainage and irrigation.

Subsidence resulting from the withdrawal of liquids has caused almost 10m of surface lowering to occur in parts of Los Angeles, while in the Central Valley of California pumping of groundwater has caused over 8m of subsidence. In low-lying areas, such lowering of the ground surface can cause increased exposure to coastal flooding. This is one of the prime reasons for the increasing flood hazard in London, Venice and Tokyo. The subsidence resulting from mining has a rather longer history. Flooded depressions, tilting homes, and deformed roads and railway lines, bear witness to its effects.

Land drainage can promote subsidence of a different type, notably in areas of organic soils. The lowering of the water table (⧫ page 204) makes peat susceptible to oxidation and deflation so that its volume decreases. At Holme Post in the English Fenlands, approximately 3·8m of subsidence occurred between 1848 and 1957.

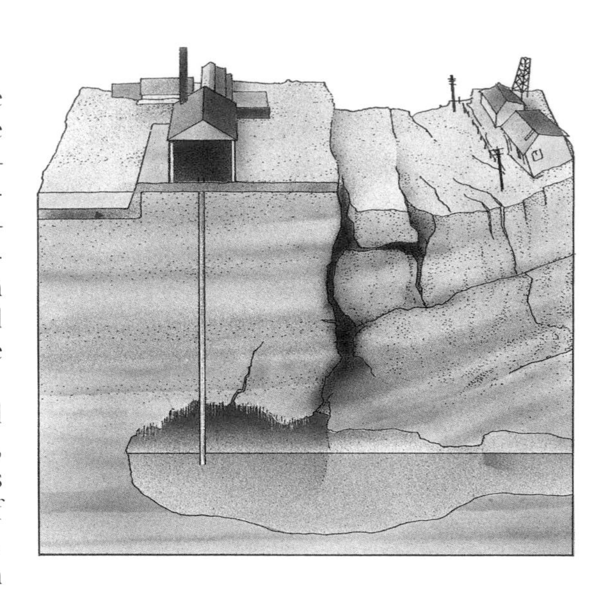

▲ ***Underground deposits of soluble minerals like salt are often extracted by drilling wells down to them, pumping down water and pumping up the salt solution so formed. The problem is that when the salt bed is dissolved away it leaves great cavities beneath the surface, and these may later lead to collapse. The salt industry in Cheshire, northern England, based on the widespread deposits of salt laid down in the arid environment of Triassic times, caused extensive damage to buildings in this way in the early years of this century.***

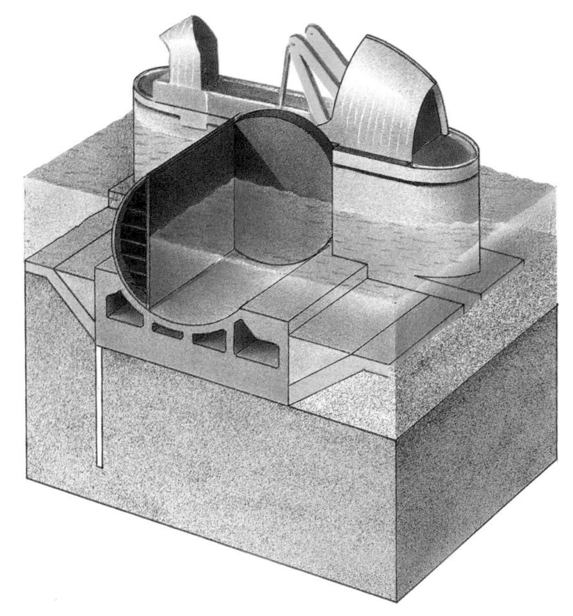

▲ ***The valley of the river Thames in the region of London has a history of flooding. This is largely due to the lowering of the land surface because of water extraction. The threat of disaster has now been alleviated by the building of the Thames barrage. This consists of a number of hemicylindrical gates that, when open, lie flush with the bottom of the river. When raised they turn upwards and form a barrier to the flow.***

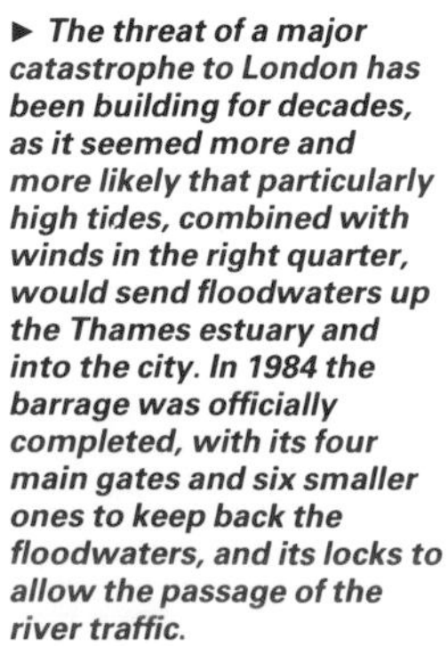

► ***The threat of a major catastrophe to London has been building for decades, as it seemed more and more likely that particularly high tides, combined with winds in the right quarter, would send floodwaters up the Thames estuary and into the city. In 1984 the barrage was officially completed, with its four main gates and six smaller ones to keep back the floodwaters, and its locks to allow the passage of the river traffic.***

▲ *The collapse of the ground surface due to melting permafrost is an ever-present danger to any human activity in cold northern areas. The Alaskan pipeline, from the oilfields in the north of the State to the ports in the south, is heated to keep the oil liquid. It is also built on stilts for part of its length so that the heat has no effect on the temperature of the soil as it crosses sensitive areas.*

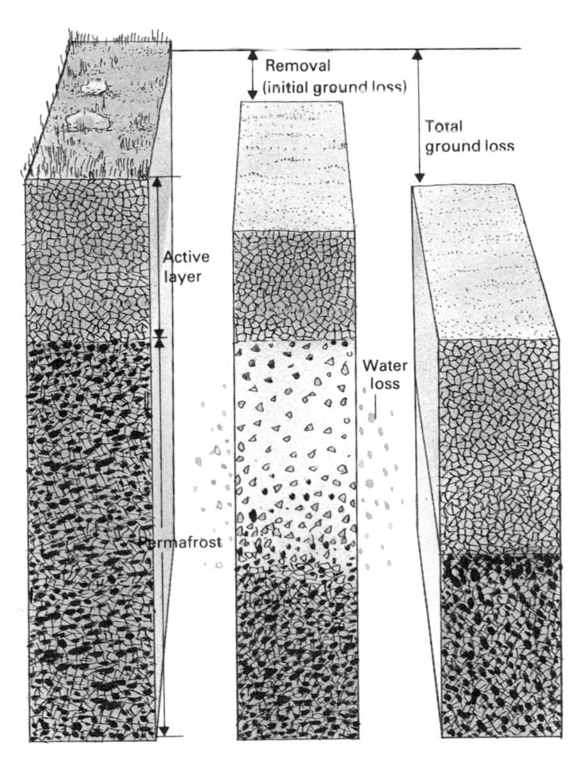

► *The permafrost, the permanently frozen soil in a tundra area, is supersaturated. That means that it holds more water than it would under normal circumstances. If the surface is warmed, or if the insulating effect of the plant cover is removed, some of the permafrost layer melts. The excess water may then run away or evaporate, decreasing the overall volume of the soil, and the land surface accordingly subsides.*

Permafrost and subsidence

In areas of the Arctic lands the subsoil, known as permafrost, is permanently frozen. Anything that causes the ice to melt can cause a phenomenon called thermokarst. If the insulating effects of a cover of forest or of peaty soil are removed, or if the temperature of the ground surface is heated by the presence of a home or an oil pipeline, the ice in the permafrost will thaw, and subsidence will occur.

The development of thermokarst subsidence can be illustrated by a simple example. Imagine an undisturbed tundra soil which has a zone of summer thawing, called the "active layer", 45cm thick. Beneath this is frozen ground which is supersaturated. It would yield, on thawing, 50 percent excess water and 50 percent saturated soil. If some of the upper ground surface (say 15cm) and its vegetation cover is removed, say by natural erosion, bare ground conditions will be produced. As a result the thickness of the layer that may thaw might increase to 60cm. As only 30cm of the original active layer remains, another 30cm of active layer must be produced by thawing the permafrost. This 60cm must be thawed since half of that will be water. Eventually the surface subsides 30cm because of melting, to give an overall depression of 45cm.

▲ **The pattern of the valley floor of the Ladakh region of the Indus attests to civilization's industry in bringing water from where it is plentiful to where it is needed. For many thousands of years the spring meltwaters from the glaciers in the Himalayas have been dammed, channeled and distributed, irrigating otherwise barren desert plains and allowing farming to feed millions of people.**

Dams, canals and reservoirs

Transporting water from areas in which it is plentiful, to areas in which it is scarce, is an important part of Man's agricultural activity. There are some landscapes of the world which are dominated by dams, canals and reservoirs. Probably the most striking example of these is the "tank" landscape of southeast India where myriads of little streams and seepages have been dammed by small earth structures to give a landscape that has been likened to a surface of vast overlapping fish scales. In the northern part of the subcontinent, in Sind and the Punjab, the landscape changes are no less striking, with the mighty snow-fed Indus and its tributaries being controlled by great embankments and barrages. The waters are distributed over thousands of square kilometers of desert by a canal network that has evolved over the past 4,000 years.

Flood control and irrigation schemes

Both for purposes of navigation and flood control Man has deliberately straightened many river channels. By removing meanders the river's course is shortened which increases the gradient and velocity of flow, so that floodwaters not only leave the basin more quickly, but also erode and deepen the channel, thereby increasing its flood capacity. The classic example is the Mississippi between Memphis, Tennessee, and Baton Rouge, Louisiana, where over a down-valley distance of 600km the length of the river has been reduced by over 270km as a result of 16 artificial cutoffs.

Some of the greatest changes have been wrought by irrigation schemes, especially in Africa and North America where about 20 percent of the total runoff is affected. One of the most apparent features of manmade dams and reservoirs is that they are becoming increasingly large. In the 1930s the Hoover Dam in the USA (221m high) was by far the tallest in the world. By the 1970s it was exceeded in height by at least thirteen others. Moreover, whereas Lake Mead, impounded by the Hoover Dam was the largest in the world in 1936, with a volume of 38 billion cubic meters of water, by the 1970s it was dwarfed. Kariba in central Africa contained 160 billion cubic meters, Egypt's Lake Nasser contained 157 billion cubic meters and Ghana's Akosombo, 148 billion cubic meters.

▼ ► The Jonglei scheme is an ambitious venture aimed at increasing the flow of the river Nile. At present the river meanders through the Sudd swamps of Sudan, where it loses much of its volume by evaporation. The aim is to bypass this and deliver more water downstream. Huge digging machines (right) are being used to cut the channel.

A particularly important consequence of the impoundment of a reservoir behind a dam is the reduction in the sediment load of the river downstream. This is well illustrated by the Nile. Until the construction of the Aswan High Dam, the later summer and autumn period of high flow was characterized by high silt concentrations of around 2,700 parts per million, but since it has been finished the silt load is rendered lower through the year as a whole and the seasonal peak reduced to just under 50 parts per million.

One of the biggest transformations of a river system that is currently taking place is the building of the Jonglei Canal on the River Nile in Sudan. It will have the biggest impact on the Nile's flow since the completion of the Aswan High Dam. However, major changes in the configuration of channels can be achieved accidentally either because of manmade changes in stream discharge or in their sediment load – both actions affect channel capacity. For example, the urbanization of a river basin results in an increase in the peak flood flows because of the runoff generated from sewers and impermeable surfaces of tarmac and tile. As a city expands the frequency of discharges which fill the channel will increase, with the result that the beds and banks of channels will be eroded so as to enlarge the channel. Conversely, downstream from a major dam the size of peak flows will be reduced and channel narrowing may occur.

▲ Not all dams are built for irrigation purposes. The Kariba Dam on the Zambesi river of East Africa, on the border between Zambia and Zimbabwe, was primarily built to provide hydro-electric power. One of the world's biggest dams, it was completed in 1958 and holds back a lake 280 km long and 32km wide. The dam itself is a massive structure, 130m high and 580m wide.

See also
Development of Soils 173-80
Landscape Features 181-96
Resources from Rocks 197-212
Engineering Geology 229-40
Manmade Earthquakes 245-48

Earthquakes are deep-seated events that occur in the depths of the Earth and so, as yet, Man has done little either to diminish or control their activity. Nonetheless the fact that he has been able inadvertently to trigger off small seismic disturbances by various engineering projects (♦ page 246) suggests that in due course it may be feasible to "defuse" earthquakes by relieving strains in the Earth's crust gradually by a series of non-destructive low-intensity earthquakes.

The most important seismicity created by Man is that associated with large reservoirs. This is probably due to the sheer pressure exerted by the mass of the water impounded, together with a change in water pressures across the contact face of faults.

Since George Perkins Marsh identified the role of Man as a geomorphological agent over a century ago, the Earth's population has increased almost five-fold. Furthermore, there have been further developments in the technology available to create change, and new areas have been exploited. Both in terms of deliberate changes and in terms of the incidental and unexpected effects of human activities, Man's role in changing the face of the Earth had been greatly magnified since the time of Marsh. Some of the changes are undoubtedly beneficial to us, but many others create hazards that we will ignore at our peril.

Controlling lava flows

In certain areas volcanic activity poses a great threat to human life and property. Unfortunately Man's ability to alter his landscape does not extend to the prevention of volcanic eruptions.

During the eruption of Helgafel in Iceland in 1973 the town of Heimaey was partly buried by erupted volcanic ash. A stream of lava threatened the harbor, but the harbor mouth was kept open by training fire hoses on the advancing lava front. This cooled and solidified the dangerous portion and diverted the flow.

Less success was recorded during the eruption of Mount Etna in 1983. Scientists tried to divert the lava flow by dynamiting part of the lava channel and letting the lava spill out in a less harmful direction. Unfortunately, the advantage gained was only temporary.

▼ Man is the most potent geomorphological agent that there is, as can be seen in this view of the coalfields of Pennsylvania. Our impact on the environment can be seen from the Moon. As the populations of the Earth increase, this impact will become greater and greater. As time goes on science will be looking for ways to lessen the damage done and still enable humanity to exist in its home world.

Engineering Geology

Geology and the construction industry – safety and economy...Stages in a project – from the desk to the structure...Tunnels – the underground problems... Foundations – footings, piles and rafts...Surface construction – cuttings and embankments...Dams – the problems with water...PERSPECTIVE...History of the science...Rock properties...Tunnel construction... Waste disposal...Dam failure...Slope stability

Engineering geology is concerned with the application of geology in the construction industry. As a consequence the practical use of engineering geology is possibly more directly visible to the layman than most other branches of geology whether it be in the form of a spectacularly engineered road through cuttings and embankments, an assured water or electricity supply from a reservoir, or a tall skyscraper. Equally the lay mind will readily appreciate the consequences of weather in creating landslides or rockfalls which closes the road, leakage which depletes the reservoir or settlement which cracks the skyscraper and puts the lift out of true. The purpose of engineering geology is to predict the geological conditions in such a manner that the structures can be designed, and built within limits of cost and time so that they will operate safely. A knowledge of the geology alone is inadequate in that the engineer needs to be able to define the behavior of the natural ground in quantitative terms. An engineering geologist must be able to translate the essentially descriptive characteristics of rocks into numbers which are a direct measure of response to changes in loading, water flow and other externally imposed events. The actual rock type, though important to the geologist, may not influence its engineering behavior. For example the properties of a fresh gabbro, strong sandstone or massive gneiss foundation to a heavy structure may be, in engineering terms, effectively identical. However, some rock types have a characteristic pattern of behavior associated with, for example, the friability of many shales and the solubility of limestones.

History of the science

Engineering geology has evolved as a subject over a period of many years. At the end of the last century a limited number of geologists became closely involved with major civil engineering projects. At this time, for example, Dr Charles Berkey was responsible for the investigation of a number of dam and tunnel projects associated with New York's water supply and Albert Heim was concerned with the interpretation of a catastrophic rock slide which occurred in Switzerland. However, it was not until the late 1920s, following the failure of the St Francis Dam in California, that geological studies became increasingly accepted as being necessarily associated with large construction projects. The development of the Tennessee Valley in the next decade involved the construction of dams, many of which were located on difficult limestone foundations, and resulted in the creation of a team of engineering geologists who were employed both in investigation and construction. Following World War II there was a rapid development of engineering geology internationally within the construction industry as the importance of accurate prediction of geological conditions was appreciated.

Engineering geology now forms a part of geotechnical engineering, geology being linked with the associated disciplines of soil mechanics and rock mechanics. The role of the engineering geologist is to identify the primary geological constraints in the project in hand, and then ensure that adequate information is provided to answer the technical problems that arise. At the present time most engineering geologists work in industry for firms who specialize in site investigation. Many are concerned with the design of the geotechnical aspects of engineering structures, and the supervision of construction. As the scale of engineering projects has increased, and the more straightforward sites have been developed, so has the complexity of the geological and geotechnical problems become progressively greater.

► Albert Heim (1849-1937), the Swiss geologist, was a pioneer of engineering geology. As early as 1878 he published a monograph on the structure of the Alps which contained a section on the stresses, the strains and the deformation of rocks, calling on both theoretical and experimental evidence. This is the kind of work that has been used as the basis for engineering geology ever since.

◄ The construction of today's modern buildings, such as these in West Berlin, must take into account the nature of the bedrock, and how it will react to the new and unnatural strains placed upon it. Engineering geology analyzes this in practical terms.

Engineers, and engineering geologists, use a slightly different terminology from academic geologists. In broad terms, the engineer divides ground conditions into rocks and soils. Rocks are hard and brittle and have a strength well in excess of the imposed engineering stresses. On the other hand, soils (engineering soils as distinct from agricultural soils) are typically represented by the weaker and more deformable soft rocks extending into weaker incoherent sands, silts and clays. This simplistic engineering classification is extended into the existence of the two separate engineering disciplines of rock mechanics and soil mechanics. In practical terms the mineral world has a total spectrum of engineering properties and behavior from the softest clay at one end to the hardest rock at the other.

Engineering construction results in a change in stress distribution within a rock or soil mass and it is convenient to consider three basic situations which can arise. Firstly a vertical load can be imposed on the ground surface as is associated with the construction of a building or a dam. Secondly, there can be a change in lateral support that will occur in an excavated road cutting, an embankment constructed on a slope or a building on an unstable natural hillside. Thirdly, removal of vertical support can occur in the underground excavation of tunnels and mines. A separate set of engineering problems arise in association with the critical issue of the behavior of groundwater which flows through the soil and land mass. Groundwater can be both a resource when pumped from wells and boreholes for irrigation and drinking water supply purposes, as well as a nuisance when it flows in an uncontrolled manner into excavations. Groundwater pressures can also significantly influence the stability of excavations and the behavior of foundations under load.

Rocks and soils provide an important source of construction materials which can be used for coarse and fine concrete aggregates, earth and rockfill for road embankments, and clays as an impermeable core for earth dams. The siting and development of local sources of such materials in quarries can play a major part in many engineering projects, particularly in more remote regions.

Stages in an engineering project

The engineering procedure which is required to build a major structure or project typically extends over many years (➧ page 232). Initially the need for a project is identified and a desk study is carried out during which all the preciously acquired information is reviewed; no physical investigation is carried out at this stage. The next phase involves an investigation which, for the larger project, is commonly divided into a series of stages. At the preliminary stage alternative sites are investigated and evaluated and a preferred option identified. The subsequent stage examines the feasibility of the scheme in more precise economic and technical terms. Finally, the design investigation establishes the ground conditions in relation to the specific location of the proposed engineering structures. Once the design is complete, it is necessary to obtain the powers to construct the scheme and raise adequate finance. The design is presented in a series of contract documents to contractors who will tender for construction. The contractor becomes responsible for the physical construction of the project while the designer supervises his work and ensures that the design criteria for the exposed conditions are adequately met. Once the scheme is completed, it will be necessary for the project to be monitored to ensure safe and economic operation.

▲ Pit props take up stresses once borne by extracted coal.

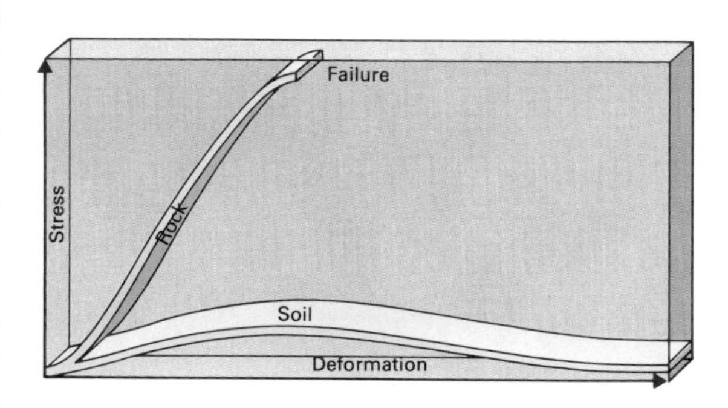

▲ Rocks deform easily at low stresses, as voids compact, but then become quite rigid until they fail at high stress. Soils, on the other hand, continue to deform at low stress and fail before any real stress can be put on them.

Engineering properties of rocks

The physical properties of rocks which are of particular relevance are deformability, strength, permeability and chemical stability. Deformability is a measure of the change in dimensions of a rock mass, under stress. If the stress is progressively increased a point will be reached when deformation continues at lower stress levels. At that point the material is said to have failed. The strength and rigidity of a particular rock type, when measured in the laboratory, is significantly greater than the equivalent measurement made on a large volume of rock "in situ". This arises from the presence of planes of weakness (or discontinuities) which may include bedding, planes, joints, fractures, faults and similar planar defects. As a consequence laboratory tests have some limitations. If deformation of clays or shales continues beyond the point of failure, a shear zone is developed with an extremely low strength. The strength of discontinuities is determined by the frictional and cohesive or interlocking effects of rough or irregular surfaces. Hence the assessment of "in situ" rock strength requires a knowledge of both the rock structure and the strength of the defect surfaces. A knowledge of strength is of particular importance in the assessment of the stability of slopes. The deformation of rocks and soils "in situ" can be determined by means of loading tests although the scale of such tests is small compared to the volume of rock which is

▲ Artificial steepening of slopes and stresses of building may lead to landslides, as in Jofeh, Amman.

► Groundwater may seep into excavations from structures undetectable from the surface.

ultimately to be loaded by the engineering structure. Consequently structures and their foundations are monitored by instruments to check the reliability of predictions.

The permeability of rock and soils is a measure of the ease with which fluids, mainly water in construction engineering, move through the material. The in situ permeability is largely determined by the presence of discontinuities and it is only in very porous materials such as coarse sandstones or gravels that intergranular flow is important. When excavations are carried out below the water table, the upper surface of permanently saturated ground, water will flow towards the excavation. As a result the prediction and control of water inflow into deep excavations or tunnels is important. The head of water behind slopes below embankments creates an uplift pressure which can lead to instability and landslides.

"In situ" stresses are important in engineering projects. Deep tunnels can encounter high stresses which are induced by the weight of overburden or locked-in tectonic stresses.

The response of a rock mass to engineering change is primarily a function of material properties, rock structure, stress and groundwater conditions and the nature of the imposed engineering change. During the investigation of any engineering project it is necessary to provide a response, in quantitative terms, to each of these issues.

From Desk Study to Report

1 Desk study

2 Remote sensing

3 Mapping

4 Geophysics

1 Before any construction work can be undertaken, the site must be thoroughly investigated to ensure that it is geologically suitable. The first phase of this is the DESK STUDY in which all previous reports of the area are examined, along with any other literature that may be relevant. At the same time the investigating team make their initial visits to the site to gain their first impressions.

2 The first observations of the site are done by REMOTE SENSING, using satellite-based imagery and aerial photography. With different scales of photography, and various scanning techniques, the investigators can obtain quick and accurate information about such factors as groundwater conditions, that would affect the project.

3 The geology of the area must be thoroughly surveyed, and this is done through conventional means of GEOLOGICAL MAPPING. The distribution of the various rock types is plotted, along with the trends of the faults and folds and other geological structures. Each rock type is sampled so that physical tests can be carried out on it.

4 The subsurface geology of the site is then investigated using GEOPHYSICS. The structures and rock-types that need to be studied are at a fairly shallow depth, and so the most appropriate geophysical techniques are resistivity surveys, in which the behavior of an electric current passing through the rock is analyzed (▸ page 212), and seismic investigation, in which the reflection and refraction of shock waves can be used to outline the subsurface structures (▸ pages 40-1). The results are then combined with the results of the geological mapping to give a complete picture of the local geology.

5 Direct sampling of the subsurface geology is important, and this is done by DRILLING. The rock samples brought up can be used to ensure that the geophysical survey has been reliable, and further geophysical work can be carried out by lowering sensors down the boreholes.

5 Drilling

***9** The end product of all this involved and lengthy work is the final REPORT. This is basically a record of all the data that have been obtained, but it also includes recommendations for any further investigations and studies that need doing and also proposals for the design and construction of the proposed engineering works. Since the accurate prediction and monitoring of ground conditions and behavior are critical to the successful completion of any engineering project, engineering geology is a very important economic application of the Earth sciences.*

***8** The next stage of the process is the stage of ANALYSIS AND DESIGN. In this all the data collected are collated, analyzed and interpreted. The results of this will dictate the final design of the engineering project. The important factors that will be taken into account here are the safety standards that will ensure that a reliable structure is built, and the economics that will enable the structure to be built for a realistic price and minimize waste.*

***7** A purely descriptive account of the geology of the site is not enough for a full survey. TESTING of the site materials, the rocks and the soils in which the proposed engineering structure is to be built, is carried out to measure their strength and suitability for the stresses and strains that will ultimately be placed upon them once the project is completed. This rigorous testing of the site materials can be done both on the site, and back in the laboratory where conditions can be closely monitored.*

***6** Drilling may not produce a full enough picture of the subsurface conditions. EXCAVATION is used to obtain direct access to the most important features of the geology of the site. Trial pits, shafts, trenches, adits and tunnels are excavated to expose the geology at depth, to sample the bedrock, and also to try out various engineering and construction methods to find which are the most appropriate and efficient for the conditions encountered.*

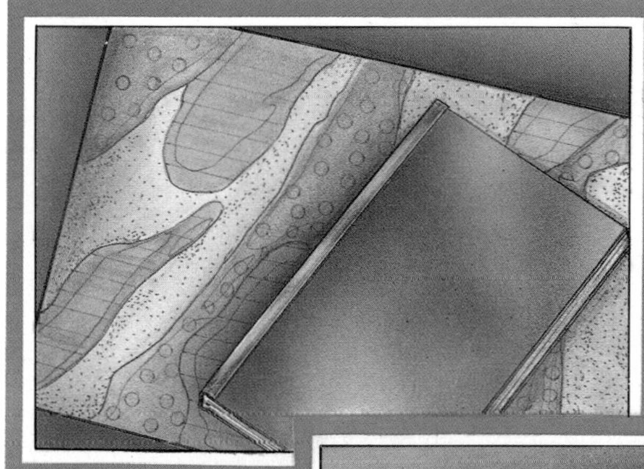

9 Report

8 Analysis

7 Laboratory work

6 Excavation

It is difficult for tunnel engineers to know what conditions await them underground

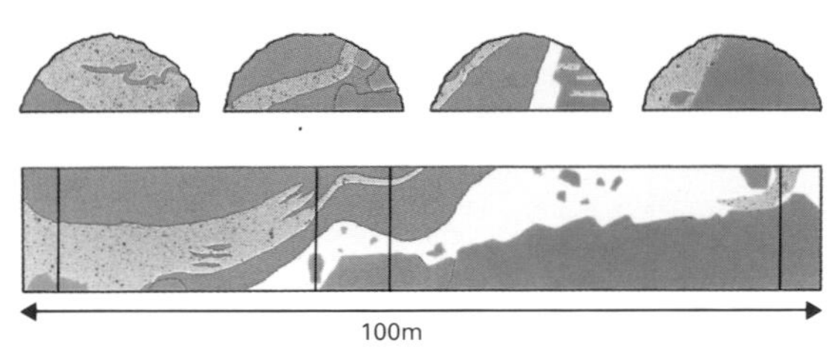

▲ ► The 10·8 km Landrucken road tunnel between Wurzburg and Fulda in West Germany, due to open in 1988, is driven through a variety of rocks types, each of which has its own properties and has to be treated differently.

▲ The cutting unit of a modern tunneling machine consists of groups of disks set in four arms about the rotational axis. The central toothed bit makes the initial cut, and the disks carry the cutting progressively outwards.

► Modern hydro-electric power stations are underground. It is important to site the machinery below the water source to make use of the head of water. These tunnels are for the Dinorwig power station in Wales.

Tunnels and tunneling

The prediction of geological conditions in underground excavations is one of the most difficult problems which occurs in engineering geology. This arises because many such excavations are at considerable depth. Many of the trans-Alpine tunnels, such as those at Simplon, St Gotthard and Mont Blanc were driven through geological conditions which had largely been predicted by surface geological mapping. The excavation of these tunnels provided new experience on both the internal geological structure of the Alps as well as the behavior of rock in deep tunnels. Tunneling has, in addition, become a specialized aspect of the construction industry with many of the engineers and workforce forming an almost elite class moving from project to project, carrying with them expertise in this challenging field.

Tunnels are put to a range of practical uses including rail and road transportation, water and sewage transfer, power development, storage and disposal, and strategic purposes. Few tunnels are less than 2 meters in diameter, which is just sufficient to provide man access, while the very largest can be 15 or 20 meters across. The typical tunnel is circular or horseshoe-shaped in cross section, with entry through a portal in a hillside or a shaft. Other excavation shapes are used for special purposes. Underground power stations built as part of hydroelectric schemes can involve a complex of cavities of different sizes and shapes. During the driving of a tunnel it is necessary to remove the rock and then to install a support system which will restrain any movements within the roof, walls or floor. The period within which a tunnel is effectively self-supporting, and before collapse begins is known as the stand-up time. In the case of the most massive rocks, such as some granites, the stand-up time is effectively infinite and no artificial support is required. However, in softer materials such as soft clays, the walls of the tunnel may squeeze in rapidly as the tunnel is advanced.

Support methods in tunnels

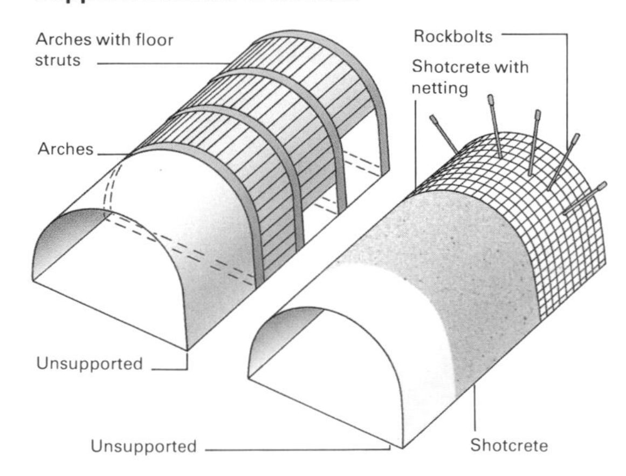

Tunnel construction

Most tunnels require a permanent lining to be installed either for safety or to improve the flow of water. During construction a temporary lining is only installed when essential. In many of the stronger rocks such as granites it is possible to avoid such a lining but in weaker rocks, fractured rocks or softer, incoherent materials a temporary lining is required.

The need for such a lining results from the redistribution of stress around the tunnel as the tunnel face advances. In shallow tunnels, roof falls can take place but as the tunnel is excavated at progressively greater depth, so the pressure increases and movement can develop in the walls and floor. Deep tunnels, driven through highly stressed rocks, can result in rock bursts from the sudden release of rock pressures. In softer rocks, such as clays, the stress associated with the weight

of overlying material can exceed the clay strength, leading to a squeeze of the clay into the tunnel. In all such situations where the tunnel sides are likely to move or collapse before the permanent lining is installed – which may not be for some years – a temporary lining is required.

Temporary linings take several forms dependent on the type of material being excavated, the stress conditions and the probability of water inflows. In the case of clays and sands, pre-cast segmental concrete units are formed into a full circular ring and then forced against the tunnel sides. In rock, the choice of lining is partly dependent on stand-up time. Traditional methods use steel arches at frequent intervals to which lagging can be added. If high stresses develop, then struts can be placed across the floor of the tunnel, converting the lining into an effective ring. Where the rock squeezes into the tunnel like clay, full circular concrete rings can be installed.

An alternative approach, the New Austrian Tunneling Method, uses shotcrete (a mixture of water, cement, sand and fine gravel), sprayed onto the rock face soon after exposure. The cement sets rapidly creating a thin skin of a weak concrete over the tunnel walls and roof. This technique is remarkably effective in preventing internal movement of the rock mass. The reason for this is not clear but probably results from the strength of the thin lining, and its effect as a seal against moisture or the drying influence of air. However, if movement does take place, then additional layers of shotcrete can be added, reinforced by netting if required. Finally rock bolts, steel rods cemented into drilled holes and anchored some depth into the rock, can be used to reinforce the rock close to the tunnel.

The method of driving a tunnel is determined by the strength of the material which is to be excavated and the associated water problems. Hard rock can be excavated by means of drilling-and-blasting, or by use of a tunneling machine. A tunneling machine can be used in massive rock but, ideally, not in the very strongest materials. A full-face machine advances by rotation of a circular steel wheel on which disks, teeth or picks are mounted resulting in advance rates of several meters per day. However, instability in the tunnel can jam the cutting mechanism and prevent free forward movement. The traditional method of rock tunneling involves the drilling of shallow boreholes into the tunnel face followed by the charging of these holes with explosives. The rates of advance of such tunnels is a few meters for each blast. One of the advantages of the drill-and-blast system is that alternative support methods can be installed dependent on the rock conditions encountered. More difficult conditions occur in softer materials such as clays and sands because the tunnel requires to be supported as excavation proceeds. In the case of clays there is the likelihood of deformation of the tunnel walls and face into the tunnel. The method of excavation adopted in softer materials commonly involves a full face tunneling machine which can both cut the face, and provide support as advance proceeds. Sands and silts flow into tunnels like granulated sugar when unsupported, especially when the ground is saturated. Under these circumstances ground improvement methods may be required, such as the injection of chemicals or compressed air into the ground.

The control of water inflow is a major factor in the design and construction of tunnels because most tunnels are driven below the water-table and so drainage of water into the tunnel is inevitable. Although much of the water inflow can be collected and controlled, catastrophic inflow associated with faults or cavernous limestones can result in flooding of tunnels causing danger and delay to construction.

▲ *The construction of high-speed roads for motor vehicles involves the removal of material to form cuttings, and the infilling of hollows to level out the grades, as here on the Medway Motorway in southeastern England.*

Structural foundations

Many of the most obvious engineering structures such as power stations, dams or skyscrapers, transmit loads into the underlying rocks. Some structures are deformable and can be built in a variety of geological situations. For example, embankments can be constructed on very soft materials, the embankment deforming as the underlying material settles. Completion is delayed sufficiently to permit settlement before, for example, a road is built on the top. In areas of mining subsidence the internal structure of low-rise buildings can be arranged so that some movement is permitted as subsidence takes place. Under these circumstances a rigid structure built of brick, or a reinforced concrete frame, could crack, settle and possibly collapse.

In the case of buildings, there are essentially three types of foundation which are available. The simplest type of foundation is a pad or strip footing which can be excavated to relatively shallow depth; the foundations of most two-storey houses are of this type. The basic geological requirement is a foundation material of good quality at shallow depth. The foundations are generally formed in pits or trenches below a broad excavation surface. If the foundation material is deformable, like clay, or there is a risk of localized subsidence, then a raft can be constructed effectively linking the footing-type foundation into a continuous plinth of reinforced concrete.

Where there is deep overburden, piles are used as foundations. A pile consists of a concrete cylinder or steel beam which carries the load from the structure to a satisfactory level at depth. Driven piles are normally hammered into softer materials until a "set" is achieved. The load of the structure is carried by a combination of friction on the sides of the pile and the support at the base. Bored piles are formed by reinforced concrete placed in large diameter boreholes. Most bored piles are end-bearing. Such piles can range in diameter from about 300mm, used for housing developments on shallow very weak overburden, to several meters in multi-storey blocks. In the case of critical structures such as nuclear power stations the foundation is generally over-designed to minimize settlement.

◄ *The foundations of Castle Peak power station in Hong Kong are strip footings, sunk into trenches excavated in granite. The foundations of the jetty in the background are formed by piles driven through soft marine muds in the top of the weathered granite.*

► *Different types of foundations are used depending on the different types of bedrock. Where the bedrock is near the surface the footings can be constructed directly on it. Where the overburden of unconsolidated material is thick, piles must be driven down to meet the bedrock. Where the overburden is very thick the building must be built on a raft to take the weight.*

Surface constructions and materials

A wide range of civil engineering activities involve the surface redistribution of hard rock from quarries, softer materials from pits or different types of waste. Such materials can be used in the manufacture of concrete, the construction of embankments for dams and roads, and in the stabilization of slopes. Production and handling of materials, therefore, form an important part of the construction industry. In road construction the most favorable route is initially selected by the calculation of the balance of excavation from cuttings and placement in embankments. Such a route will identify special problems, such as a potential bridge crossing, landslide area or tunnel location through a mountain ridge. The material excavated from cuttings will usually be re-used in embankments. Soft materials, such as wet sands, soft clays and organic deposits are unsuitable for the foundation of road embankments and are disposed of in spoil tips. Once the embankment has been formed the road is constructed from a thick layer of broken rock, followed by progressively thinner layers of finer aggregate and finally the running surface of concrete or asphalt.

Most engineering projects require sources of high quality rock derived from quarries, or gravels and sands extracted from pits for the production of concrete. In addition, other sources of material are required, including sands for filters, coarse aggregates for drainage layers, rockfill for the shoulders of dams, impermeable material for dam cores and large rock blocks for protection of slopes against wave attack in reservoirs, or in the sea. Many civil engineering projects require new sources of construction material to be identified and developed quickly; such sources may have a maximum life of two or more years. In developed areas, however, sources of material may be already available from existing working quarries. Concrete aggregates require stable, strong rocks which will not react with the cement. Granites, massive limestones and sandstones, and gneisses are generally suitable. However, rocks or alluvial gravels which contain fine-grained silica, micaceous minerals and clay, and some volcanic rocks, are liable to alkali reactions leading to deterioration of the concrete.

Foundation types

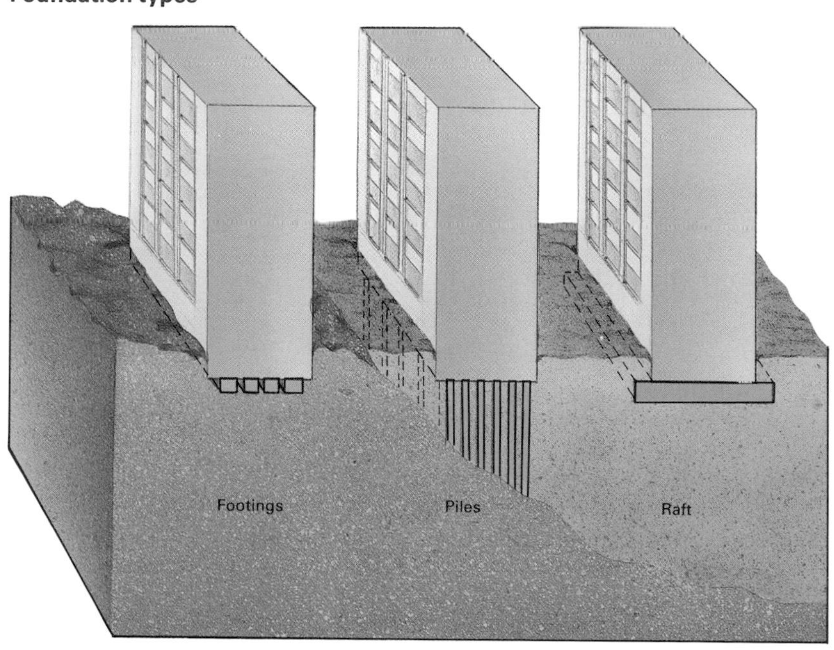

Disposal of toxic waste

▲ *Toxic waste must be contained in an area where it will not contaminate the ground, the water or the air round about. The most favorable site is in an impermeable layer like clay.*

Waste disposal

The disposal of waste materials, including household waste, toxic chemicals or radioactive chemicals, requires the provision of an engineered facility which will minimize or prevent pollution. Shallow disposal in existing or abandoned excavations is currently used for the disposal of large volume wastes. The major risk in such circumstances is associated with the development of pollution of groundwater flowing through the waste. Ideally such disposal is by containment in a clay foundation through which the groundwater moves slowly, or in geological situations where the water table is below the base of the waste and cannot reach it.

In the case of toxic or radioactive waste absolute containment is required for minimum periods of time, often measurable in terms of thousands of years, in order to ensure that the waste does not enter the biological cycle and affect living things. Such disposal requires engineered facilities in conditions where groundwater movement is extremely slow and there is a minimal risk of geological disturbance, such as faulting and earthquakes.

The cheapest commercial source of electricity is from hydro-electric schemes, made possible by dams

Dams and their failure

*Problems can arise where there are unstable slopes on the reservoir flanks. On 3 October 1963 part of mount Toc, on the southern side of the Vaiont reservoir, slid into the reservoir pushing a wave about 200m high across the 267m arch dam. The dam withstood this sudden water load, but major devastation occurred in the Longarone valley downstream, where five villages were swept away and 2,600 people perished (**▸ page 222**).*

*In the case of larger reservoirs, or those in seismically active areas, there is a risk that earthquake activity will be induced or enhanced. The cause of such induced seismicity is not fully understood but it is believed to be triggered by changes in water pressure at depth, associated with reservoir filling (**▸ pages 245-8**).*

Some of the most critical geological conditions associated with foundations occur during dam construction. A failure can result in the uncontrolled release of water, potentially leading to loss of life and devastation of wide areas. Hence dams are regarded as major targets during times of war. The Malpasset dam, which failed on 2 December 1959, resulted in a major re-evaluation of the interaction between concrete dams and their foundations. Malpasset dam was a 67m arch dam built on micaceous schists in southern France. The dam failure resulted in the collapse of the left wing of the arch, a wedge of rock below the dam foundation level being removed. Part of the dam was left in the valley floor, and on the right bank. Observations on the parts of the dam that remained after the failure indicated that a downstream displacement had taken place, the greatest movement being on the left bank. Subsequent investigation revealed that the rock mass on the left bank was more deformable than that on the right and in fact was about one thirtieth the deformability of concrete. It is possible that this situation contributed to unequal movement in the arch as the water load developed during a rising reservoir level.

▲ Dams, such as the Tarbela Dam in Pakistan, have their own engineering problems. The bedrock has to cope with both the weight of the structure and the pressure of the water.

▼ The Malpasset Dam failed when the pressure of water on the dam scooped out a wedge of rock downstream. The wedge was bounded by a fault and a joint.

The Malpasset Dam

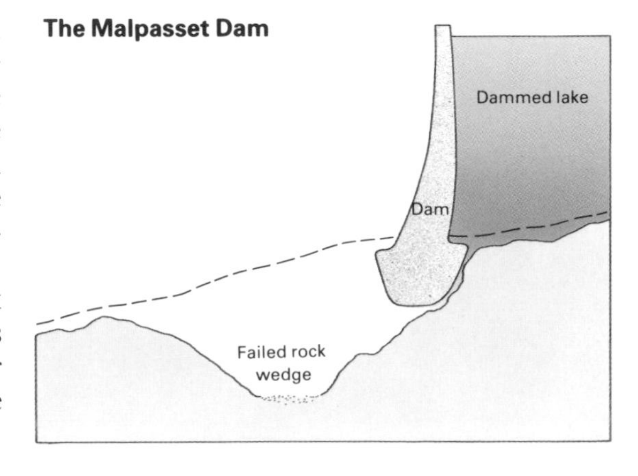

Dams and dam construction

The choice of dam to be constructed at a particular site is determined by the foundation conditions, and also the availability of construction materials.

When the foundation is of adequate quality concrete dams can be constructed. These dams require that settlement is kept to a minimum. The most critical type of concrete dam is the arch dam in which the water load in the reservoir is transmitted through the thin concrete arch into the rock abutments at the valley sides, as well as into the foundation. The advantage of such a dam arises from economy in the use of concrete which can be an expensive factor in construction in remote sites. Embankment dams can be built on deep, deformable overburden, on weak rock foundations or in sites where suitable materials are available and concrete costs would be high.

Inevitably, the load of any dam, of whatever type, is significant and can influence a large volume of the underlying foundation. This requires careful monitoring by instrumentation during and after construction ensuring that the design criteria are not exceeded and the scheme operates safely.

▲ ► *The Latiyan Dam is founded on a complex and potentially weak bedrock of sandstone, shale, quartzite and limestone. It is also cut by a fault. The base of the dam slopes upstream to give a more even transmission of stress.*

▼ *The Monteynard arch dam in France is built in a narrow limestone gorge which provides strong side abutments for the arch.*

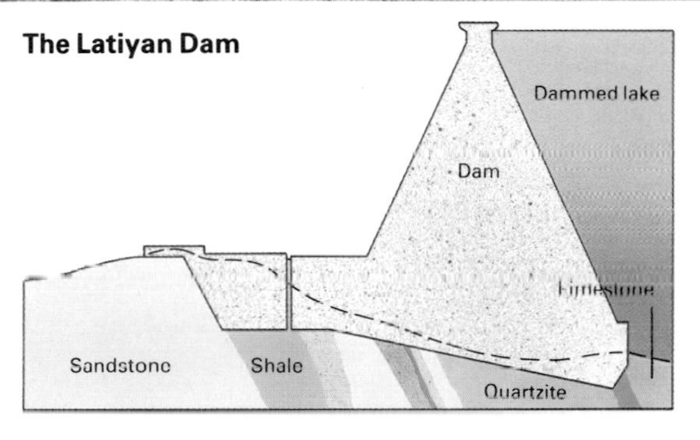

Siting and constructing a dam

The construction of dams results in the impounding of water in reservoirs which needs to be retained by both the dam and the natural conditions in the reservoir basin. The watertightness of reservoirs is an important factor which requires detailed geological study particularly where deep buried channels, limestones and similar permeable rocks, or abandoned mine workings are present. The potential loss of water at the dam site, or in the reservoir basin, can be prevented or controlled by the use of grouting in which a cement-water mixture is injected into the rock or other materials through which leakage is anticipated.

The Latiyan dam is a 112m buttress structure constructed on Devonian quartzitic sandstones and shales and located in northern Iran. The selection of the site of the dam was controlled by the shape of the valley (which made it economical) and the widespread presence of landslipped terrain on the valley sides. On the left bank of the site the buttress dam is keyed into the hill side within a deep excavation in folded sandstone. The right bank is founded on a steep hillside upstream.

In general circumstances this geological configuration is good because the strata are at angles to the loading imposed from the dam into the foundations. However, much of the downstream area of the foundation is composed of deformable shales which were too weak to carry the imposed stresses from the dam. This geological situation has been overcome by constructing a thrust block on the downstream side of the dam crossing the bands of shales. A dam is one of the most dangerous things that can be constructed on the wrong bedrock.

See also
Faulting and Folding 153-6
Landscape Features 181-96
Resources from Rocks 197-212
Changing the Landscape 213-28
Predicting Natural Disasters 241-4
Manmade Earthquakes 245-8

▲ Any attempt to cut into a natural slope, or to modify it in any way, results in an imbalance of the gravity stresses involved. When this is not taken into account the result may be a landslip in which the natural topography tries to reassert itself. This cutting was excavated in metamorphic rocks for a road in the United Arab Emirates. A major rock-slide developed which blocked the road. The cutting has now been stabilized by flattening the sideslopes so that they do not form such a precarious angle.

Slope stability

Most engineering projects require the evaluation of slope stability in connection with the engineering of new, excavated slopes, or construction on landslipped hillsides. The mechanism of failure of slopes is closely associated with the nature of the material which forms the slope. The development of instability results from the removal of land from the toe of the slope by undercutting, the addition of load at the top of the slope or a rise in water pressure within the slope. The stabilization of slopes relies upon the artificial countering of such conditions by engineering methods.

Slopes in clays are notoriously associated with landsliding. The typical landslide is associated with a rotational movement of the slipped block, so that the existing ground surface at the top of the slope subsides forming a back-scar while the toe extends forward (◆ page 185). The plane of failure is formed by a curved slip surface along which the clay has become polished and grooved by the movement. Such rotational slides may be composed of a series of slipped blocks forming a multiple slide complex. If the resulting slope becomes degraded by weathering and further slope movements, the hillside tends to have a characteristic undulating form. Excavation at the toe of a single or multiple slide or the construction of an embankment on top can lead to collapse.

During periods of heavy rainfall the water table will rise and this will lead to increased water pressures in the slope and consequential movement. Many such slopes are in a state of quasi-equilibrium moving continuously but irregularly. If more water is available to the sliding material either from precipitation or from springs discharging onto the slope, a softened mud particularly in sandy or silty materials can be formed which will move down the slope in a thin sheet as mudflow. Heavy rainstorms can lead to catastrophic landslides, such as the tip slide in colliery waste at Aberfan on 2 October 1966 (◆ page 222), and such events are regular occurrences in the tropics during monsoon periods.

The stabilization of such landslides can be carried out by means of slope flattening, thereby removing load from the top of the slope, the addition of a supporting bank at the toe of the slide, and the removal of water by drainage through boreholes and gravel-filled trenches.

Rock slopes fail by a combination of gravitationally-induced falls arising from undercutting, toppling of slabs and columns, translational failure on planes of weakness such as weak bedding planes or large-scale creep. The difference between such movements and those in weaker rock materials is that the pre-existing defects, such as joints, faults and bedding planes determine the shape of the failed block, and have a significant controlling influence on the mechanism of failure. Many failures in high rock slopes develop a considerable acceleration during descent which, if the foot of the slope is on level ground, results in the formation of a high velocity rock flow extending for a considerable distance from the toe of the slope. The very physical size of many rock slopes makes their stabilization a difficult venture.

Predicting Natural Disasters

22

Earthquakes and their human and social impact – the death that can strike without warning...The benefits of prediction – evacuation before the event... Levels of prediction – where earthquakes can strike, when they can strike, what phenomena can be observed that may anticipate them?...PERSPECTIVE... Where prediction has been successful...The prediction of volcanic eruptions

In 1906 one of the most famous (or notorious) earthquakes in history took place in San Francisco, killing about 700 people and causing about 400 million dollars worth of damage. Horrendous though this disaster was, the effects of the shock were much less than they would be if an earthquake of the same size were to occur in the same city today. It is impossible to say how many people would die in a modern earthquake-ravaged San Francisco; but the number would almost certainly be in the order of 200,000 and the property damage could exceed $10,000 million.

In South America, the Mediterranean area, the Soviet Union, Japan, China and numerous other places, many millions of people are seriously threatened by earthquakes and more than 10,000 of them will actually die in seismic disasters in an average year. To be able to predict exactly when and where an earthquake is to take place would therefore save many lives. Unfortunately, despite considerable research over the past few decades, there is still no generally applicable method of earthquake prediction and none is remotely in sight.

Of course, even if accurate predictions were possible, it would not prevent property damage. In the absence of the perfect earthquake-proof building, the only way of doing that would be to transfer thousands of cities, towns and villages away from the seismic zones. Even of that were practicable, however, it is unlikely to happen. The chief benefit of having a viable method of earthquake prediction would be that people could save themselves if they knew that a large earthquake would definitely strike in a specified place at a specified time.

◄ **The Tangshan earthquake of 27 July 1976, that destroyed the city and killed people in Peking 150 kilometers away, highlights the uncertain state of modern earthquake prediction. It struck without warning at a time when the Chinese were having some success at predicting earthquakes elsewhere in their country**

Successful prediction

Although no universally applicable method of earthquake prediction has yet been discovered, a few individual earthquakes have been successfully forecast, partly by good luck, partly by hard work and partly because the earthquakes concerned gave better-than-usual premonitory signals. The best known example is the magnitude 7·3 shock that took place in Hiacheng, China, in 1975. Only 1,328 people died in this disaster even though more than 8 million people were living in the soon-to-be-stricken area. The Chinese had managed to predict the earthquake from a wide variety of scientific measurements (such as ground level, ground tilts, water levels in wells, magnetic field and gravity) as well as from what scientists in the west then regarded as "non-scientific" information (for example, the behavior of animals). There was therefore time to evacuate all the buildings in the immediate danger area. As 90 percent of the buildings in Haicheng were destroyed, more than a million people may otherwise have been killed. The Haicheng success occurred just as China was re-opening its doors to the west, and the immediate reaction of western seismologists was that the Chinese must have solved the prediction problem. This belief was strengthened when in 1976 several earthquakes with magnitudes up to 7·2 were also successfully predicted in the Sungpan-Pingwu region. Unfortunately the Chinese failed to anticipate the even larger (magnitude 7·8) earthquake in Tangshan where more than 280,000 people died later in 1976. There have also been numerous failures since. The Chinese have evidently not yet discovered a general method with a high success rate.

Another partial success story comes from Japan where, during the long swarm of earthquakes that struck the Matsushiro region during 1965-6, seismologists were able to combine data from a number of sources to predict successfully the timing of some of the peaks of activity later in the swarm. Unfortunately, the tourist business complained bitterly that the published predictions were ruining its trade.

The problem of issuing warnings to the public is an important one. Even if a general earthquake prediction technique were to be found, it is unlikely that it would be perfect. It is inevitable, therefore, that there would be false alarms. This actually happened in China in 1974, and several people were killed during the emergency activities, leading to complaints to the authorities. In the United States, however, where resort to legal action is common, a scientist who issued a false warning could, however good his intentions, soon find himself in deep and expensive legal waters. Some Californian seismologists have already had a taste of this and, as a result, are being much less forthcoming about their prediction data even when their expertise might be beneficial to the public.

No one will now guarantee that a complete solution to earthquake prediction in even possible

Levels of prediction

It is possible to envision earthquake prediction at a series of levels of ever higher sophistication. The first and most fundamental level is that of simply being able to specify in very general terms *where* an earthquake is most likely to occur. Most of the world's seismic events are known to take place along comparatively narrow bands coincident with lithospheric plate boundaries (♦ page 25); and so it follows that prediction at this most basic level is possible, if not very useful.

The next level up is that of being able to determine where, *within* the Earth's seismic bands, the chances of a large earthquake occurring soon are greatest. Earthquakes are the result of sudden releases of strain that has built up over many years. Once the strain has been released in a particular locality, it takes time to accumulate again; and so it is unlikely that a large earthquake will recur there soon after. Conversely, where a large earthquake has not taken place recently, it may very often be presumed that the strain is still building up to a seismic release. Zones with long histories of major seismic activity but within which there have been no major seismic events within the past 30 years are known as "seismic gaps". They represent areas where large earthquakes in the near future are most probable. This does not mean that large earthquakes *will* occur there, but only that they are statistically most likely to do so. Nevertheless, seismic gaps are considered to be particularly dangerous areas in which to live and work. The major seismic gaps have already been identified; and to that extent it can be said that prediction at the second level is now possible.

The third and most sophisticated level of prediction is that of being able to specify the precise time of an impending earthquake in any given area or even within a single well-studied area. This is the capability that has proved elusive, although a few earthquakes have in fact been successfully anticipated.

There are many phenomena that might be used in principle to detect an impending earthquake, and some have been made to work in practice. For example, it is possible that pre-earthquake effects in the upper crust could produce changes in ground level that could be detected by repeated leveling surveys. Likewise, there might be premonitory tilting of the ground that could be detected using tiltmeters. Deformation of either kind might also lead to changes in tide or sea levels that could be measured by tide gauges or similar instruments. Stresses set up in the crust prior to an earthquake might produce changes in the velocities of seismic waves in the local rocks; and these changes could perhaps be detected by measuring the velocities of the waves from explosions generated specially for the purpose or from small earthquakes in the region. Stressed rocks are also known sometimes to release anomalous quantities of radon gas, generate electric currents and undergo magnetic changes, all of which might be detected by sensitive enough instruments.

In fact, these phenomena and others have all been observed (but not all at once) before some earthquakes. There are two problems, however. One is that there are many earthquakes for which premonitory phenomena have not been observed. Second, when precursory effects are observed, they frequently fail to fall into any well defined pattern that can be used to specify the time of an impending event.

The result is that, although research on prediction has been taking place in Japan since early this century, and in the USA, the USSR and China since at least the early 1960s, scientists appear to be no nearer discovering a universally applicable method.

▲ *When Mexico City was devastated by an earthquake on 19 September 1985, the epicenter was 400km away. The earthquake itself took place on a destructive plate margin, in an area due to have such an earthquake. Mexico City was so badly hit because it was built on unconsolidated lake-bed sediments that amplified the shock waves. Several thousand died, but many more survived buried in wreckage for days.*

▲ *Relatively few (64) died in the San Fernando Valley earthquake – it happened early in the morning with few people about. Prior evacuation would limit earthquake casualties.*

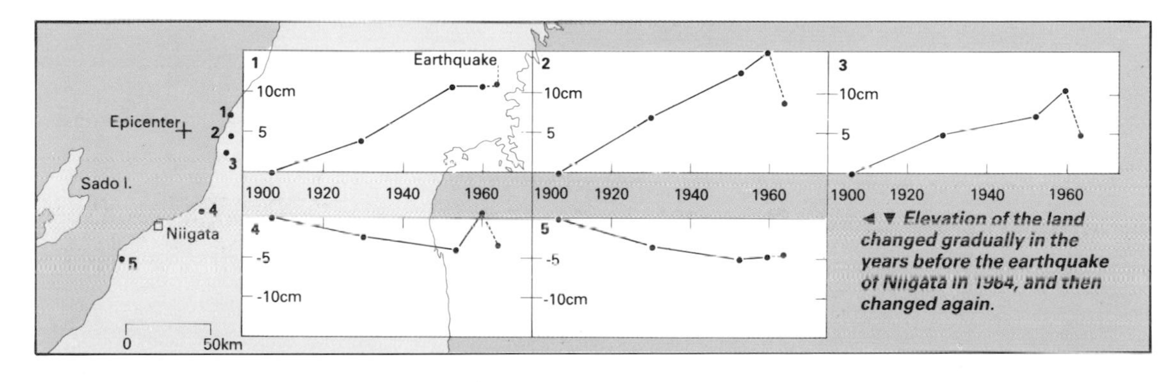

◄ ▼ *Elevation of the land changed gradually in the years before the earthquake of Niigata in 1964, and then changed again.*

▼ *An orange glow in the sky sometimes appears at the time of earthquakes, as here in Matsushiro in 1966. If ground electrical levels are responsible they may be used in prediction.*

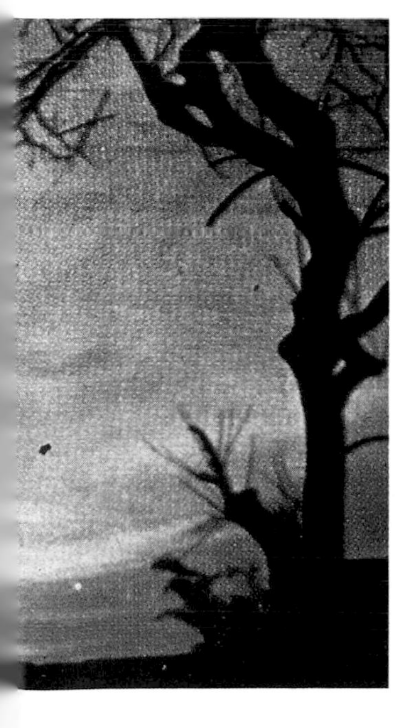

See also
The Dynamic Earth 9-24
Earthquakes and Volcanoes 25-32
Engineering Geology 229-40

▲ In 1976 La Soufrière volcano on Guadeloupe island showed all the signs of impending eruption – 75,000 inhabitants were evacuated but the threatened eruption did not occur. Although eruptions are easier to predict than earthquakes, predictions are not always accurate.

Predicting volcanic eruptions

Seismic and volcanic activity differ in obvious ways. An earthquake is a sudden release of invisible energy, although the effects are often all too visible. A volcanic eruption, by contrast, involves the emission of something tangible, whether it be flowing lava, gases or rock fragments. Despite these differences, trying to predict a volcanic eruption is in many ways similar to trying to predict a seismic shock. It is again necessary to look for premonitory changes in ground level, ground tilt, gravity, magnetism and seismicity. The chief difference is that, in the volcanic case, historical data play a more important role. Each volcano is unique; and when it comes to assessing its future behavior, there is no substitute for experience of the volcano's idiosyncrasies, gleaned from many years of concentrated observation.

Some volcanologists believe that most volcanic eruptions could be predicted if only adequate money were available to provide monitoring equipment for all active and dormant volcanoes and to enable scientists to build up the necessary historical knowledge. They may or may not be right; but as Haroun Tazieff, the famous French volcanologist, has pointed out, predicting the outbreak of volcanic activity is seldom useful. For one thing, it is not usual for large numbers of people to live in close proximity to a volcano; the existence of the volcano is often sufficient warning of the danger in itself. Secondly, the onset of volcanic activity is usually only dangerous when it takes the form of a huge, unexpected explosion. This, too, is rare; most explosions occur only after a first phase of activity that is comparatively mild.

The danger from volcanoes therefore lies less in the outbreak of activity than in the subsequent course of events. What people in volcanic areas really need to know is whether, once it has started, the activity will proceed in such a way as to present danger and hence the need to evacuate. Predicting the development of activity is often more difficult than predicting its onset, and it is here that detailed knowledge of a volcano's earlier behavior can be crucial. Indeed, so difficult is the task that, as with earthquakes, expensive false alarms are only to be expected. The classic case is that of the volcano on Guadeloupe. In 1976 what turned out to be a false prediction led to the evacuation of 75,000 people for 15 weeks at a cost of more than $500 million.

Manmade Earthquakes

Lake Mead earthquakes – the first observation of induced seismicity...Reservoirs around the world – sites of artificially triggered earthquakes...Earthquake control – the feasibility and practicality of reducing the stresses in earthquake regions...PERSPECTIVE... Earthquakes induced by naturally by rivers...The Denver earthquakes...The Rangely experiment in earthquake induction

Water engineers in the United States began to fill Lake Mead, the reservoir behind the newly-constructed Hoover Dam (then called the Boulder Dam), in 1935. The job took six years. In the meantime, seismometers had been installed near the dam in 1938 and had immediately begun to register hundreds of small earthquakes a year, mostly with magnitudes lower than 3·0. Something strange seemed to be happening. During the 15 years prior to the construction of the dam, the few inhabitants had felt no earthquakes at all in what had long been considered an aseismic region. Yet in 1936 they felt 21 and in 1937 there were 116. In 1939, for example, there were 128 felt earthquakes and 663 more that were detectable by instruments.

The seismologist D.S. Carder suggested a few years later that the earthquakes were being induced by the reservoir, although he could offer no conclusive proof. The seismometers had not been in place before the impounding of the water had begun, and so there was no truly scientific "before and after" comparison. Moreover, the world was at war and people had their minds on more important things. The case of Lake Mead was largely forgotten and was seen, by the few who did remember it, as a mere academic curiosity.

Earthquakes induced by rivers

In 1963 the American geologist L.D. McGinnis discovered that the frequency of small earthquakes in the Mississippi Valley between Cairo, Illinois and Memphis, Tennessee corresponded closely with the rate of change in the level of the water in the Mississippi River itself. There are significantly more earthquakes when the water is rising or falling, such as during times of flood, than when the level is nearly stationary.

In 1981 the American geologist I.G. Wong also reported such a correlation involving microearthquakes (earthquakes in the magnitude range 1·0-3·0) in the vicinity of the Colorado River in southeast Utah.

Some earthquakes must therefore be induced by water-level changes. They are not strictly manmade, for the rivers would be rising and falling in Man's absence due to seasonal rainfall; but they do emphasize the role of water loads in inducing seismic activity.

Where water-level changes caused by irrigation are concerned, however, Man is more clearly involved. In 1975 the American geologist D.L. Turcotte noticed that in the early 1950s there had been an increase in the frequency of small earthquakes in the Columbia Plateau region of Washington. This coincides with extensive irrigation that was introduced there during the early 1950s and which since then has raised the water table by up to 200m in places.

▼ Earthquakes induced by the build-up of water pressure in reservoirs are quite frequent events. They do not necessarily occur along the zones of seismicity in the Earth's crust.

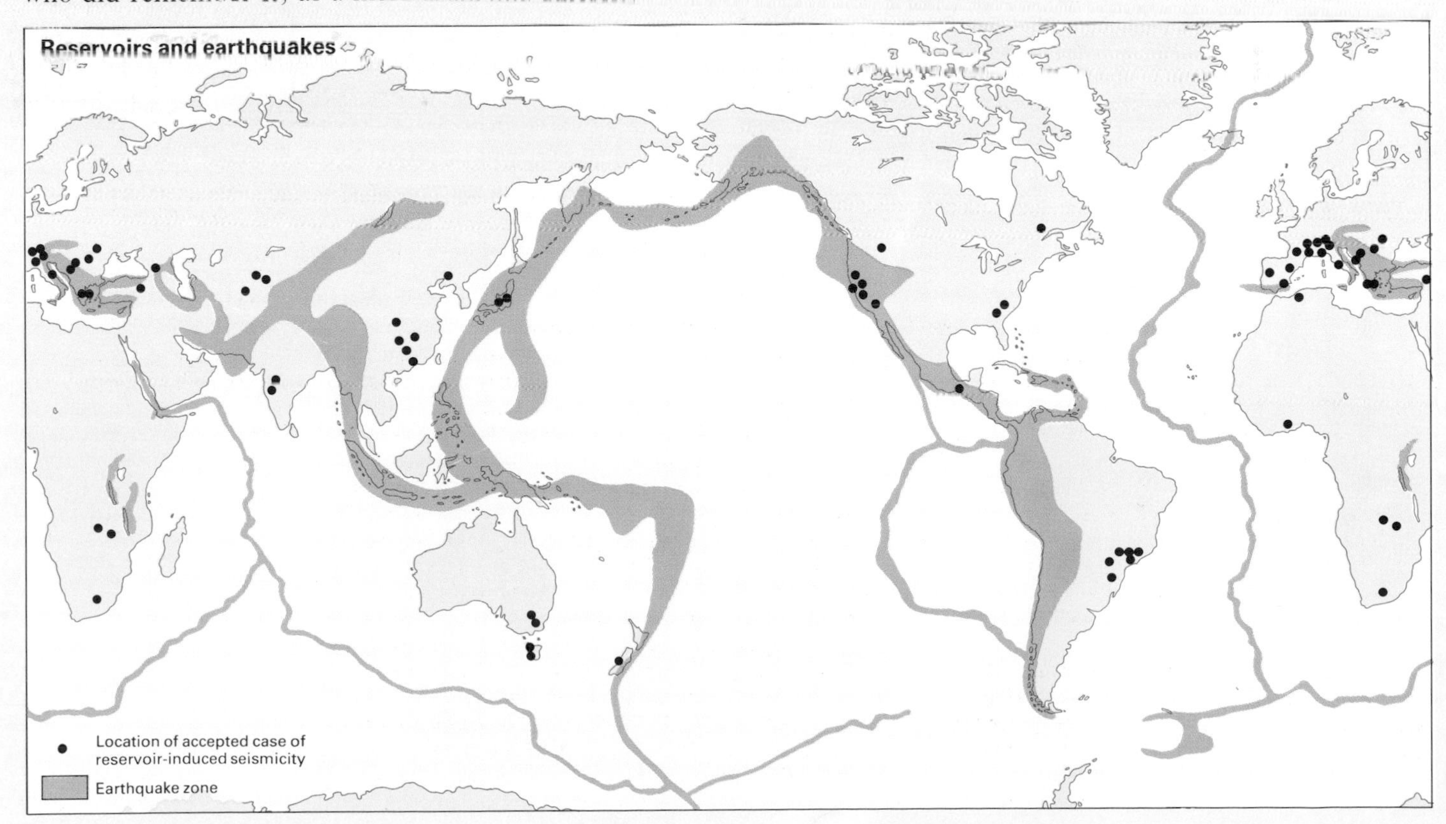

The Denver earthquakes

*Up to 1962 the Denver area of Colorado in the United States was practically earthquake free. The only shock that had even been recorded there was one of intensity VII on the Modified Mercalli Scale (**♦ page 26**) on 7 November 1882. Yet between April 1962 and September 1965 the Denver region was struck by no fewer than 710 small earthquakes in the magnitude range 0·7-4·3. The mystery was solved in November 1965 by a local consulting geologist, David Evans, who pointed the finger at the US Army's Rocky Mountain Arsenal about 19km to the northeast of the city.*

Since 1942 the Army had been experimenting at the Arsenal with chemical warfare materials. When their waste was found to be contaminating the local groundwater supply and endangering crops, the Army decided to get rid of it by injecting it under pressure into a 2,000m deep well drilled specially for the purpose. Injection began on March 1962 and continued to 30 September 1963 at an average rate of 5·5 million US gallons a month. On 17 September 1964 injection was restarted under gravity (with no pressure) at an average rate of 2·0 million US gallons a month. Pressure was again applied from 6 April 1965 onwards, when the rate of disposal was increased to 4·0 million US gallons a month; and in June 1965 the pressure and rate were increased further.

The earthquakes began about a month after the injection was first started. During the first period to September 1963, earthquakes averaged almost 20 a month with one month having more than 40. When no fluid was being injected the shocks dropped to only a few a month. When injection was resumed and the pressure increased the frequency of earthquakes rose to more than 80 a month.

The Army disclaimed responsibility at first, but when the evidence became overwhelming they ceased the injection early in 1966. Seismic activity tailed off until by the end of the 1960s it was negligible. But there was an unexpected burst of activity in 1967 during which three earthquakes of magnitude greater than 5 took place – bigger than any that had occurred while fluid injection was in progress. This remains unexplained, although it shows that the relationship between the time of injection and the time of earthquake activity is a complex one.

By the end of the 1960s geologists were regarding Lake Mead as one of the earliest known examples of a fairly common phenomenon, namely, the generation of earthquakes by large reservoirs. By the mid-1960s induced seismicity was suspected at lakes in Greece, in France, and at a number of other sites, but in none of these cases were seismometers installed before the start of impounding the water. The evidence for induced earthquakes was strong, but circumstantial.

What finally convinced most skeptics was the tragedy at the Koyna Reservoir in India. This new, very large reservoir was filled between 1962 and 1965. The first earthquakes close by were felt in 1963 and thereafter the shocks became bigger and more frequent until on 10 December 1967 an earthquake of magnitude 6·4 killed 177 people, injured about 2,000 others and caused extensive damage. There could be little doubt that the shocks were reservoir-induced in this case, for Koyna lies on a Precambrian shield, one of the most stable and least seismic types of terrain in the world.

By the early 1980s about a hundred reservoirs were thought to have induced earthquakes, although the cause of the phenomenon is still by no means fully understood. The obvious possibility is that the huge mass of water in a large reservoir puts stress on the rocks beneath and thereby causes pre-existing faults to slip, but calculations suggest that reservoir-induced stresses are actually very small compared to the natural stresses already present. A more likely hypothesis is that as water seeps into the rocks surrounding a reservoir it increases the pressure in the pores and hence triggers the release of strain already present. Manmade earthquakes can also be induced by the injection of fluid into the ground.

Large underground nuclear explosions are usually followed for days or even weeks by a series of earthquakes with magnitudes smaller than that of the original explosion. A nuclear explosion is not itself a manmade earthquake, since it is not related to any natural process. The aftershocks are true earthquakes, however, because they are due to local releases of natural strain. Moreover, these adjustments probably take place in much the same way as the subsurface adjustments that occur beneath filling reservoirs, by a change in the distribution of fluid and local increase in pore pressures.

All nuclear explosions with an equivalent magnitude of at least 5 produce aftershocks for at least a day. By contrast some large reservoirs induce no earthquakes. The geological state of the rocks in the vicinity is crucial, and a large reservoir will only lead to earthquakes if the local rocks already contain faults.

◄ ► Denver, Colorado (left) is as unlikely a place as any to experience earthquakes, yet between 1962 and 1965 a number of tremors shook the area (right above). It was eventually discovered that the frequency of earthquakes was directly related to the volume of fluid disposed of in the nearby Rocky Mountain Arsenal well (right below).

▲ ► *Lake Mead, behind the Hoover Dam, is one of the earliest examples of an area in which earthquakes have been induced by the filling of a reservoir with water. Earthquakes were recorded there as soon as the lake began to fill in 1935, and continued until the end of the 1940s. The whole area in the vicinity of the lake settled, in places by as much as 12cm.*

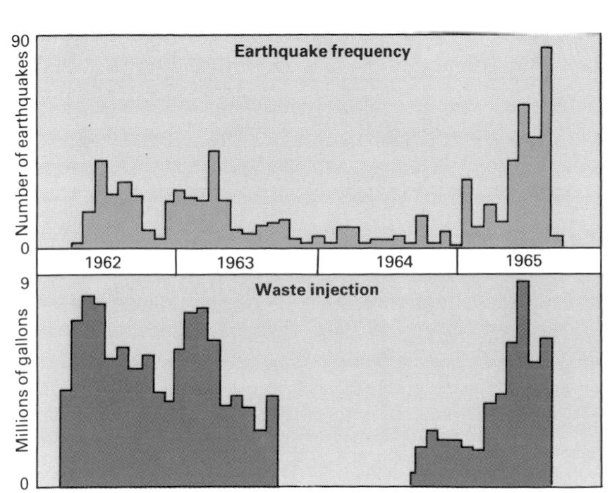

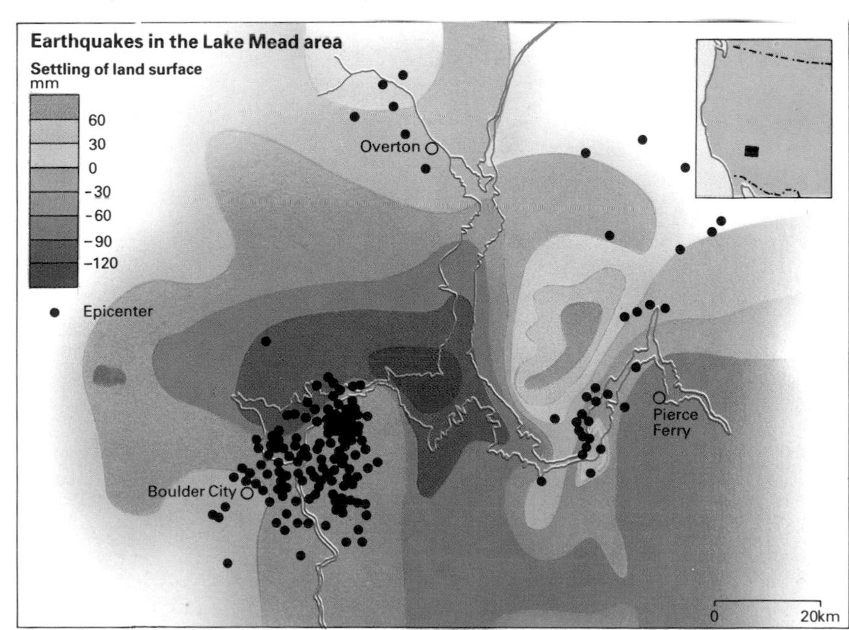

See also
Earthquakes and Volcanoes 25-32
Faulting and Folding 153-6
Changing the Landscape 213-28
Engineering Geology 229-40
Predicting Natural Disasters 241-4

The Rangely experiment

Once geologists had become convinced that the earthquakes generated in the vicinity of reservoirs, fluid injection wells and nuclear explosions were manmade, they decided to carry out a controlled experiment at the Rangely oilfield in Colorado. The oil company concerned had already been injecting water into the field since 1958 to increase underground pressure and enhance the recovery of oil. Moreover, seismometers had been installed in the area in 1962 and had been registering small earthquakes, although it was not clear whether they were due to water injection or whether the region was slightly seismic.

In October 1969, geophysicists from the US Geological Survey installed a new dense array of seismometers nearby. Then in November 1970 the normal injection of water by the oil company was stopped and pumps reduced the pressure. By April 1971 the frequency of earthquakes had decreased by a factor of about 20 to fewer than 10 shocks a month. Water injection was restarted in June 1971 and the frequency of earthquakes rose again significantly, but not until late 1972.

The experiment clearly demonstrated that earthquakes could be started and stopped simply by injecting water. Unfortunately, it also showed that the timing of the increase and decrease in the number of earthquakes could not be controlled very accurately – discouraging ideas of earthquake control by fluid injection.

The discovery of manmade earthquakes led to the exciting suggestion in the early 1970s that they might be used to mitigate the effects of big shallow earthquakes and perhaps even abolish them altogether. The idea behind such earthquake modification or earthquake control, as it came to be called, was that by injecting fluids into a highly-stressed fault zone it should be possible to induce many small man-made earthquakes and thus release in short harmless bursts the strain that would otherwise build up and be released in one big earthquake.

Unfortunately, the idea was hopelessly optimistic. For one thing, many small manmade earthquakes would be needed. Because the earthquake magnitude scale is logarithmic with an energy factor of about 30 between each division, it would require about $30\times30\times30\times30 = 810{,}000$ harmless shocks of magnitude 4·0 to disperse the energy of a potentially great earthquake of magnitude 8·0.

A much more serious problem, however, is that it has proved impossible so far to exert any fine control over the frequency and timing of manmade earthquakes. Such events can be "switched on" and "switched off", but it is not yet possible to determine precisely when the switching will occur. To attempt earthquake modification in a highly-populated seismic zone – which is where it is most needed – would therefore present an unwarranted risk with severe social, economic and legal implications. Moreover, even if a perfect modification system could be devised for one region, it would not necessarily work in another where the geology is quite different.

Earthquake modification may ultimately prove possible, but not without many more years, and perhaps even decades, of research.

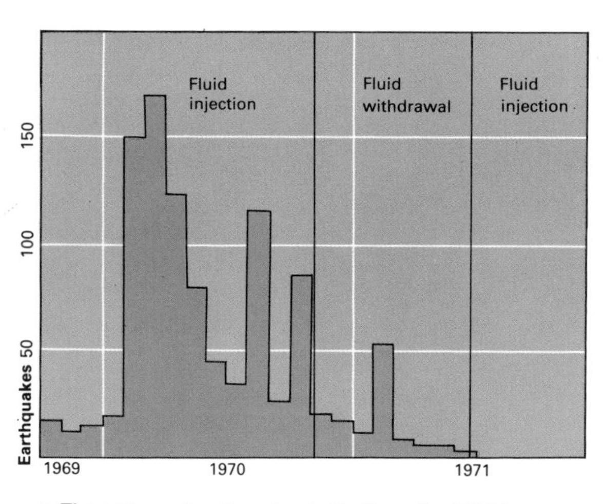

▲ ***The pattern of earthquakes in the Rangely oil field.***

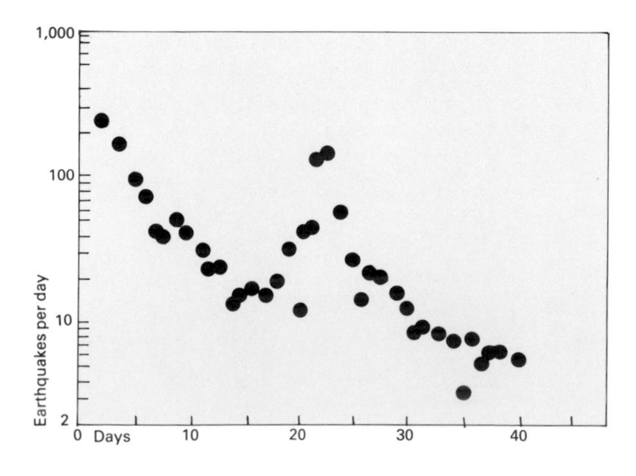

◄ ***Following the nuclear explosion BENHAM there were a number of earthquakes in the area.***

▲ ***An underground nuclear explosion is not, in itself, an earthquake, yet it may induce them.***

Credits

Key to abbreviations: GSF Geoscience Features Ltd. Wye; OSF Oxford Scientific Films, Long Hanborough; SPL Science Photo Library, London. b bottom; bl bottom left; br bottom right; c center; cl center left; cr center right; t top; tl top left; tr top right.

1 Vautier-de Nanxe, Paris **2-3** P.W. Bading/Zefa **4-5** Erwin Christian/Zefa 897 David Gallent **8** Vautier-de Nanxe **9** Bodleian Library, Oxford **10tr** Bildarchiv Preussischer Kulturbesitz **10** Francis Danby: The Flood, Tate Gallery, London **11** British Museum of Natural History **12tr** Black Hills Institute of Geological Research **14b** M.S. Price/Natural Science Photos **16** GSF **20** R.D. Ballard/Woods Hole, Mass. **25** NOAA **26tr** Picturepoint **26br** Dr. John S. Shelton, California **27** National Air Survey Center **29** BBC Hulton Picture Library/Bettman Archive **30bl**, br GSF **31** Photri **32tr** Mansell Collection, London **32l** Ann Ronan Picture Library **39** NASA/Photri/Zefa **40t** Joseph F. Viesti/Susan Griggs Agency **41c** Larry D. Brown/Cornell University **41b** Consortium for Continental Reflection Profiling: L. Brown, L. Serpa, T. Setzer/J. Oliver, S. Kaufman, R. Lillie, D. Steiner, Geology, II, 1983 **42bl** Philip Daly, Bath **42t** Geodaetisk Institut **45tr** Wellcome Institute for the History of Medicine **49t** Robert Harding Picture Library **51** Timothy O'Keefe/Bruce Coleman Ltd. **55t** Vautier-de Nanxe, Paris **59** Kitt Peak National Observatory/Cerro Tololo Inter American Observatory **60-61t** British Museum of Natural History **60b** Paul Brierly **62bl**, br A. C. Waltham **63** © De Beers Consolidated Mines Ltd. **64tr** Paul Brierley bl, tl, GSF c, Imitor **65** GSF **66tl**, tr, c, Imitor b, GSF **67** GSF **68** Philip Daly, Bath **69** Planetarium Armagh, © AAT **71** NASA **73** Scala, Florence **74tr** GSF, c, G. R. Roberts, New Zealand **75** GSF **76tl** Sinclair Stammers/SPL br, Geological Survey, London **77** Geological Museum/BMNH **82** A.C. Waltham **83** Canada Center for Remote Sensing **84tr** Dr. R. Goldring, University of Reading **85t** Ronald Templeton/OSF c, R.C.L. Wilson/RIDA 86t Prof. W.S.F. Kidd **87tl** GSF tr Institute of Geological Science, London, c, Ken Lucas/Seaphot **88tl** GSF b, Biofotos/Heather Angel **89bl** Institute of Geological Science, tr, Geological Society **93** GSF **96tl** GSF, bl, RIDA **97br** Prof. N. Alvarez/SPL **100t** Michael Holford, Loughton, br, Novosti Press Agency **102-3** National Air Survey Corp. **107** Images Colour Library, Leeds **108** Dr. J. Decker **109** Prof. S. Warren Carey, Tasmania **110** Geoslides, London **112** Dr.J.E. Guest/Woodmansterne **113** OSF 118-119 GSF **120** Aerofilms **122** A. C. Waltham **123** GSF **126** Dr. R. Legeckis/SPL **128t**, b Seaphot **129tr**, cr, NASA **129** © SF/Daily Telegraph **130t**, b, Zefa **131** Bruce Coleman Ltd. **131** Abril/Zefa **138t** Jan Griffier: Frostfair. Guildhall Art Gallery, City of London/ Bridgeman Art Library **139tr** NASA/SPL **139bl** NASA **140** National Society for Clean Air **141** GSF **142** Maurice Nimmo, Haverfordwest **143tl** Biophoto Associates, Leeds, cl,tc,cc Maurice Nimmo, tr, GSF **146t** GSF, ct,cb, G. R. Roberts, b, Maurice Nimmo **147** Maurice Nimmo **148t** Maurice Nimmo, br, Andrew Lawson **149t** GSF, bl,Geoslides **150** br, Maurice Nimmo **151tl** GSF cl,bl,br Maurice Nimmo, tr, Biophoto Associates **152** Maurice Nimmo **153** Dr. J.S. Shelton **154tr** Dr. J.S. Shelton, b, Maurice Nimmo **155** Maurice Nimmo **156** Dr. J. S. Shelton **158** Earth Satellite Corp./SPL **159t** Sinclair Stammers/SPL, c, David Bayliss/RIDA **160tl** Garo Nalbadian, Jerusalem **161** Robert Harding Picture Library **162** US Geological Survey/John S. Shelton **163** Aerofilms **164tl** University of Sheffield, br, Dougal Dixon **165tr** Dr. J.S. Shelton, b,National Portrait Gallery/National Photographic Record, c from Bagnold: *Blown Sand and Desert Dunes p. 36* 166 Bruce Coleman Ltd. **167** K. Weck/Zefa **168** A. C. Waltham **170tl** Institute of Geological Science, bl, OSF, **171** Heather Angel **172** Deep Sea Drilling Project/Scripps Institute of Oceanography **173** Biophoto Associates, Leeds **174tl** Heather Angel, bl, Catherine Ellis, Dept. of Zoology, University of Hull/SPL **175bl** Dr. John A. Catt, br, M.J. Thomas/Frank Lane Agency **180** Zefa **181** G. R. Roberts **183** G.R. Roberts **184-5** Dr. J. S. Shelton **186** c GSF b, Zefa **187** Landsat/Eros Data Center **188t** A. Post/U.S. Geological Survey **188b** G. R. Roberts **189** Dr. J. S. Shelton **192** Dr. J. S. Shelton **193t** G.R. Roberts **193b** Dougal Dixon **194l** G.R. Roberts, r, GSF **195** GSF **196** Zefa **197** Robert Harding Picture Library **198** Inco Ltd. **199** Robert Harding **200t** David Alan Harvey/Susan Griggs, c, Seaphot, **201t** Robert Harding **203t** National Coal Board, b, Pfaff/Zefa **204** Sonia Halliday and Laura Lushington **206** Bruce Coleman Ltd. **207** US Geological Survey, b, Jerry Hardman-Jones, Leeds **208t** Susan Griggs Agency, r, GSF **209t** David Davies/Bruce Coleman Ltd., c, Tony Morrison **210t** SPL, b, Seismograph Service (England) Ltd. **211tr** SPL, cr, Travel Photo International **212** Shuttle Imaging Radar Jet Propulsion Laboratory Pasadena **213** Photri **214bl** Picturepoint Ltd., br Clive Sawyer/Zefa, **215t** Ronald Sheridan, b, J. Allen Cash **216** K. Goebel/Zefa, tr,Janet and Colin Bord **217t** Western Evening Herald/photo: Hugh Baddeley Productions, b, Camera Press **218r** Landsat, bl,KLM Aerocarto **219** Dr. Georg Gerster/John Hillelson Agency **220** BBC Hulton Picture Library/Bettman Archive **221t** Camerapix Hutchison, b, J. Hobday/ Natural Science Photos **222** Tor Eigeland/Susan Griggs Agency **223** Colorific **224** Airviews, Manchester **225** The British Petroleum Co. plc **226** Robert Harding Picture Library **227tr** Charnock, br, Peter Fraenkel **228** Dr. J.S. Shelton **229bl** Starfoto/Zefa, br, Eidgenössische Technische Hochschule, Zurich **230** National Coal Board **231-234** J. L. Knill **235** Central Electricity Generating Board **236tl** Aerofilms, bl, J. L. Knill **238tl** Robert Harding Picture Library **239-240** J. L. Knill **241** Dr. R. Adams, International Seismological Centre, Newbury, Berks **242t** Frank Spooner Pictures, b, U.S. Geological Society **243t** J. Eyerman/Colorific, b John Topham Picture Library **244** Frank Spooner Picture Library **246b** Robert Harding Picture Library **247** Michael Freeman/Bruce Coleman Ltd. **248** Photri

Additional picture research Mary Fane

Cartographic editors Nicholas Harris and Zoë Goodwin

Cartographic production Location Map Services; Alan Mais, Lovell Johns Ltd

Indexer Susan Harris

Typesetters Peter MacDonald and Ron Barrow

Artists Robert Burns; Kai Choi; Chris Forsey; Alan Hollingbery; Kevin Maddison; Julia Osorno; Colin Salmon; Mick Saunders.

Further Reading

General

Ager, D.V. *The Nature of the Stratigraphical Record* (2nd edition – Macmillan)
Bolt, B.A. *Inside the Earth* (W.H. Freeman)
Brownlow, A.H. *Geochemistry* (Prentice-Hall)
Coates, D.R. *Geology and Society* (Chapman and Hall)
Eicher, D.L., McAlester, A.L. and Rottman, M.L. *The History of the Earth's Crust* (Prentice-Hall)
Elder, J. *The Bowels of the Earth* (Oxford University Press)
Goudie, A. *The Human Impact* (Basil Blackwell)
Goudie, A. and Gardner, R. *Discovering Landscape in England and Wales* (Allen and Unwin)
Harland, W.B., Cox, A.V., Llewellyn, P.G., Pickton, C.A.G., Smith, A.G. and Walters, R. *A Geologic Time Scale* (Cambridge University Press)
Levin, H.L. *The Earth Through Time* (Saunders)
Owen, H.G. *Atlas of Continental Displacement, 200 Ma-Present* (Cambridge University Press)
Park, R.G. *Foundations of Structural Geology* (Blackie)
Press, F. and Siever, R. *Earth* (3rd edition – W.H. Freeman)
van Andel, Tj.H. *New Views on an Old Planet: Continental Drift and the History of the Earth* (Cambridge University Press)

Specific Topics

Bell, F.G. *Fundamentals of Engineering Geology* (Butterworths)
Berry, L.G., Mason, B. and Dietrich, R.V. *Mineralogy* (2nd edition – W.H. Freeman)
Blong, R.J. *Volcanic Hazards* (Academic Press)
Bolt, B.A. *Earthquakes: A Primer* (W.H. Freeman)
Bonneau, M. and Souchier, B. *Constituents and Properties of Soils* (Academic Press)
Bullard, F.M. *Volcanoes of the Earth* (University of Texas Press)
Carey, S.W. *The Expanding Earth* (University of Tasmania)
Dent, D. and Young, A. *Soil Survey and Land Evaluation* (Allen and Unwin)
Francis, P. *Volcanoes* (Penguin Books)
Gregory, K.J. and Walling, D. *Man and Environmental Processes* (Dawson)
Gribbin, J. *Climatic Change* (Cambridge University Press)
Gribbin, J. *Future Weather* (Penguin Books)
Imbrie, J. and Imbrie, K.P. *Ice Ages* (Macmillan)
Knill, J. *Industrial Geology* (Oxford University Press)
Lamb, H.H. *Climate, History and the Modern World* (Methuen)
McElhinny, M.W. *Palaeomagnetism and Plate Tectonics* (Cambridge University Press)
Menard, H.W. *Ocean Science* (W.H. Freeman)
Merrill, R.T. and McElhinny, M.W. *The Earth's Magnetic Field* (Academic Press)
Nield, E.W. and Tucker, V.C.T. *Palaeontology: An Introduction* (Pergamon)
Ollier, C.D. *Weathering* (Longman)
Reading, H.G. *Sedimentary Environments and Facies* (Blackwell Scientific Publications)
Sawkins, F.J. *Metal Deposits in Relation to Plate Tectonics* (Springer-Verlag)
Skinner, B.J. *Earth Resources* (Prentice-Hall)
The Times Atlas of the Oceans (Times Books)
van Andel, Tj.H. *Science at Sea – Tales of an Old Ocean* (W.H. Freeman)
Walker, J.G.C. *Evolution of the Atmosphere* (Macmillan)

Glossary

Abyssal
Relating to the oceanic deeps, ususally below 3,000m.
Aftershock
An EARTHQUAKE that follows a larger earthquake and originates from the same at a nearby FOCUS.
Anaerobic
Existing without oxygen.
Anticline
A FOLD in which the BEDS arch upwards at the axis.
Aquifer
A BED of ROCK capable of holding water.
Asthenosphere
The layer of the Earth immediately below the LITHOSPHERE, representing a zone of weakness that allows the movement of PLATES above it.
Basalt
A fine-grained IGNEOUS ROCK, low in silica and characterized by its dark color and its ability to produce LAVA flows.
Bed
Layer of SEDIMENTARY ROCK which is clear enough for it to be distinguished from the layers above and below.
Calcareous
Containing calcium carbonate, as in limestone.
Carbonate
A compound containing CO_3. MINERAL carbonates are typical of limestones.
Clastic
Referring to a sedimentary rock formed from fragments broken from previously existing rocks.
Clay
Very fine sediment or SEDIMENTARY ROCK having particle sizes of less that .004mm and a plastic consistency.
Continent
A large landmass consisting of ROCKS that are distinct from those of the OCEAN floor. The term covers the dry land areas and the surrounding shallow seas.
Convection
A movement in which hot fluid, being of low density, rises above cold fluid which moves in to take its place.
Core
The central mass of the Earth, comprising an inner solid part and an outer liquid part, about 3,470km in radius, and probably consisting largely of iron.
Country rock
The ROCK in which a geological reaction takes place.
Crust
The outer layer of the Earth. It consists of two types - continental, which is rich in silica (about 40km thick) and oceanic, which is poor in silica (about 10km thick).
Crystal
The regular shape produced by the natural formation of a substance. Bounded by flat faces, the shape reflects the internal arrangement of the constituents.
Delta
A buildup of sediment at the mouth of a river, in a series of sandbanks and bars that effectively split up the river into a number of channels.
Diagenesis
The process by which a sediment becomes a SEDIMENTARY ROCK.
Dike (dyke)
An intrusion of igneous rock whose surface is different from that of adjoining material. Dikes are usually vertical.
Earthquake
Shock waves produced by the movement of rocks along a FAULT within the Earth's CRUST or upper MANTLE.
Epicenter
The point on the Earth's surface directly above the FOCUS of an EARTHQUAKE.
Epicontinental
At the edge of a CONTINENT, usually referring to a shallow sea that overlies the continental shelf.
Erosion
The process by which an exposed part of the Earth's surface is broken down.
Evaporite
A ROCK produced by the precipitation of dissolved salts from a body of water that is drying up.
Fault
A crack in the ROCKS along which two rock masses are seen to have moved.
Focus
The point from which EARTHQUAKE waves emanate.
Fold
A SEDIMENTARY ROCK structure in which the BEDS are seen to have been bent.
Fossil
The remains of a once-living organism preserved in rock.
Fossilization
The natural process by which a living organism is turned into a FOSSIL. Usually only the hard parts survive, like sharks' teeth in marine clays.
Glacier
A mass of ice, produced by compacted snow, that moves in a particular direction.
Gondwana
The southern supercontinent composed of the present-day continents of South America, Africa, India, Madagascar, Australia and Antarctica. It began to break up about 200 million years ago.
Greenhouse effect
An increase in atmospheric temperature caused by an increased volume of carbon dioxide in the atmosphere. This absorbs infrared radiation reflected from the Earth's surface.
Gutenberg discontinuity
The boundary between the Earth's MANTLE andCORE.
Half-life
The time taken for a radioactive isotope to decay to half of its original quantity. This is constant for each radioactive isotope and can be used to date MINERALS and ROCKS containing radioactive isotopes.
Horst
A block that has been uplifted between two FAULTS.
Hot spot
An area of increased thermal activity in the MANTLE.
Humus
The partially decayed organic material in the SOIL.
Ice age
A period of time in which temperatures are lower and glaciers spread over a significant portion of the Earth.
Igneous
Solidifying of a ROCK from a molten state.
Ionosphere
A division of the atmosphere between 70km and several hundred km in height in which the air is ionized - the atoms are split up into free electrons and positively charged particles.
Island arc
A chain of volcanic islands produced by a SUBDUCTION ZONE and formed by the side of an oceanic TRENCH. The arc shape is produced because of the geometry of a more-or-less straight plate edge cutting into the curve of the Earth's sphere.
Isostasy
The theoretical buoyancy balance of different parts of the Earth's CRUST as though they were floating on a denser substratum. This results in lighter material rising above denser regions.
Isthsmus
A narrow strip of land connecting two larger land areas, such as the Isthsmus of Panama.
Joint
A crack in the ROCKS along which there has been no movement.
Lateral continuity
The concept that suggests that if two BEDS of SEDIMENTARY ROCK are obviously identical in age and physical characteristics, but are some distance from one another, then they are likely to be parts of the same bed with the intervening portions eroded away.
Laurasia
The northern SUPERCONTINENT comprising the present day continents of North America, Europe and Asia. The northern part of PANGEA.
Lava
Molten ROCK produced from a VOLCANO at fissure.
Leaching
The removal of materials in solution from ROCK or SOIL by percolating groundwater.
Lithosphere
The region corresponding to the CRUST and the rigid part of the MANTLE above the ASTHENOSPHERE. The lithosphere constitutes the plates of PLATE TECTONICS.
Magma
Molten ROCK originating in the CRUST or MANTLE.
Mantle
The silica-rich layer that constitutes the bulk of the Earth. It lies between the CORE and the CRUST.
Mass movements
Movement of a portion of the Earth's surface as a single piece, as in a landslide or a slump.
Mesosphere
In geophysics, the part of the MANTLE beneath the ASTHENOSPHERE.
In meteorology, the layer of atmosphere between about 50 and 80km above the ground. In the mesosphere the air temperature reaches its lowest - about -120°C.
Metamorphism
The change of one ROCK type into another by great heat or pressure without that rock's passing through a molten phase.
Mineral
A naturally formed inorganic substance having a precise chemical formula and often a crystalline shape. Minerals are the building bricks of ROCKS.
Moho
Abbreviation for the Mohorovičić discontinuity - the boundary between the CRUST and the MANTLE.
Nappe
A large-scale FOLD or series of folds in which the limbs are thrown over and buried like folds in tumbled cloth.
Oasis
A fertile spot in a desert area where water is available.
Ocean
The great body of salt water that covers more than two-thirds of the Earth's surface, particularly that part that lies over what is known as the oceanic CRUST.
Ooze
Unconsolidated sediment of organic origin on the OCEAN floor.
Ophiolite
Sequence of rocks consisting of basaltic DIKES, LAVAS and deep-sea sediments that were once part of the oceanic CRUST.
Ore
A material containing a metal that can be extracted in economic quantities.
Overburden
The natural material overlying any useful deposit of economic MINERAL.
Paleomagnetism
The properties of the Earth's magnetic field at a time in the past revealed by examining the alignment of magnetic particles in rocks formed at that time.
Pangea
The SUPERCONTINENT that existed between about 250 and 200 million years ago and comprising all the continental masses. It can be thought of as a combination of GONDWANA and LAURASIA.
Panthalassa
The OCEAN that covered the portion of the Earth not occupied by PANGEA.
Permafrost
A condition in which a deep layer of SOIL cannot thaw out during the summer, despite the thawing of the layers above. The result is a poorly-drained summer landscape typical of northern Canada and USSR.
Plate tectonics
The study of the movement of the LITHOSPHERIC plates over the surface of the globe. This is the mechanism that moves the CONTINENTS and opens and closes the OCEANS.
Precambrian
The whole span of geological time between the formation of the Earth and the beginning of the Cambrian Period about 590 million years ago.
Pyroclastic
ROCK fragment blasted out of a VOLCANO.
Radiation
Any form of energy, particularly heat or light, that can be transmitted through a medium without having an effect on that medium. For example, the sun's rays may warm a rock surface but not the atmosphere that it passed through.
Reef
A ridge of rocks lying just beneath the surface of the water, usually built up from the skeletons of corals.

Resistivity
The property of a substance that determines its resistence to an electrical current passing through it.
Rift valley
A valley formed by the subsidence of a block between two parallel FAULTS.
Rock
Any naturally-formed mass of MINERAL matter. To a geologist, even unconsolidated sand represents rock.
Saltation
The movement of a particle along a surface in a series of jumps, propelled by a current.
Scarp
A slope or a cliff that cuts across the bedding planes of a rock face.
Sedimentary
To do with the deposition of loose material by natural processes. The formation of a sandbank in a river is a sedimentary process.
Sedimentation
The deposition of loose material, such as sand, mud or gravel on the bed of a river, a desert basin or the bottom of the sea.
Shield
A block of CONTINENT consisting mostly of PRE-CAMBRIAN rocks, contorted and worn flat, that has been resistant to deformation for a long time.
Smoker
An eruption of hot, chemical-laden water from an oceanic ridge. The thick suspension of material looks like billowing smoke hence the name.
Soil
In geology and geography, the loose, unconsolidated layer that overlies most ROCKS of the CONTINENTS. It consists of broken and chemically altered MINERAL material from the underlying rocks, and partially decayed organic material from plant life. In engineering, any loose unconsolidated material.
Strata
The various layers in which a SEDIMENTARY ROCK is formed. The singular "stratum" is rarely used.
Stratosphere
The part of the Earth's ATMOSPHERE that lies between the TROPOSPHERE and the MESOSPHERE. In effect this is between about 12 and 50km above the ground.
Strike
The direction or bearing of a horizontal outcrop of a dipping BED. For example, in a dipping bed of rock protruding above water level, the line along which the upper surface of the bed meets the water is the line of strike. It is perpendicular to the "dip" of the bed.
Subduction zone
A destructive plate margin, where one plate of LITHOSPHERE is being drawn down beneath another. The surface expression of a subduction zone is an oceanic TRENCH and this is accompanied by either an arc of volcanic islands or a volcanic fold mountain region on an adjacent CONTINENT.
Supercontinent
A large CONTINENT produced by the fusion of several smaller continents. The classic example from the past is PANGEA, consisting of all the known continents. A modern example is Eurasia – made up of Europe and Asia, joined along the Urals, and India, joined to Asia by the Himalayas.
Superposition
The principle that states that, in any undisturbed sequence of SEDIMENTARY ROCKS, the oldest is at the bottom and the newest at the top. This is one of the basic concepts of STRATIGRAPHY.
Syncline
A FOLD in which the BEDS sag downward in the axis. The opposite of an ANTICLINE.
Trench
A deep valley found on the OCEAN floor along a SUBDUCTION ZONE. It is the result of the downward movement of one oceanic plate as it is pulled beneath another. Trenches can be up to 11km deep and are the deepest parts of the oceans.
Troposphere
The lower portion of the Earth's ATMOSPHERE, up to approximately 12km in height. This is the densest part of the atmosphere, containing most of the mass, and it is the most important, supporting all the weather processes.
Turbidity current
A mass of particles in suspension that moves as a dense fluid. These are common on continental slopes, bringing unconsolidated material from the continental shelf down to the ABYSSAL plains. In appearance a turbidity current looks like a billowing cloud.
Unconformity
A break in the sequence of SEDIMENTARY ROCKS. It is caused by one sequence being uplifted, being eroded flat, and then having another sequence deposited upon it. An unconformity is usually taken to represent dry land at some time in the geological past.
Volatile
Any substance that is easily evaporated away. Volatiles in MAGMA include water and carbon dioxide.
Volcano
The surface expression of the eruption of MAGMA through the Earth's crust. This may take the form of a fountain of LAVA, or a mountain built up of successive layers of ash and lava flows.

Derivations

As in the other sciences, Earth science uses many technical terms derived from Greek and Latin. However, there are many Earth science terms that are derived from other languages as well. The reason is that a great number of geological and geographical features are mainly found in particular areas, and in these areas there has always been a word for them in the local tongue. The local words have thus slipped into the nomenclature, and are applied to these same features wherever they are found. Very often the official name for some feature is taken directly from the place name of a particular example of that feature. Some notable ones are given here.

ARABIC
Erg
Sandy desert.
Hammada
Rocky desert.

DUTCH
Inselberg
The product of onion-skin weathering in an arid landscape, producing rounded isolated hills on an otherwise flat plain. They are particularly common in the former Dutch colonies in southern Africa.

ENGLISH (idiomatic)
Clint, Grike
Yorkshire terms for exposed blocks and their intervening gullies, the results of solution weathering of limestone areas.
Lias
Literally "layers". The lowermost sequences of the Jurassic, consisting of markedly alternating beds of limestone and shale.

FRENCH
Arête
Sharp ridge, peak or horn formed by erosion where two glaciers meet.
Cirque
Bowl-shaped, steep-sided hollow in rock formed through erosion by ice.
Crevasse
Deep crack in a glacier
Moraine
Sand, rocks or other debris carried along by a glacier and left as a heap once the glacier has melted.
Névé
Hard-packed mountain snow that has been converted into granular ice in a mountain glacier and has dug out CIRQUES.
Roche moutonnée
Rocks that resist the erosion of a glacier and become polished.

HAWAIIAN
Aa, Pahoehoe
Types of basaltic lava flows, common on Hawaii.
Pele's hair, Pele's tears
Fine volcanic material ejected from Hawaiian-type eruptions, named after the Hawaiian goddess of fire.

ICELANDIC
Geyser
A fountain of hot ground-water in volcanic regions. Named after a particular example.
Thufer
Small hillocks caused by frost-heaving.

INUIT (eskimo)
Nunatak
A mountain protruding through an ice-sheet.
Pingo
A tundra landscape feature in which a mass of soil is forced up into a hill by the expansion of ice beneath.

ITALIAN
Caldera
Literally "a bowl". A vast crater formed by the collapse of a volcanic peak into an empty magma chamber below, like that on Mount Vesuvius.
Lapilli
Literally "small stones". Small lumps of volcanic ash ejected from a Strombolian-type volcano.
Tombolo
A sandspit that connects an island to the mainland. Named after one such about 120km north of Rome.
Volcano
Derived from the proper name Vulcano, a volcanic island in the Mediterranean. This, in turn is derived from Vulcan, the Roman god of fire.

NORWEGIAN
Fjord
A steep-sided inlet formed by the inundation of a U-shaped valley.

RUSSIAN
Tundra
Literally "a marsh". Extent of swampy land, common in Siberia, underlain by permafrost and on which only low plants can grow.

SERBO-CROAT
Karst
A bare arid landscape formed by the solution erosion of limestone. The proper name of the area of the Adriatic coast where it is typical.

SPANISH
Butte, Cuesta, Mesa
Types of hill formed by arid erosion in the same area.
Playa
Arid area of inland drainage, with temporary lakes in the wet season, such as those in the former Spanish-speaking areas of western North America.
Ria
An inlet of the sea formed by a drowned valley, like those to the south of Cape Finisterre in Spain.
Salina
A PLAYA region with salt deposits.

Index

Page numbers in *italics* refer to illustrations, captions or side text.

D

E

F

G

H

I

J

K

L

M

N

T

U

W

Y

Z